Protein Metabolism and Homeostasis in Aging

ADVANCES IN EXPERIMENTAL MEDICINE AND BIOLOGY

Recent Volumes in this Series

Volume 686
RARE DISEASES EPIDEMIOLOGY
Edited by Manuel Posada de la Paz and Stephen C. Groft

Volume 687
BCL-2 PROTEIN FAMILY: ESSENTIAL REGULATORS OF CELL DEATH
Edited by Claudio Hetz

Volume 688
SPHINGOLIPIDS AS SIGNALING AND REGULATORY MOLECULES
Edited by Charles Chalfant and Maurizio Del Poeta

Volume 689
HOX GENES: STUDIES FROM THE 20TH TO THE 21ST CENTURY
Edited by Jean S. Deutsch

Volume 690
THE RENIN-ANGIOTENSIN SYSTEM: CURRENT RESEARCH PROGRESS IN THE PANCREAS
Edited by Po Sing Leung

Volume 691
ADVANCES IN TNF FAMILY RESEARCH
Edited by David Wallach

Volume 692
NEUROPEPTIDE SYSTEMS AS TARGETS FOR PARASITE AND PEST CONTROL
Edited by Timothy G. Geary and Aaron G. Maule

Volume 693
POST-TRANSCRIPTIONAL REGULATION BY STAR PROTEINS
Edited by Talila Volk and Karen Artzt

Volume 694
PROTEIN METABOLISM AND HOMEOSTASIS IN AGING
Edited by Nektarios Tavernarakis

Protein Metabolism and Homeostasis in Aging

Edited by

Nektarios Tavernarakis, PhD
Institute of Molecular Biology and Biotechnology (IMBB)
Foundation for Research and Technology Hellas (FORTH)
Heraklion, Crete, Greece

Springer Science+Business Media, LLC

Landes Bioscience

Springer Science+Business Media, LLC
Landes Bioscience

Printed in the USA.

Springer Science+Business Media, LLC, 233 Spring Street, New York, New York 10013, USA
http://www.springer.com

Please address all inquiries to the publishers:
Landes Bioscience, 1002 West Avenue, Austin, Texas 78701, USA
Phone: 512/ 637 6050; FAX: 512/ 637 6079
http://www.landesbioscience.com

The chapters in this book are available in the Madame Curie Bioscience Database.
http://www.landesbioscience.com/curie

Protein Metabolism and Homeostasis in Aging, edited by Nektarios Tavernarakis. Landes Bioscience / Springer Science+Business Media, LLC dual imprint / Springer series: Advances in Experimental Medicine and Biology.

ISBN: 978-1-4419-7001-5

Library of Congress Cataloging-in-Publication Data

Protein metabolism and homeostasis in aging / edited by Nektarios Tavernarakis.
p. ; cm. -- (Advances in experimental medicine and biology ; v. 694)
Includes bibliographical references and index.
ISBN 978-1-4419-7001-5
1. Proteins--Metabolism. 2. Cells--Aging. 3. Aging--Physiological aspects. 4. Homeostasis. I. Tavernarakis, Nektarios. II. Series: Advances in experimental medicine and biology ; v. 694. 0065-2598
[DNLM: 1. Cell Aging--physiology. 2. Aging--physiology. 3. Caenorhabditis elegans--genetics. 4. Caenorhabditis elegans Proteins--metabolism. 5. Homeostasis--physiology. 6. Models, Biological. W1 AD559 v.694 2010 / QU 375 P967 2010]
QP551.P69715 2010
612'.01575--dc22

2010021376

DEDICATION

To my father

PREFACE

Aging is loosely defined as the accumulation of changes in an organism over time. At the cellular level such changes are distinct and multidimensional: DNA replication ceases, cells stop dividing, they become senescent and eventually die. DNA metabolism and chromosomal maintenance, together with protein metabolism are critical in the aging process. The focus of this book is on the role of protein metabolism and homeostasis in aging. An overview is provided of the current knowledge in the area, including protein synthesis, accuracy and repair, post-translational modifications, degradation and turnover, and how they define and influence aging. The chapters mainly focus on well-characterised factors and pathways, but new areas are also presented, where associations with aging are just being elucidated by current experimental data.

Protein turnover, the balance between protein synthesis and protein degradation are carefully maintained in healthy cells. Chapters 1 and 2 illustrate that aging cells are characterised by alterations in the rate, level and accuracy of protein synthesis compared to young ones, and that mRNA translation, essential for cell growth and survival, is controlled at multiple levels. The theory that growth and somatic maintenance are believed to be antagonistic processes is described in Chapter 3: inhibition of protein synthesis results in decreased rates of growth and development, but also confers an extension of lifespan, as shown for example by the effects of dietary restriction in various models organisms. Quality control mechanisms ensure misfolded or damaged proteins are remodelled or repaired, but when this fails proteins are targeted to go through degradation in order to avoid untimely cell death. The ubiquitin/proteasome system keeps cells clear of abnormal, damaged or denatured proteins (Chapter 4), while autophagy degrades long-lived proteins and small organelles (Chapter 5); compromised activity of both processes has been tightly correlated to aging. Accumulation of damaged or misfolded proteins within cells has been associated with human age-related, neurodegenerative diseases. The paradoxical situation of autophagy up-regulation in models of premature aging is also discussed in Chapter 6.

The insulin/IGF-1, TGFβ, TOR and p38/ MAP kinase signalling pathways play a part in regulating protein turnover and have been linked to aging through a number of their components, discussed in Chapter 7. Inhibition of the insulin/IGF and TOR pathways results in lifespan extension in worms, as detailed in Chapter 8, similarly with other longevity pathways, including dietary intake and mitochondrial function. The role of mitochondria in protein quality control and the influence of reactive oxygen species in aging are presented in detail in Chapter 9. Chapter 10 discusses that different types of stress, intracellular, oncogenic and environmental, such as food and space restrictions, oxidative stress, temperature fluctuations and accumulation of damaged proteins, have been shown to induce premature aging and/or senescence through mechanisms independent of telomere shortening. In Chapter 11 the free radical and oxidative stress theories of aging are portrayed to link such stress factors to the occurrence of aging through the function of mitochondria, the activity of detoxifying enzymes and degradation pathways and their effects on protein turnover, while resistance to stress has been directly associated to lifespan extension and delayed aging in model organisms. Stress is also a major inducer of the sumoylation pathway, a post-translational protein modification that, together with substrate interactions with other ubiquitin-like proteins, show differential activity in aging tissues and has recently been linked to the onset of cellular senescence; these pathways are presented in detail in Chapters 12 and 13. The critical importance of maintaining cell homeostasis is evident in every chapter in this book, but is presented in more detail in Chapter 14, especially focusing on hormone signalling in response to environmental cues. Hormones and cytokines that affect muscle homeostasis during aging are presented in Chapter 15. The main pathways that take part in skeletal muscle atrophy and regeneration are illustrated, followed by a description of current gene and cell therapies to rescue muscle atrophy and wasting. The volume concludes with Chapter 16, where common techniques used in protein metabolism and homeostasis research are presented and critically reviewed.

We would like to thank all authors of this book for their excellent contributions, and hope that the reader will enjoy the chapters and be inspired to further their knowledge in this ever-expanding and exciting field. As the advances of modern technology and medicine have significantly raised life expectancy, it is becoming ever more important to gain a deeper insight and understanding of the mechanisms that influence the aging process. Such knowledge is an essential prerequisite for the development of effective strategies to increase health span and quality of life for the elderly.

Artemisia M. Andreou, PhD
Nektarios Tavernarakis, PhD
March 2010, Heraklion, Greece

ABOUT THE EDITOR...

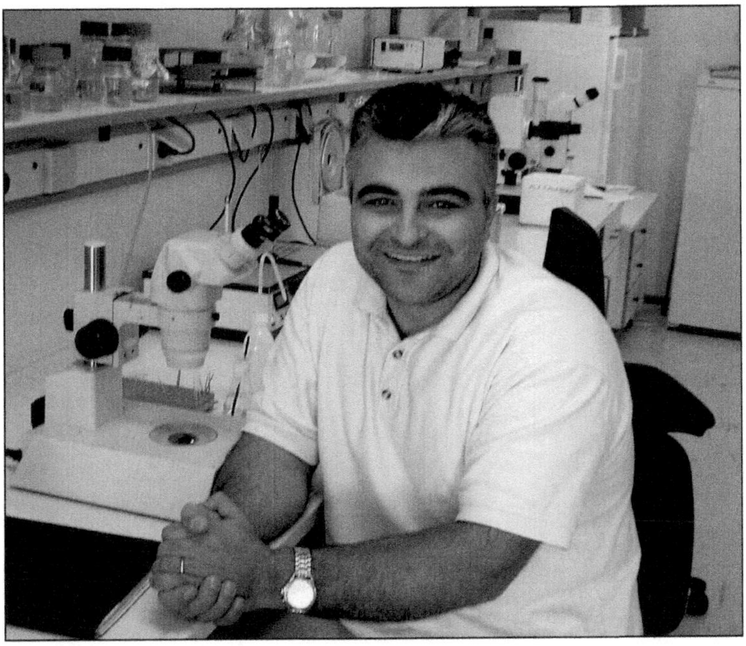

NEKTARIOS TAVERNARAKIS is a Research Director (Professor) at the Institute of Molecular Biology and Biotechnology in Heraklion, Crete, Greece, heading the *Caenorhabditis elegans* molecular genetics laboratory. He earned his PhD degree at the University of Crete studying gene expression regulation in yeast and trained in *C. elegans* genetics and molecular biology at Rutgers University, New Jersey, USA. His research focuses on studies of neuronal function and dysfunction, using the nematode *Caenorhabditis elegans* as a model organism. His main interests are the molecular mechanisms of necrotic cell death in neurodegeneration and senescent decline, the molecular mechanisms of sensory transduction and integration by the nervous system, the interplay between cellular metabolism and aging, and the development of novel genetic tools for *C. elegans* research. He is the recipient of a European Research Council (ERC) Advanced Investigator grant award, a European Molecular Biology Organisation (EMBO) Young Investigator award, an International Human Frontier in Science Program Organization (HFSPO) long-term award, the Bodossaki Foundation Scientific Prize for Medicine and Biology, the Alexander von Humboldt Foundation, Friedrich Wilhelm Bessel research award. Professor Tavernarakis is a member of EMBO.

PARTICIPANTS

Joy Alcedo
Friedrich Miescher Institute
for Biomedical Research
Basel
Switzerland

Artemisia M. Andreou
Institute of Molecular Biology
and Biotechnology (IMBB)
Foundation for Research and
Technology Hellas (FORTH)
Heraklion, Crete
Greece

Emmanuelle Boilan
University of Namur
Research Unit on Cellular Biology
Namur
Belgium

Louise Boisen
Department of Molecular Biology
University of Aarhus
Aarhus
Denmark

Bart P. Braeckman
Department of Biology
Ghent University
Ghent
Belgium

Queelim Ch'ng
MRC Centre for Developmental
Neurobiology
King's College
London
UK

Devon Chandler-Brown
Department of Pathology
University of Washington
Seattle, Washington
USA

Niki Chondrogianni
National Hellenic Research Foundation
Institute of Biological Research
and Biotechnology
Athens
Greece

Florence Debacq-Chainiaux
University of Namur
Research Unit on Cellular Biology
Namur
Belgium

Jérémie Dedessus Le Moutier
University of Namur
Research Unit on Cellular Biology
Namur
Belgium

Krisztina Takács-Vellai
Department of Genetics
Eötvös Loránd University
Budapest
Hungary

Nektarios Tavernarakis
Institute of Molecular Biology and Biotechnology (IMBB)
Foundation for Research and Technology Hellas (FORTH)
Heraklion, Crete
Greece

Olivier Toussaint
University of Namur
Research Unit on Cellular Biology
Namur
Belgium

Jacques R. Vanfleteren
Department of Biology
Ghent University
Ghent
Belgium

Tibor Vellai
Department of Genetics
Eötvös Loránd University
Budapest
Hungary

Manlio Vinciguerra
Mouse Biology Unit
European Molecular Biology Laboratory (EMBL)
Rome
Italy

Geoffroy Weemaels
University of Namur
Research Unit on Cellular Biology
Namur
Belgium

Andrea C. Weil
Johann Wolfgang Goethe University
Faculty for Biosciences and Cluster of Excellence Macromolecular Complexes
Institute of Molecular Biosciences
Frankfurt
Germany

CONTENTS

1. SYNTHESIS, MODIFICATION AND TURNOVER OF PROTEINS DURING AGING 1

Suresh I.S. Rattan

Abstract........ 1
Introduction........ 1
Efficiency and Accuracy of Protein Synthesis during Aging........ 1
Altered Protein Synthesis during Aging........ 2
Post-Translational Modifications during Aging........ 4
Protein Turnover during Aging........ 9
Conclusion........ 9

2. REGULATION OF mRNA TRANSLATION AS A CONSERVED MECHANISM OF LONGEVITY CONTROL........ 14

Ranjana Mehta, Devon Chandler-Brown, Fresnida J. Ramos, Lara S. Shamieh and Matt Kaeberlein

Abstract........ 14
Introduction........ 14
Genome Scale Longevity Screens in Yeast and Nematodes........ 16
mRNA Translation is a Public Determinant of Longevity........ 17
Is DR Mediated by Reduced mRNA Translation?........ 17
Possible Mechanisms for How Translation Influences Aging........ 20
Does mRNA Translation Modulate Aging in Mammals?........ 23
Conclusion........ 25

3. PROTEIN SYNTHESIS AND THE ANTAGONISTIC PLEIOTROPY HYPOTHESIS OF AGING........ 30

Pankaj Kapahi

Abstract........ 30
Evolution of Aging........ 30
Insulin-Like Signaling (ILS)........ 31

TOR Pathway 31
Protein Synthesis 32
Direct Screens to Identify Genes That Antagonistically Regulate Growth and Longevity 32
Dietary Restriction (DR), Protein Synthesis and Antagonistic Pleiotropy 33
Mechanism of Lifespan Extension by Inhibition of Protein Synthesis 34
Conclusion 35

4. PROTEASOME FUNCTION DETERMINES CELLULAR HOMEOSTASIS AND THE RATE OF AGING 38

Niki Chondrogianni and Efstathios S. Gonos

Abstract 38
Protein Homeostasis and Aging: Which Are the Key Players? 38
An Introduction to the Proteasome Biology 39
Proteasome during Aging 41
Proteasome Activation: Is There a Way to Restore Proteasome Function? 42
Conclusion 44

5. AUTOPHAGY AND LONGEVITY: LESSONS FROM *C. ELEGANS* 47

Kailiang Jia and Beth Levine

Abstract 47
Introduction 47
DAF-2 Insulin/IGF-1-Like Signaling 48
Dietary Restriction 52
Mitochondrial Activity 52
Autophagy 53
Autophagy and *C. elegans* Longevity Pathways 54
Conclusion 57

6. AUTOPHAGY AND AGING: LESSONS FROM PROGERIA MODELS 61

Guillermo Mariño, Alvaro F. Fernández and Carlos López-Otín

Abstract 61
Introduction 61
Autophagy and Physiological Aging 62
Autophagy and Premature Aging 64
Conclusion 66

7. REGULATION OF PROTEIN TURNOVER BY LONGEVITY PATHWAYS 69

Tibor Vellai and Krisztina Takács-Vellai

Abstract 69
Protein Metabolism and Aging 69

Longevity Pathways That Promote Protein Synthesis 74
Interactions between Molecular Mechanisms Involved in Protein Synthesis and Degradation 77
Conclusion 78

8. PROTEIN METABOLISM AND LIFESPAN IN *CAENORHABDITIS ELEGANS* 81

Geert Depuydt, Jacques R. Vanfleteren and Bart P. Braeckman

Abstract 81
Introduction 81
Dietary Restriction, TOR Signaling and Protein Homeostasis 83
Reduced Protein Synthesis Extends Lifespan 84
Translation Initiation 85
A Model for Translation Inhibition Induced Longevity 94
HSF-1 Mediated Defense against Proteotoxicity 96
Autophagy 97
Proteasome Function in Proteotoxicity and Longevity 101
Conclusion 101

9. MITOCHONDRIAL PROTEIN QUALITY CONTROL SYSTEMS IN AGING AND DISEASE 108

Karin Luce, Andrea C. Weil and Heinz D. Osiewacz

Abstract 108
Introduction 108
Mitochondrial Chaperones Are Necessary for Regulated Mitochondrial PQC 110
Role of the Mitochondrial Proteases in Maintaining Mitochondrial Functions 112
Role of the Membrane-Bound AAA Proteases on Diseases, Apoptosis and Aging 117
Mitochondrial Lon Protease Activity and Aging 118
Conclusion 120

10. P38^{MAPK} IN THE SENESCENCE OF HUMAN AND MURINE FIBROBLASTS 126

Florence Debacq-Chainiaux, Emmanuelle Boilan, Jérémie Dedessus Le Moutier, Geoffroy Weemaels and Olivier Toussaint

Abstract 126
Introduction 126
Senescence Is the Hardest Word to Say 126
The Role of DNA Damage Checkpoint Genes in Senescence 127
Signal Transduction and Gene Expression in SIPS: Central Role of p38^{MAPK} 127
TGF-β1 and p38^{MAPK} in H_2O_2- and UVB-Induced SIPS 129
p38^{MAPK}, p53 and Rb 131
Role of Caveolin-1 in Cellular Senescence and Interplay with p38^{MAPK} 131
Premature Senescence as an Anti-Oncogenic Defense Mechanism 132
Signaling Pathway Mediating Ras-Induced Premature Senescence—The Tumor Suppressing Function of p38^{MAPK} 132
Conclusion: The Next Steps… 133

11. PROTEIN HOMEOSTASIS IN MODELS OF AGING AND AGE-RELATED CONFORMATIONAL DISEASE................ 138

Elise A. Kikis, Tali Gidalevitz and Richard I. Morimoto

Abstract.. 138
Protein Folding Problem in Aging.. 138
Invertebrate Models of Late Onset Conformational Diseases......................... 145
Failure of Homeostasis in Conformational Disease of Aging.......................... 148
Modifiers of Conformational Disease... 149
Conclusion 153

12. ROLES FOR SUMO MODIFICATION DURING SENESCENCE...... 160

Artemisia M. Andreou and Nektarios Tavernarakis

Abstract.. 160
Introduction to the SUMO Modification System 160
Cellular Senescence... 163
SUMO and Senescence 164
Conclusion 167

13. POST-TRANSLATIONAL MODIFICATION OF CELLULAR PROTEINS BY UBIQUITIN AND UBIQUITIN-LIKE MOLECULES: ROLE IN CELLULAR SENESCENCE AND AGING.. 172

Johannes Grillari, Regina Grillari-Voglauer and Pidder Jansen-Dürr

Abstract.. 172
Introduction... 172
Role of Ubiquitination, SUMOylation and ISG15 in Cellular Senescence................ 174
Ubiquitination in Tissues during Aging... 182
Segmental Progeroid Syndromes and the Ubiquitin System 185
Role of Ubiquitinylation and SUMOylation in Aging of Short-Lived Model Organisms... 186
Conclusion 186

14. SENSORY INFLUENCE ON HOMEOSTASIS AND LIFESPAN: MOLECULES AND CIRCUITS... 197

Joy Alcedo, Wolfgang Maier and Queelim Ch'ng

Abstract.. 197
Introduction... 197
How External Sensory Cues Influence Homeostasis...................................... 198
Molecular Mechanisms Linking Sensory Transduction and Hormonal Outputs...... 198
The Circuitry Underlying the Processing of Sensory Information 201
Sensory Influence on Lifespan... 204
Conclusion: Connections Between the Sensory Influence on Homeostasis and Lifespan? 206

15. REGULATION OF MUSCLE ATROPHY IN AGING AND DISEASE 211

Manlio Vinciguerra, Antonio Musaro and Nadia Rosenthal

Abstract 211
Muscle Atrophy and Wasting Diseases 211
Molecular Mechanisms of Muscle Atrophy and Wasting 212
Muscle Cytoskeletal Disorders and Atrophy 216
Cellular and Molecular Bases of Muscle Repair 217
Gene and Cell Therapies to Rescue Muscle Atrophy and Wasting 223
Conclusion 226

16. CONFRONTING CELLULAR HETEROGENEITY IN STUDIES OF PROTEIN METABOLISM AND HOMEOSTASIS IN AGING RESEARCH 234

Louise Boisen and Peter Kristensen

Abstract 234
Analysing Aging 234
Antibody Phage Display as a Discovery Tool 236
Conclusion 241

INDEX 245

ACKNOWLEDGEMENTS

I am grateful to Dr. Artemisia M. Andreou for her valuable help in editing this book.

CHAPTER 1

Synthesis, Modification and Turnover of Proteins during Aging

Suresh I.S. Rattan*

Abstract

Alterations in the rate and extent of protein synthesis, accuracy, post-translational modifications and turnover are among the main molecular characteristics of aging. A decline in the cellular capacity through proteasomal and lysosomal pathways to recognize and preferentially degrade damaged proteins leads to the accumulation of abnormal proteins during aging. The consequent increase in molecular heterogeneity and impaired functioning of proteins is the basis of several age-related pathologies, such as cataracts, sarcopenia and neurodegerative diseases. Understanding the proteomic spectrum and its functional implications during aging can facilitate developing effective means of intervention, prevention and therapy of aging and age-related diseases.

Introduction

A decline in the rate of total protein synthesis is one of the most commonly observed age-associated biochemical changes in cells, tissues, organs and organisms, including human beings. The implications and consequences of slower rates of protein synthesis are manifold in the context of aging and age-related pathology. Although there is a considerable variability among different tissues and cell types in the extent of decline (varying from 20% to 80%), the fact remains that the bulk protein synthesis slows down during aging.[1,2] However, it should be pointed out that age-related slowing down of bulk protein synthesis does not mean that the synthesis of each and every protein becomes slower uniformly during aging. Furthermore, even though bulk protein synthesis slows down with age, total protein content of the cell generally increases because of an accumulation of abnormal proteins during aging. Age-related changes in protein synthesis are regulated both at the transcriptional and pretranslational levels in terms of the availability of individual mRNA species for translation,[2] and at the translational and post-translational levels in terms of alterations in the components of the protein synthetic machinery and the pattern of postsynthetic modifications that determine the activity, specificity and stability of a protein.

The aim of this chapter is to provide an overview of the regulation and misregulation, synthesis, modifications and turnover of proteins, including the role of protein errors and accumulation of abnormal proteins during aging.

Efficiency and Accuracy of Protein Synthesis during Aging

Eukaryotic protein synthesis is a highly complex process, which requires about 200 small and large components to function effectively and accurately in order to translate one mRNA molecule, while using large quantities of cellular energy. There are three major components of the

*Suresh Rattan—Laboratory of Cellular Aging, Department of Molecular Biology, Aarhus University, DK8000 Aarhus—C, Denmark. Email: rattan@mb.au.dk

Protein Metabolism and Homeostasis in Aging, edited by Nektarios Tavernarakis.

translational apparatus: (1) the translational particle, the ribosome; (2) the amino acid transfer system or charging system; and (3) the translational factors.[3] The protein-synthesizing apparatus is highly organised and its macromolecular components are not freely diffusible within cells. The rate and accuracy of protein synthesis can be critical for aging. Since, the error frequency of amino acid misincorporation is generally considered to be quite high (10^{-3} to 10^{-4}) as compared with nucleotide misincorporation, the role of protein error feedback in aging has been a widely discussed issue.[4,5]

So far, no direct estimates of protein error levels in any aging system have been made primarily due to the lack of appropriate methods to determine spontaneous levels of error in a normal situation. However, several indirect estimates of the accuracy of translation in cell-free extracts, using synthetic templates or natural mRNAs have been made. Studies performed on animal tissues, such as chick brain, mouse liver and rat brain, liver and kidney, did not reveal any major age-related differences in the capacity and accuracy of ribosomes to translate poly(U) in cell-free extracts.[6] However, these attempts to estimate the error frequencies during in vitro translation of poly(U) template were inconclusive because the error frequencies encountered in the assays were several times greater than the estimates of natural error frequencies. Another indirect method that has been used to detect misincorporation of amino acids during aging is the method of two-dimensional (2D) gel electrophoresis of proteins, but no age-related increase in amino acid misincorporation affecting the net charge on proteins was observed in histidine-starved human fibroblasts and in nematodes. In contrast to this, using mRNA of CcTMV coat protein for translation by cell extracts prepared from young and old human fibroblasts, a seven-fold increase in cysteine misincorporation during cellular aging has been observed.[7,8] These studies also showed that an aminoglycoside antibiotic paromomycin (Pm), which is known to reduce ribosomal accuracy during translation in vivo and in vitro, induces more errors in the translation of CcTMV coat protein mRNA by cell extracts prepared from senescent human fibroblasts than those from young cells. Indirect evidence that indicates the role of protein errors in cellular aging can be drawn from studies on the increase in the sensitivity of human fibroblasts to the life-shortening and aging-inducing effects of Pm and another aminoglycoside antibiotic G418.[9,10] Further evidence in support of the role of errors comes from experiments which showed that an induction and increase in protein errors can accelerate aging in human cells and bacteria.[1,2,4,11,12] Similarly, an increase in the accuracy of protein synthesis can slow aging and increase the lifespan in fungi.[13,14] Although a global "error catastrophe" as a cause of aging due to errors in each and every macromolecule is considered unlikely, it is not ruled out that some kind of errors in various components of protein synthetic machinery, including tRNA charging, may have long-term effects on cellular stability and survival.

Altered Protein Synthesis during Aging

The translational process can be envisaged to proceed in three steps—initiation, elongation and termination, followed by post-translational modifications, including folding, which give the protein a functional tertiary structure. The initiation step is considered to be the main target for the regulation of protein synthesis during cell cycle, growth, development, hormonal response and under stress conditions including heat shock, irradiation and starvation.[15,16] With respect to aging, however, the rate of initiation appears to remain unaltered. For example, using in vitro assays, the conversion of isolated 40S and 60S ribosomal subunits into the 80S initiation complex has been reported to decrease by less than 15% in old *Drosophila*, rat liver and kidney and mouse liver and kidney. On the other hand, since polysomal fraction of the ribosomes decreases during aging, it implies that the activity of an anti-ribosomal-association factor eIF-3 may increase during aging. The activity of eIF-2, which is required for the formation of the ternary complex of Met-tRNAi, GTP and eIF-2, has been reported to decrease in rat tissues during development and aging. Attenuation of hypusine formation on eIF-5A during senescence of human diploid fibroblasts has been reported.[17] Recently, a novel role of eIF-5A in protein chain elongation has been reported,[18] which may also be important in relation to aging. Similar studies on other eIFs and in other aging systems are yet to be performed and it is necessary that detailed studies on eIFs are also undertaken in the context of aging and the question of the regulation of protein synthesis at the level of initiation is reinvestigated.

Several studies have been performed on age-related changes in the number of ribosomes, thermal stability, binding to aminoacyl-tRNA, the level of ribosomal proteins and rRNAs, sensitivity to aminoglycoside antibiotics and the fidelity of ribosomes.[6] Although there is a slight decrease in the number of ribosomes in old animals, this does not appear to be a rate-limiting factor for protein synthesis due to ribosomal abundance in cells. Instead, several studies indicate that the biochemical and biophysical changes in ribosomal characteristics may be more important for translational regulation during aging. For example, the ability of aged ribosomes to translate synthetic poly(U) or natural globin mRNA decreases significantly. A decrease in the translational capacity of ribosomes has also been observed in rodent tissues such as muscle, brain, liver, lens, testis and parotid gland and in various organs of *Drosophila*.[19,20]

The reasons for the functional changes observed in aging ribosomes are not known at present. Some attempts have been made to study the effect of aging on rRNAs and ribosomal proteins. Although a three-fold increase in the content of rRNA has been reported in late passage senescent human fibroblasts, it is not clear whether the quantity and quality of individual rRNA species undergo alterations during aging and what effect such a change might have on ribosome function. Similarly, although an increase in mRNA levels for ribosomal protein L7 has been reported in aged human fibroblasts and rat preadipocytes, there are no differences in the electrophoretic patterns of the ribosomal proteins in young and old *Drosophila* and mouse liver.[6]

The formation of the 80S initiation complex is followed by the repetitive cyclic event of peptide chain elongation, a series of reactions catalysed by elongation factors (EFs; also abbreviated as eEFs).[21,22] Various estimates of the elongation rates in eukaryotic cells give a value in the range of 3 to 6 amino acids incorporated per ribosome per second, which is several times slower than the prokaryotic elongation rate of 15 to 18 amino acids incorporated per second.[23] With regard to aging, a slowing-down of the elongation phase of protein synthesis has been suggested to be crucial in bringing about the age-related decline in total protein synthesis. This is because a decline of up to 80% in the rate of protein elongation has been reported by estimating the rate of phenylalanyl-tRNA binding to ribosomes in poly(U)-translating cell-free extracts from old *Drosophila*, nematodes and rodent organs.[24] In vivo, a two-fold decrease in the rate of polypeptide chain elongation in old WAG albino rat liver and brain cortex has been reported. Similarly, a decline of 31% in the rate of protein elongation in the livers of male Sprague-Dawley rats has been reported, by measuring the rate of polypeptide chain assembly which was 5.7 amino acids per second in young animals and was 4.5 amino acids per second in old animals.[25] However, these estimates of protein elongation rates have been made for "average" size proteins. It will be important to see if there is differential regulation of protein elongation rates for different proteins during aging.

The elongation of polypeptide chain is mediated by 2 elongation factors, eEF-1 and eEF-2 in eukaryotes (a third factor, EF-3, is reported only in yeast), which are highly conserved during evolution.[21] The activity of eEF-1 declines with age in rat livers and *Drosophila* and the drop parallels the decrease in protein synthesis.[6,20] This decline in the activity of eEF-1 has been correlated only to EF-1A as no changes were observed in the EF-1beta-gamma-mediated activity. Using more specific cell-free stoichiometric and catalytic assays, a 35-45% decrease in the activity and amounts of active eEF-1A has been reported for serially passaged senescent human fibroblasts, old mouse and rat livers and brains.[6,20,26]

In the case of eEF-2 that catalyses the translocation of peptidyl-tRNA on the ribosome during the elongation cycle, conflicting data are available regarding the changes during aging. For example, a lack of difference in the rate of translocation has been observed during the translation of poly(U) by cell-free extracts prepared from young and old *Drosophila* and from rodent organs.[27,28] Similarly, although the proportion of heat-labile eEF-2 increases during aging, the specific activity of eEF-2 purified from old rat and mouse liver remains unchanged.[29] In contrast, a decline of more than 60% in the amount of active eEF-2 has been reported during aging of human fibroblasts in culture, measured by determining the content of diphtheria toxin-mediated ADP-ribosylatable eEF-2 in cell lysates.[30] However, using the same assay, no age-related change in the amount of ADP-ribosylatable eEF-2 was detected in rat livers.[31] Increased fragmentation of eEF-2 due to oxidation has also been reported in old rat livers.[32] Acidic variants of eEF-2 in isolated rat heart and cultured cardiomyocytes have also been reported.[33] Further studies are required to determine if there are any qualitative and quantitative changes in eEF-2 at the levels of transcription, translation and post-translational modifications and how such changes are related with the regulation of protein synthesis during aging.

The cycle of peptide chain elongation continues until one of the three stop codons is reached. There is no aa-tRNA complementary to these codons and instead a termination factor or a release factor (RF) binds to the ribosome and induces the hydrolysis of both the aminoacyl linkage and the GTP, releasing the completed polypeptide chain from the ribosome. Studies in aging *Drosophila* and old rat livers and kidneys have shown that the release of ribosome bound N-formylmethionine, a measure of the rate of translation termination, was not affected with age.[28] Direct estimates of the activity of the termination factor during aging have not been yet made.

Post-Translational Modifications during Aging

Accurate translation of mRNA, followed by appropriate modifications of the polypeptide chain, is essential for its normal folding, targeting and specificity. A misregulation in any of these steps can have far reaching biological consequences, including its effects on cell growth, division and survival. A large number of post-translational modifications of proteins have been described that determine the activity, stability, specificity, transportability and lifespan of a protein. Several of these modifications are highly specific and regulated involving various enzymatic pathways, but there are several nonenzymatic modifications of proteins which occur stochastically but depend on the ternary structures.[34,35]

Phosphorylation

Phosphorylation of serine, threonine and tyrosine residues is one of the best studied modifications of proteins. The coordinated activities of protein kinases, which catalyse phosphorylation and of protein phosphatases which catalyse dephosphorylation, regulate several biological processes, including protein synthesis, cell division, signal transduction, cell growth and development.[36] Altered pattern of protein phosphorylation may be one of the reasons for age-related alterations in protein function and activity and can be a major cause of the failure of homeodynamics and aging. For example, inhibition of DNA synthesis and the loss of proliferative capacity is the ultimate characteristic of normal diploid cells undergoing aging in vitro.[37] Although several putative inhibitors of DNA synthesis have been identified in senescent cells, little is known about the mechanisms of action and the regulation of activity of these inhibitors. It is possible that the activity of several of these inhibitors is regulated by phosphorylation. For example, several studies have shown age-related alterations in cell-cycle-regulated gene expression of various genes such as *c-fos*, *c-jun*, *JunB*, *c-myc*, *c*-Ha-*ras*, *p53*, *cdc2*, *cycA*, *cycB*, *cycD* and retinoblastoma gene *RB1*. Although phosphorylation is involved in regulating the activities of the gene products of almost all these genes, a decrease in phosphorylated cyclin E and Cdk2 and failure to phosphorylate *RB1* gene product $p110^{Rb}$ and *cdc2* product $p34^{cdc2}$ during cellular aging have been reported at present.[38] It will be important to find out if there are age-related alterations in the phosphorylation state of other cell cycle related gene products, proteins involved in DNA and RNA synthesis, including various transcription factors.[39,40]

Various components of the protein synthetic apparatus undergo phosphorylation and dephosphorylation and thus regulate the rates of protein synthesis.[23] For example, phosphorylation of eIF-2 correlates with inhibition of initiation reactions and consequently the inhibition of protein synthesis. Conditions like starvation, heat shock and viral infection, which inhibit the initiation of protein synthesis, induce the phosphorylation of eIF-2 in various cells. Stimuli such as insulin and phorbol esters modulate the phosphorylation of eIF-3, eIF-4B and eIF-4F by activating various protein kinases.[23] Since the activity of eIF-2 has been reported to decrease during aging, it is possible that the phosphorylation status of eIF-2 also changes during aging. However, no studies are available on age-related changes in the phosphorylation pattern of initiation factors.

At the level of protein elongation, the phosphorylation of elongation factors eEF-1A and eEF-2 appears to be involved in regulating their activities. It will be important to see whether the age-related decline in the activities of elongation factors is accompanied by a parallel change in the extent of phosphorylation of these enzymes. Incidentally, it has been reported that there is an increase in the levels of phosphorylated eEF-1 and eEF-2 during mitosis when minimal protein synthesis occurs. Furthermore, there is indirect evidence that alterations in the phosphorylation and dephosphorylation of eEF-2 due to changes in the activities of eEF-2-specific protein kinase III,[41] and PP2A phosphatase[42] may affect the rates of protein synthesis during aging in rat livers.

Phosphorylation also occurs in other proteins that participate in the translational process. For example, the regulatory role of phosphorylation of aa-tRNA synthetase in protein synthesis has been suggested.[43] However, to what extent the decline in the activity and the accumulation of heat-labile aa-tRNA synthetases reported in studies performed on various organs of aging mice and rats is related to their phosphorylation is not known. Furthermore, since the phosphorylation of the S6 ribosomal protein correlates with the activation of protein synthesis, failure to phosphorylate S6 protein in senescent human fibroblasts in response to serum,[44] can be one of the reasons for the decline in the rate of protein synthesis observed during aging.

Pathways of intracellular signal transduction depend on sequential phosphorylation and dephosphorylation of a wide variety of proteins. All phosphorylation reactions result from the action of one or more kinases and the ratio between two interconvertible forms of kinases (active and inactive) acts as a control mechanism for many cellular functions. Studies performed on aging cells have not shown any deficiency in the amount, activity or ability of protein kinase-C (PKC) to elicit signalling pathway.[45] There is also evidence that senescent human fibroblasts retain their ability to phosphorylate proteins in the PKC signal transduction pathway. It appears that the PKCs are largely unaltered in fibroblasts, although the body of information about phosphorylation mechanisms is still very limited.[39]

Growth factor receptors for EGF, FGF, PDGF, insulin, glucocorticoids and several other hormones also possess protein kinase activity. Therefore, deficiencies in the phosphorylation process of receptors would be a logical explanation for the age-related decline of responsiveness to hormonal action and growth stimulation. However, there is no age-related decline in the autophosphorylation activity of various growth factor receptors.[46,47] Similarly, most of the PKC-mediated pathways of intracellular signal transduction in response to various mitogens including phorbol esters appear to remain unaltered in senescent fibroblasts.[46-48] However, a decline in both serine/threonine- and tyrosine-specific protein kinase signals after activation has been observed in the case of T-lymphocytes in aging mice.[49] Similarly, alterations in MAPK phosphorylation have been observed in rodents,[50] while PKC phowsphorylation changes are seen in human pathological aging, including Alzheimer's disease and neurodegenerative processes.[51].

Oxidation

It is often observed that inactive and abnormal proteins accumulate in old cells and tissues. This increased amount of debris in the cytoplasm can be inhibitory for cell growth and normal metabolism and thus contribute towards failure of homeostasis. One of the reasons for the inactivation of enzymes can be their oxidative modification by oxygen free radicals and by mixed-function oxidation (MFO) systems or metal catalyzed oxidation (MCO) systems. Since some amino acid residues, particularly proline, arginine and lysine, are oxidized to carbonyl derivatives, the amount of carbonyl content of proteins has been used as an estimate of protein oxidation during aging.[52-56]

An increase in the levels of oxidatively modified proteins has been reported in old human erythrocytes of higher density and in cultured human fibroblasts from normal old donors and from individuals suffering from progeria and Werner's syndrome.[57] Similarly, there was a two-fold increase in the protein carbonyl content of brain proteins of retired breeder Mongolian gerbils, which was reversed by treatment with the spin-trapping compound N-*tert*-butyl-phenylnitrone.[58]

An age-related increase in the carbonyl content has also been reported in houseflies, fruitflies, nematodes and mouse organs.[59-63]

The loss of activity of 6-phosphogluconate dehydrogenase and liver malic enzyme during aging is related to the loss of lysine and histidine residues by oxidation.[64] Oxidation of a cysteine residue in glyceraldehyde-3-phosphate dehydrogenase may be responsible for its inactivation during aging in rat muscles.[65] It has also been reported that the concentration of the oxidation products of human lens proteins and skin collagen increases along with the accumulation of oxidative forms of alpha-crystallin in patients with age-related cataract.[66] However, the content of *ortho*-tyrosine and dityrosine, formed by the oxidation of phenylalanine and tyrosine, respectively, did not increase in the aging human lens.[67] Structural alterations introduced into proteins by oxidation can lead to aggregation, fragmentation, denaturation and distortion of secondary and tertiary structure, thereby increasing the proteolytic susceptibility of oxidized proteins. Furthermore, toxic products of carbonyl modifications can react with other macromolecules and affect various metabolic processes.

Generally it is thought that there are no systems for the repair of oxidatively damaged proteins and these must be degraded to avoid their accumulation. However, at least one kind of protein repair system has been identified. For example, oxidatively-induced disulphide bridges in cysteine and methionine lead to the formation of sulfenic acid, sulfinic acid and methionine sulphoxide which can accumulate during aging. These damages can be repaired by the methyl-sulfoxide reductase (MSR) system comprising of at least 3 Msr enzymes whose activity is decreased during aging.[68,69] Senile graying of human hair due to increased oxidative stress is accompanied by a reduction of the activities of MSR system.[70] In contrast to this, overexpression of one of the Msr genes increases the lifespan of Drosophila and improves their stress tolerance to oxidative damage-inducing agents.[71]

Glycation

Glycation is one of the most prevalent covalent modifications in which the free amino groups of proteins react with glucose to form a ketoamine called Amadori product. This is followed by a sequence of further reactions and rearrangements producing the so-called advanced glycosylation end products (AGEs).[72,73] Most commonly, it is the long-lived structural proteins such as lens crystallins, collagen and basement membrane proteins which are more susceptible to glycation. In the case of skin, vimentin has been found to be as the main protein becoming glycated during aging.[74] The glycated proteins are then more prone to form crosslinks with other proteins, leading to structural and functional alterations.[5]

An increase in the levels of glycated proteins during aging has been observed in a wide variety of systems. For example, there is an increase in the level of glycated lysine residues of rat sciatic nerve, aorta and skin collagen during aging.[75] There is an increase in the glycation of human collagen and osteocalcin during aging.[76] The formation and the accumulation of the AGEs are implicated in the physiology and pathology of senescence. It has been observed that pentosidine (cross-linked glycated lysine and arginine), carboxylmethyllysine (CML, glycated and oxidated proteins) and pyrroline increase with age in normal and diabetic humans.[77] By using AGE-specific antibodies, an AGE-modified form of human hemoglobin has been identified, whose levels increase during aging and in patients with diabetes-induced hyperglycemia.[78]

Deamidation, Racemization and Isomerization

Age-related changes in the catalytic activity, heat stability, affinity for substrate and other physical characteristics, such as the conformation of proteins may also be due to the charge change introduced by conversion of a neutral amide group to an acidic group by deamidation. Spontaneous deamidation of asparaginyl and glutaminyl residues of several proteins has been related with the observed accumulation of their inactive and heat labile isoforms during aging.[65] The sequential deamidation of two asparagine residues of triphosphate isomerase is responsible for the differences of the isoenzymes present in aging cells and tissues, such as bovine eye lens and human skin fibroblasts from old donors and patients with progeria and Werner's syndrome.[79] Deamidation of glucose-6-phosphate isomerase produces the variant of the enzyme that accumulates in aging bovine lenses.[80]

The interconversion of optical isoforms of amino acids, called racemization, has been reported to increase during aging. For example, the concentration of D-aspartate in protein hydrolysates from human teeth, erythrocytes and eye lens increases with age.[81] Racemization of tyrosine has been reported to occur in the aging brunescent human cataract lenses.[82] The spontaneous prolyl *cis-trans* isomerization in proteins that may cause some of the so-called spontaneous conformational changes has been implicated in the age-related decline in the activity of certain enzymes. However, no definitive examples of enzymes undergoing this kind of post-translational modification during aging are available. It is also not known to what extent the conformational changes associated with old rat muscle phosphoglycerate kinase, enolase and other enzymes are associated with racemization and isomerization.

ADP-Ribosylation

The structure and function of many proteins such as nuclear proteins topoisomerase I, DNA ligase II, endonuclease, histones H1, H2B and H4, DNA polymerases and cytoplasmic proteins adenyl cyclase and elongation factor eEF-2 is modulated by ADP-ribosylation. ADP-ribosylation of proteins is involved in various cellular processes such as maintenance of chromatin structure, DNA repair, protein synthesis, cell differentiation and cell transformation.[83,84]

Indirect evidence suggests that poly-ADP-ribosylation of proteins may decrease during aging because the activity of poly(ADP-ribose)polymerase (PARP) decreases in aging human fibroblasts both as a function of donor age and during serial passaging in vitro.[85] Similarly, the direct relationship observed between maximum lifespan of a species and the activity of PARP in mononuclear leukocytes of 13 mammalian species indicates its important role in aging and longevity.[83,86]

One cytoplasmic protein that can be specifically ribosylated by diphtheria toxin and exotoxin A is the protein elongation factor eEF-2. ADP-ribosylation of the diphthamide (modified histidine 715) residue of eEF-2 results in the complete abolition of its catalytic activity.[21] There is evidence that increased ADP-ribosylation of eEF-2 is correlated with cellular aging. For example, the amount of eEF-2 that can be ADP-ribosylated in the presence of diphtheria toxin in cell-free extracts decreases significantly during aging of human fibroblasts in culture.[30,32]

Methylation

Methylation of nitrogens of arginine, lysine and histidine and carboxyls of glutamate and aspartate residues is a widely observed post-translational modification that is involved in many cellular functions. Although most of our present understanding regarding the significance of protein methylation has come from studies on bacterial chemotaxis, muscle contraction, electron transport, processing of pituitary hormones and gene expression, its role in aging is beginning to emerge.[87]

Proteins whose activities are increased by methylation include alcohol dehydrogenase, histones, ribosomal proteins, cytochrome C, eEF-1A, myosin, myelin and rhodopsin. Of these, decreased methylation of histones has been reported in livers and brains of aging rats. On the other hand, there is no difference in the extent of methylation of newly synthesized histones during cellular aging of human fibroblasts in culture. Studies on the levels of methylated histidine, arginine and lysine of myosin isolated from the leg muscles of aging rats, mice and hamsters showed unchanged levels of histidine, decreased levels of arginine and trimethyllysine and increased levels of monomethyllysine.[6,88]

During the aging of erythrocytes, there is an increase in the number of methyl groups per molecule of band 2.1 (ankyrin) and band 3 protein, which correlates with increased membrane rigidity of erythrocytes during aging.[89] Similarly, there is a several fold increase in the number of methyl acceptor proteins in the eye lenses from aged humans and persons suffering from cataract.[90] The number of carboxylmethylatable sites of cerebral membrane-bound proteins also increases in rat brain during aging.[91] At present, age-related changes in the methylation of other proteins such as ribosomal proteins, calmodulin, cytochrome C and myosin have not been studied. It is clear that protein methylation is involved in diverse functions including protein synthesis and turnover and that it should be studied thoroughly in relation to the process of aging.

Proteolytic Processing

Many newly synthesised proteins undergo post-translational proteolytic processing by which certain conformational restraint on the inactive precursor is released and a biologically active protein is generated. Several inactive precursors of enzymes called zymogens, precursors of growth factors, peptide and protein hormones such as insulin, precursors of extracellular matrix and many other secretory proteins including various proteases such as collagenase, undergo proteolytic processing. There are no systematic studies performed on age-related changes in post-translational proteolytic processing of any proteins. However, there is some evidence that alterations in proteolytic processing may be one of the reasons for the appearance or disappearance of certain proteins during aging. For example, the appearance of the "senescent cell antigen" on the surface of a wide variety of aging cells is derived from the proteolysis of band 3 protein.[92] The exposure of senescent cell-specific epitopes on fibronectin,[93] may also be due to altered proteolytic processing. Progressive proteolysis of a 90 kDa protein, Tp-90 terminin , into Tp-60 and Tp-30 terminin in senescent cells and in cells committed to apoptosis has been reported.[94]

Altered proteolytic cleavage of the beta-amyloid precursor protein is well known to play an important role in the pathogenesis of Alzheimer's disease.[95,96] Increased proteolysis of a conformationally more labile single-chain form of the lysosomal protease cathepsin B has been suggested as a reason for the age-related decline in its activity during aging of human fibroblasts.[97] Similarly, alterations in the activity of collagenase during aging of human fibroblasts has been suggested to be due to structural and catalytic changes.[98]

Other Modifications

In addition to the types of post-translational modifications mentioned above, there are some other modifications that determine the structure and function of various proteins and may have a role to play during aging. For example, the incorporation of ethanolamine into protein elongation factor eEF-1A may be involved in determining its stability and interaction with intracellular membranes.[99] Whether this modification has any role in the regulation of the activity of eEF-1A during aging is not known at present. Similarly, the protein initiation factor eIF-5A contains an unusual amino acid, hypusine, which is synthesized post-translationally as a result of a series of enzymatically catalysed alterations of a lysine residue.[100] Since the absence of hypusine in eIF-5A blocks the initiation of protein synthesis, reduced activity of a hypusine synthase enzyme has been reported in senescent human fibroblasts.[17]

Detyrosination of microtubules affecting the cytoskeletal organization and many other cellular functions, may also be important during aging. Furthermore, the roles of chaperones in protein folding and conformational organization are yet to be studied in relation to the aging process. There is some evidence that both the pentose-mediated protein crosslinking and transglutaminase-mediated crosslinking of proteins is involved in aging. For example, there is a high correlation between pentosidine protein crosslinks and pigmentation in senescent and cataract affected human lens.[101] Similarly, an increase in transglutaminase activity during cellular apoptosis, differentiation and aging of human keratinocytes indicates an important role of this modification in the process of aging.[102]

Protein tyrosine sulfation is another post-translational modification that may have significance in protein alteration during aging because it is involved in determining the biological activity of neuropeptides and the intracellular transportation of a secretory protein.[103] Similarly, prenylation, the covalent attachment of isoprenoid lipids on cysteine-rich proteins, is involved in the regulation of the activity of some proto-oncogenic ras proteins and the nuclear lamins A and B.[104] These studies have indicated a critical role for prenylation in the regulation of oncogenesis, nuclear structure, signal transduction and cell cycle progression, functions very much related with the causative aspects of aging. There is an age-dependent decrease in the activity of prenyltransferases in the rat liver, which may account for the changes in the synthesis and turnover of mevalonate pathway lipids, including cholestrol, ubiquinone and dolichol.[105]

Protein Turnover during Aging

Efficient macromolecular turnover is integral to the normal function and survival of a biological system. Although there are large variations in the rates of degradation of individual proteins, it is generally observed that overall protein turnover slows down during aging.[1,20] The physiological consequences of decreased protein turnover include the accumulation of altered and abnormal proteins in the cell, an altered pattern of post-translational modifications due to increased dwell time and a disruption of the organisation of the cytoskeleton and extracellular matrix.[5]

Age-related decline in protein turnover is generally due to a decrease in the proteolytic activity of various lysosomal and cytoplasmic proteases. Molecular details of various pathways of protein degradation, such as the proteasome-mediated, ubiquitin-mediated and the lysosome-mediated pathways, are now being studied in relation to aging.[106-108] For example, ubiquitin marking of proteins for degradation and ubiquitin-mediated proteolysis did not decline in aging human fibroblasts and no change in the levels of ubiquitin mRNA and ubiquitin pools was detected.[109] Significantly reduced proteasomal activities during aging, with or without a parallel decrease in the amount of proteasomal components, have been reported for human fibroblasts, keratinocytes and other systems.[69,110,111] The exact reasons for the age-related loss of proteasomal activities are still not known and may include the accumulation of abnormal proteasomal subunits, damaged by oxidative and glycoxidative pathways.[112] Similarly, reduced lysosomal activity may be due to altered pH and overcrowding of the lysosomes with abnormal and damaged macromolecules such as age-pigments or lipofuscin.[113,114]

Other reasons for age-related changes in the activities of various proteases leading to a decrease in the rate of protein turnover include slower transcription, reduced rates of (protein) synthesis and altered pattern of postsynthetic modifications, as discussed above. Furthermore, there is evidence that certain inhibitors of various proteases, such as tissue inhibitor of metalloproteinases,[115] and trypsin inhibitor,[116] had increased levels of expression and activity during aging of human fibroblasts. This results in a decrease in protease activity, thus leading to decreased protein degradation during aging.

Conclusion

The synthesis, modifications and turnover of proteins are interdependent processes that practically set a limit on the efficiency of genetic information transfer from coded molecules to functional molecules. Therefore, analysing the synthesis of proteins, their modifications, which determine their activity, stability and specificity and turnover of inactive and altered proteins is central to understanding aging. Only a complete understanding of the proteomic spectrum and its functional implications with respect to normal aging can facilitate the development of effective means of intervention, prevention and therapy of aging and age-related diseases.[117,118]

References

1. Rattan SIS. Synthesis, modifications and turnover of proteins during aging. Exp Gerontol 1996; 31:33-47.
2. Rattan SIS. Transcriptional and translational dysregulation during aging. In: von Zglinicki T, ed. Aging at the Molecular Level. Dordrecht: Kluwer Acad Publ, 2003:179-91.
3. Abbott CM, Proud CG. Translational factors: in sickness and in health. Trends Biochem Sci 2004; 29:25-31.
4. Holliday R. The current status of the protein error theory of aging. Exp Gerontol 1996; 31:449-52.
5. Hipkiss A. Accumulation of altered proteins and ageing: causes and effects. Exp Gerontol 2006; 41:464-73.
6. Rattan SIS. Translation and post-translational modifications during aging. In: Macieira-Coelho A, ed. Molecular Basis of Aging. Boca Raton, Florida: CRC Press, 1995:389-420.
7. Luce MC, Bunn CL. Altered sensitivity of protein synthesis to paromomycin in extracts from aging human diploid fibroblasts. Exp Gerontol 1987; 22:165-77.
8. Luce MC, Bunn CL. Decreased accuracy of protein synthesis in extracts from aging human diploid fibroblasts. Exp Gerontol 1989; 24:113-25.
9. Holliday R, Rattan SIS. Evidence that paromomycin induces premature ageing in human fibroblasts. Monogr Devl Biol 1984; 17:221-33.

10. Buchanan JH, Stevens A, Sidhu J. Aminoglycoside antibiotic treatment of human fibroblasts: intracellular accumulation, molecular changes and the loss of ribosomal accuracy. Eur J Cell Biol 1987; 43:141-7.
11. Nyström T. Translational fidelity, protein oxidation and senescence: lessons from bacteria. Ageing Res Rev 2002; 1:693-703.
12. Nyström T. Aging in bacteria. Curr Opin Microbiol 2002; 5:596-601.
13. Silar P, Rossignol M, Haedens V et al. Deletion and dosage modulation of the eEF1A gene in Podospora anserina: effect on the life cycle. Biogerontology 2000; 1:47-54.
14. Holbrook MA, Menninger JR. Erythromycin slows aging of Saccharomyces cerevisiae. J Gerontol Biol Sci 2002; 57A:B29-B36.
15. Dever TE. Translation initiation: adept at adapting. TIBS 1999; 24:398-403.
16. Hershey JWB, Merrick WC. The pathway and mechanism of initiation of protein synthesis. In: Sonenberg N, Hershey JWB, Mathews MB, eds. Translational Control of Gene Expression. New York: Cold Spring Harbor Laboratory Press, 2000:33-88.
17. Chen ZP, Chen KY. Dramatic attenuation of hypusine formation on eukaryotic initiation factor 5A during senescence of IMR-90 human diploid fibroblasts. J Cell Physiol 1997; 170:248-54.
18. Saini P, Eyler DE, Green R et al. Hypusine-containing protein eIF5A promotes translation elongation. Nature 2009; 459:118-21.
19. Ward W, Richardson A. Effect of age on liver protein synthesis and degradation. Hepatol 1991; 14:935-48.
20. Van Remmen H, Ward WF, Sabia RV et al. Gene expression and protein degradation. In: Masoro E, editor. Handbook of Physiology: Aging. Oxford University Press, 1995:171-234.
21. Riis B, Rattan SIS, Clark BFC et al. Eukaryotic protein elongation factors. TIBS 1990; 15:420-4.
22. Andersen GR, Nissen P, Nyborg J. Elongation factors in protein biosynthesis. Trends Biochem Sci 2003; 28:434-41.
23. Merrick WC. Mechanism and regulation of eukaryotic protein synthesis. Microbiol Rev 1992; 56:291-315.
24. Richardson A, Semsei I. Effect of aging on translation and transcription. Rev Biol Res Aging 1987; 3:467-83.
25. Merry BJ, Holehan AM. Effect of age and restricted feeding on polypeptide chain assembly kinetics in liver protein synthesis in vivo. Mech Ageing Develop 1991; 58:139-50.
26. Rattan SIS. Regulation of protein synthesis during ageing. Eur J Gerontol 1992; 1:128-36.
27. Webster GC. Protein synthesis in aging organisms. In: Sohal RS, Birnbaum LS, Cutler RG, editors. Molecular Biology of Aging: Gene Stability and Gene Expression. New York: Raven Press; 1985:263-89.
28. Webster GC. Effect of aging on the components of the protein synthesis system. In: Collatz KG, Sohal RS, eds. Insect Aging. Berlin: Springer-Verlag, 1986:207-16.
29. Takahashi R, Mori N, Goto S. Accumulation of heat-labile elongation factor 2 in the liver of mice and rats. Exp Gerontol 1985; 20:325-31.
30. Riis B, Rattan SIS, Derventzi A et al. Reduced levels of ADP-ribosylatable elongation factor-2 in aged and SV40-transformed human cells. FEBS Lett 1990; 266:45-7.
31. Rattan SIS, Ward WF, Glenting M et al. Dietary calorie restriction does not affect the levels of protein elongation factors in rat livers during ageing. Mech Ageing Develop 1991; 58:85-91.
32. Parrado J, Bougria M, Ayala A et al. Effects of aging on the various steps of protein synthesis: fragmentation of elongation factor 2. Free Rad Biol Med 1999; 26:362-70.
33. Jäger M, Holtz J, Redpath NT et al. The ageing heart: influence of cellular and tissue ageing on total content and distribution of the variants of elongation factor-2. Mech Ageing Dev 2002; 123:1305-19.
34. Soskic V, Groebe K, Schrattenholz A. Nonenzymatic post-translational protein modifications in ageing. Exp Gerontol 2008; 43:247-57.
35. Stefani M. Protein folding and misfolding, relevance to disease and biological function. In: Smith HJ, Simons C, Seewell RDE, eds. Protein Misfolding in Neurodegenerative Diseases: Mechanisms and Therapeutic Strategies. Boca Raton: CRC Press, 2008:2-66.
36. Dephoure N, Zhou C, Villén J et al. A quantitative atlas of mitotic phosphorylation. Proc Natl Acad Sci USA 2008; 105:10762-7.
37. Rattan SIS. Cellular senescence in vitro. Encyclopedia of Life Sciences 2008; doi:10.1002/9780470015902.a0002567.pub2.
38. Stein GH, Dulic V. Origins of G1 arrest in senescent human fibroblasts. BioEssays 1995; 17:537-43.
39. Tresini M, Lorenzini A, Torres C et al. Modulation of replicative senescence of diploid human cells by nuclear ERK signaling. J Biol Chem 2007; 282:4136-51.
40. Sedding DG. FoxO transcription factors in oxidative stress response and ageing—a new fork on the way to longevity? Biol Chem 2008; 389:279-83.

41. Riis B, Rattan SIS, Palmquist K et al. Elongation factor 2-specific calcium and calmodulin dependent protein kinase III activity in rat livers varies with age and calorie restriction. Biochem Biophys Res Commun 1993; 192:1210-6.
42. Riis B, Rattan SIS, Palmquist K et al. Dephosphorylation of the phosphorylated elongation factor-2 in the livers of calorie-restricted and freely-fed rats during ageing. Biochem Mol Biol Int 1995; 35:855-9.
43. Meinnel T, Mechulam Y, Blanquet S. Aminoacyl-tRNA synthetases: occurrence, structure and function. In: Söll D, RajBhandary UL, eds. tRNA: Structure, Biosynthesis and Function. Washington D.C.: ASM Press, 1995:251-92.
44. Kihara F, Ninomyia-Tsuji J, Ishibashi S et al. Failure in S6 protein phosphorylation by serum stimulation of senescent human diploid fibroblasts, TIG-1. Mech Ageing Dev 1986; 20:305-13.
45. Blumenthal EJ, Miller ACK, Stein GH et al. Serine/threonine protein kinases and calcium-dependent protease in senescent IMR-90 fibroblasts. Mech Ageing Dev 1993; 72:13-24.
46. De Tata V, Ptasznik A, Cristofalo VJ. Effect of tumor promoter phorbol 12-myristate 13-acetate (PMA) on proliferation of young and senescent WI-38 human diploid fibroblasts. Exp Cell Res 1993; 205:261-9.
47. Farber A, Chang C, Sell C et al. Failure of senescent human fibroblasts to express the insulin-like growth factor-1 gene. J Biol Chem 1993; 268:17883-8.
48. Derventzi A, Rattan SIS, Clark BFC. Phorbol ester PMA stimulates protein synthesis and increases the levels of active elongation factors EF-1a and EF-2 in ageing human fibroblasts. Mech Ageing Dev 1993; 69:193-205.
49. Miller RA. Aging and immune function: cellular and biochemical analyses. Exp Gerontol 1994; 29:21-35.
50. Pardo VG, Facchinetti MM, Curino A et al. Age-related alteration of 1alpha,25(OH)(2)-vitamin D (3)-dependent activation of p38 MAPK in rat intestinal cells. Biogerontology 2007; 8:13-24.
51. Battaini F, Govoni S, Trabucchi M. Protein kinase C signal transmission during aging. In: Macieira-Coelho A, ed. Molecular Basis of Aging. Boca Raton: CRC Press, 1995:269-91.
52. Levine RL. Carbonyl modified proteins in cellular regulation, aging and disease. Free Rad Biol Med 2002; 32:790-6.
53. Dukan S, Farewell A, Ballesteros M et al. Protein oxidation in response to increased transcriptional or translational errors. Proc Natl Acad Sci USA 2000; 97:5746-9.
54. Grune T. Oxidative stress, aging and the proteasomal system. Biogerontology 2000; 1:31-40.
55. Cloos PA, Christgau S. Post-translational modifications of proteins: implications for aging, antigen recognition and autoimmunity. Biogerontology 2004; 5:139-58.
56. Perez VI, Buffenstein R, Masamsetti V et al. Protein stability and resistance to oxidative stress are determinants of longevity in the longest-living rodent, the naked mole-rat. Proc Natl Acad Sci USA 2009.
57. Verbeke P, Clark BFC, Rattan SIS. Reduced levels of oxidized and glycoxidized proteins in human fibroblasts exposed to repeated mild heat shock during serial passaging in vitro. Free Rad Biol Med 2001; 31:1593-602.
58. Carney JM, Starke-Reed PE, Oliver CN et al. Reversal of age-related increase in brain protein oxidation, decrease in enzyme activity and loss in temporal and spatial memory by chronic administration of the spin-trapping compound N-tert-butyl-alpha-phenylnitrone. Proc Natl Acad Sci USA 1991; 88:3633-6.
59. Sohal RS, Agarwal S, Dubey A et al. Protein oxidative damage is associated with life expectancy of houseflies. Proc Natl Acad Sci USA 1993; 90:7255-9.
60. Sohal RS, Ku H-H, Agarwal S. Biochemical correlates of longevity in two closely related rodent species. Biochem Biophys Res Commun 1993; 196:7-11.
61. Beal MF. Oxidatively modified proteins in aging and disease. Free Rad Biol Med 2002; 32:797-803.
62. Goto S, Nakamura A, Radak Z et al. Carbonylated proteins in aging and excercise: immunoblot approaches. Mech Ageing Dev 1999; 107:245-53.
63. Yasuda K, Adachi H, Fujiwara Y et al. Protein carbonyl accumulation in aging dauer formation-defective (daf) mutants of Caenorhabditis elegans. J Gerontol Biol Sci 1999; 54A:B47-B51.
64. Gordillo E, Ayala A, Bautista J et al. Implication of lysine residues in the loss of enzymatic activity in rat liver 6-phosphogluconate dehydrogenase found in aging. J Biol Chem 1989; 264:17024-8.
65. Gafni A. Age-related effects in enzyme metabolism and catalysis. Rev Biol Res Aging 1990; 4:315-36.
66. Stadtman ER. Protein oxidation and aging. Science 1992; 257:1220-4.
67. Wells-Knecht MC, Huggins TG, Dyer DG et al. Oxidized amino acids in lens protein with age. Measurement of o-tyrosine and dityrosine in the aging human lens. J Biol Chem 1993; 268:12348-52.
68. Mary J, Vougier S, Picot CR et al. Enzymatic reactions involved in the repair of oxidized proteins. Exp Gerontol 2004; 39:1117-23.

69. Petropoulos I, Conconi M, Wang X et al. Increase of oxidatively modified protein is associated with a decrease of proteasome activity and content in aging epidermal cells. J Gerontol Biol Sci 2000; 55A:B220-B7.
70. Wood JM, Decker H, Hartmann H et al. Senile hair graying: H_2O_2-mediated oxidative stress affects human hair color by blunting methionine sulfoxide repair. FASEB J 2009.
71. Ruan H, Tang XD, Chen ML et al. High-quality life extension by the enzyme peptide methionine sulfoxide reductase. Proc Natl Acad Sci USA. 2002; 99:2748-53.
72. Meli M, Frey J, Perier C. Native protein glycoxidation and aging. J Nutr Health Aging 2003; 7:263-6.
73. Ramasamy R, Yan SF, Schmidt AM. Methylglyoxal comes of AGE. Cell 2006; 124:258-60.
74. Kueper T, Grune T, Prahl S et al. Vimentin is the specific target in skin glycation. Structural prerequisites, functional consequences and role in skin aging. J Biol Chem 2007; 282:23427-36.
75. Oimomi M, Maeda Y, Hata F et al. A study of the age-related acceleration of glycation of tissue proteins in rats. J Gerontol 1988; 43:B98-101.
76. Miksík I, Deyl Z. Changes in the amount of e-hexosyllysine, UV absorbance and fluorescence of collagen with age in different animal species. J Gerontol 1991; 46:B111-6.
77. Lee AT, Cerami A. Role of glycation in aging. Ann NY Acad Sci 1992; 663:63-70.
78. Makita Z, Vlassara H, Rayfield E et al. Hemoglobin-AGE: a circulating marker of advanced glycosylation. Science 1992; 258:651-3.
79. Gracy RW, Yüksel KÜ, Chapman ML et al. Impaired protein degradation may account for the accumulation of "abnormal" proteins in aging cells. In: Adelman RC, Dekker EE, eds. Modifications of Proteins during Aging. New York: Alan R. Liss, 1985:1-18.
80. Cini JK, Gracy RW. Molecular basis of the isozyme of bovine glucose-6-phosphate isomerase. Arch Biochem Biophys 1986; 249:500-5.
81. Brunauer LS, Clarke S. Age-dependent accumulation of protein residues which can be hydrolyzed to D-aspartic acid in human erythrocytes. J Biol Chem 1986; 261:12538-43.
82. Luthra M, Ranganathan D, Ranganathan S et al. Racemization of tyrosine in the insoluble protein fraction of brunescent aging human lenses. J Biol Chem 1994; 269:22678-82.
83. Beneke S, Alvarez-Gonzalez R, Bürkle A. Comparative characterization of poly(ADP-ribose) polymerase-1 from two mammalian species with different life span. Exp Gerontol 2000; 35:989-1002.
84. Bürkle A. Physiology and pathophysiology of poly(ADP-ribosyl)ation. BioEssays 2001; 23:795-806.
85. Dell'Orco RT, Anderson LE. Decline of poly(ADP-ribosyl)ation during in vitro senescence in human diploid fibroblasts. J Cell Physiol 1991; 146:216-21.
86. Grube K, Bürkle A. Poly(ADP-ribose) polymerase activity in mononuclear leukocytes of 13 mammalian species correlates with species-specific life span. Proc Natl Acad Sci USA 1992; 89:11759-63.
87. McBride AE, Silver PA. State of the Arg: protein methylation at arginine comes of age. Cell 2001; 106:5-8.
88. Rattan SIS, Derventzi A, Clark BFC. Protein synthesis, post-translational modifications and aging. Ann NY Acad Sci 1992; 663:48-62.
89. Mays-Hoopes LL. Macromolecular methylation during aging. Rev Biol Res Aging 1985; 2:361-93.
90. McFadden PN, Clarke S. Protein carboxyl methyltransferase and methyl acceptor proteins in aging and cataractus tissue of the human eye lens. Mech Ageing Develop 1986; 34:91-105.
91. Sellinger OZ, Kramer CM, Conger A et al. The carboxylmethylation of cerebral membrane-bound proteins increases with age. Mech Ageing Develop 1988; 43:161-73.
92. Kay MMB. Molecular aging of membrane molecules and cellular removal. In: Goldstein AL, ed. Biomedical Advances in Aging. New York: Plenum Press, 1990:147-61.
93. Porter MB, Pereira-Smith OM, Smith JR. Common senescent cell-specific antibody epitopes on fibronectin in species and cells of varied origin. J Cell Physiol 1992; 150:545-51.
94. Hébert L, Pandey S, Wang E. Commitment to cell death is signaled by the appearance of a terminin protein of 30 kDa. Exp Cell Res 1994; 210:10-8.
95. Selkoe DJ. Aging brain, aging mind. Sci Amer 1992; 267:135-42.
96. Esler WP, Wolfe MS. A portrait of Alzheimer secretases—new features and familiar faces. Science 2001; 293:1449-54.
97. DiPaolo BR, Pignolo RJ, Cristofalo VJ. Overexpression of the two-chain form of cathepsin B in senescent WI-38 cells. Exp Cell Res 1992; 201:500-5.
98. Sottile J, Mann DM, Diemer V et al. Regulation of collagenase and collagenase mRNA production in early- and late-passage human diploid fibroblasts. J Cell Physiol 1989; 138:281-90.
99. Whiteheart SW, Shenbagamurthi P, Chen L et al. Murine elongation factor 1a (EF-1a) is post-translationally modified by novel amide-linked ethanolamine-phosphoglycerol moieties. J Biol Chem 1989; 264:14334-41.

100. Park MH, Wolff EC, Folk JE. Is hypusine essential for eukaryotic cell proliferation? TIBS 1993; 18:475-9.
101. Nagaraj RH, Sell DR, Prabhakaram M et al. High correlation between pentosidine protein crosslinks and pigmentation implicates ascorbate oxidation in human lens senescence and cataractogenesis. Proc Natl Acad Sci USA 1991; 88:10257-61.
102. Norsgaard H, Clark BFC, Rattan SIS. Distinction between differentiation and senescence and the absence of increased apoptosis in human keratinocytes undergoing cellular aging in vitro. Exp Gerontol 1996; 31:563-70.
103. Huttner WB. Protein tyrosine sulfation. TIBS 1987; 12:361-3.
104. Marshall CJ. Protein prenylation: a mediator of protein-protein interactions. Science 1993; 259:1865-6.
105. Thelin A, Runquist M, Ericsson J et al. Age-dependent changes in rat liver prenyltransferases. Mech Ageing Dev 1994; 76:165-76.
106. Merker K, Grune T. Proteolysis of oxidised proteins and cellular senescence. Exp Gerontol 2000; 35:779-86.
107. Gaczynska M, Osmulski PA, Ward WF. Caretaker or undertaker? The role of the proteasome in aging. Mech Ageing Dev 2001; 122:235-54.
108. Shringaarpure R, Davies KJA. Protein turnover by the proteasome in aging and disease. Free Rad Biol Med 2002; 32:1084-9.
109. Pan J-X, Short SR, Goff SA et al. Ubiquitin pools, ubiquitin mRNA levels and ubiquitin-mediated proteolysis in aging human fibroblasts. Exp Gerontol 1993; 28:39-49.
110. Bulteau AL, Petropoulos I, Friguet B. Age-related alterations of proteasome structure and function in aging epidermis. Exp Gerontol 2000; 35:767-77.
111. Brégégére F, Milner Y, Friguet B. The ubiquitin-proteasome system at the crossroads of stress-response and ageing pathways: a handle for skin care? Aging Res Rev 2006; 5:60-90.
112. Carrard G, Bulteau AL, Petropoulos I et al. Impairment of proteasome structure and function in aging. Int J Biochem Cell Biol 2002; 34:1461-74.
113. Terman A, Kurz T, Gustafsson B et al. Lysosomal labilization. IUBMB LIfe 2006; 58:531-9.
114. Terman A, Gustafsson B, Brunk UT. Autophagy, organelles and ageing. J Pathol 2007; 211:134-43.
115. Wick M, Bürger C, Brüsselbach S et al. A novel member of human tissue inhibitor of metalloproteinases (TIMP) gene family is regulated during G1 progression, mitogenic stimulation, differentiation and senescence. J Biol Chem 1994; 269:18953-60.
116. Hearn MG, Edland SD, Ogburn CE et al. Trypsin inhibitor activities of fibroblasts increase with age of donor and are unaltered in familial Alzheimer's disease. Exp Gerontol 1994; 29:611-23.
117. Rattan SIS. Theories of biological aging: genes, proteins and free radicals. Free Rad Res 2006; 40:1230-8.
118. Rattan SIS. Increased molecular damage and heterogeneity as the basis of aging. Biol Chem 2008; 389:267-72.

Chapter 2

Regulation of mRNA Translation as a Conserved Mechanism of Longevity Control

Ranjana Mehta, Devon Chandler-Brown, Fresnida J. Ramos, Lara S. Shamieh and Matt Kaeberlein*

Abstract

Appropriate regulation of mRNA translation is essential for growth and survival and the pathways that regulate mRNA translation have been highly conserved throughout eukaryotic evolution. Translation is controlled by a complex set of mechanisms acting at multiple levels, ranging from global protein synthesis to individual mRNAs. Recently, several mutations that perturb regulation of mRNA translation have also been found to increase longevity in three model organisms: the budding yeast *Saccharomyces cerevisiae*, the nematode *Caenorhabditis elegans* and the fruit fly *Drosophila melanogaster*. Many of these translation control factors can be mapped to a single pathway downstream of the nutrient responsive target of rapamycin (TOR) kinase. In this chapter, we will review the data suggesting that mRNA translation is an evolutionarily conserved modifier of longevity and discuss potential mechanisms by which mRNA translation could influence aging and age-associated disease in different species.

Introduction

Aging is influenced by a complex interaction between genes and environment. Many different types of damage have been suggested to play a causal role in aging, including oxidation of macromolecules (DNA, RNA, lipids and proteins), cell senescence resulting from telomere shortening, apoptotic cell death, advanced glycation end-products and protein misfolding/aggregation.[1-7] Despite many decades of study, however, the relative contribution of each of these processes to aging and how this translates into the multitude of age-associated phenotypes seen in people remains largely unknown.

Many of the primary advances in our understanding of basic mechanisms of aging over the past several years have come from studies in invertebrate organisms. Three organisms, in particular, have been particularly informative regarding the genetic and environmental determinants of longevity: the budding yeast *Saccharomyces cerevisiae*, the nematode *Caenorhabditis elegans* and the fruit fly *Drosophila melanogaster*. Studies from these organisms have demonstrated the existence of at least a handful of "public" mechanisms of longevity control—presumably related to aspects of aging that are conserved across evolutionarily divergent species.[8] These include the longevity modifying functions of insulin-like signaling (ILS) pathways in nematodes and flies, sirtuin protein deacetylases in all three organisms and the target of rapamycin (TOR) kinases in all three organisms.

*Corresponding Author: Matt Kaeberlein—Department of Pathology, University of Washington, Seattle, Washington USA. Email: kaeber@u.washington.edu

Protein Metabolism and Homeostasis in Aging, edited by Nektarios Tavernarakis.

The best-characterized "public" mechanism of longevity control is dietary restriction (DR). DR can be defined as a reduction in nutrient availability in the absence of malnutrition and has been observed to increase life span in a variety of organisms, including yeast, nematodes, flies, mice and rats.[9,10] Although it remains unclear whether DR also promotes longevity in humans, initial indications are that DR improves health and survival in primates.[11-13a] The genetic pathway(s) by which DR slows aging has been studied extensively in invertebrate organisms. Data from both yeast and nematodes suggest an important role for TOR signaling downstream of DR in these organisms.[14,15] TOR activity is reduced by DR, a reduction in TOR activity is sufficient to increase life span and epistasis studies have placed TOR and DR in the same pathway with respect to longevity.

TOR signaling modulates multiple cellular processes, including protein degradation by autophagy, mitochondrial metabolism and mRNA translation.[16] An important question, therefore, is which of these TOR regulated processes are most relevant for longevity control in response to nutrient and growth factor abundance. It has been demonstrated that autophagy is necessary for life span extension from DR or reduced TOR signaling in worms; however, induction of autophagy does not appear to be sufficient to increase life span on its own.[17-19] In contrast, mutations that modulate mRNA translation have been found to be sufficient to increase life span in yeast, nematodes and flies, raising the possibility that DR promotes longevity, at least in part, via a conserved longevity pathway involving TOR and TOR-regulated translation factors (Fig. 1).[20]

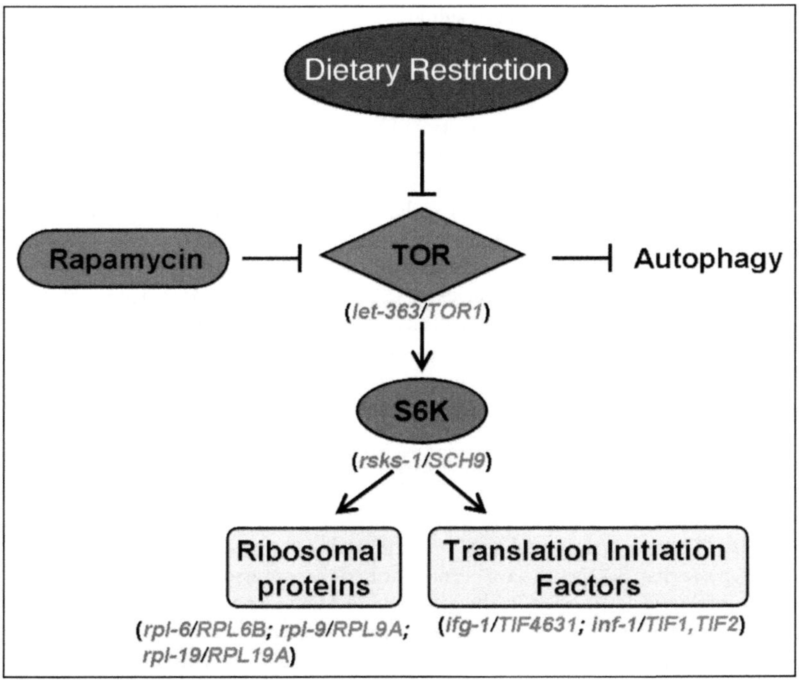

Figure 1. A conserved longevity pathway involving TOR signaling and protein translation. Components of the TOR signaling pathway have been shown to modulate longevity in yeast, nematodes and flies and may represent a mechanism for lifespan extension in response to dietary restriction. Many downstream targets of TOR signaling, including ribosomal proteins and translation initiation factors, have a conserved role in life span determination. Conserved aging genes highlight the role of the TOR signaling pathway in modulating longevity. Homolog gene pairs that modulate aging in both yeast and nematodes are shown in parentheses (nematode genes are on the left and yeast genes are on the right). All enclosed factors are known to affect aging in invertebrate organisms.

This hypothesis and the evidence supporting an important role for mRNA translation in aging across evolutionarily divergent species are discussed in detail below.

Genome Scale Longevity Screens in Yeast and Nematodes

With the sequencing of the complete genome of *S. cerevisiae*, it was possible to create a set of comprehensive single gene deletion mutants that span every known open reading frame in the yeast genome. The yeast ORF deletion collection contains haploid *MAT***a** and *MAT*α mating type strains, as well as homozygous and heterozygous diploids, in an isogenic background.[21] Two distinct aging paradigms have been developed in yeast, referred to as replicative and chronological aging and the deletion collection has been screened to identify single-gene deletions that increase life span in both aging models.[22,23]

Replicative life span (RLS) measures age as the number of mitotic divisions a single cell can undergo.[24] *Saccharomyces cerevisiae* is a budding yeast, meaning that it mitoses asymmetrically, a characteristic conferring two particular implications: a mother-daughter pair can be identified and the mother cell, which retains most of the cell volume and its contents, also retains a majority of the age associated damage, such as damaged proteins and extrachromosomal rDNA circles (ERCs). Replicative age is monitored by micromanipulation of the daughter cells away from the mother cell and RLS is defined by the total number of daughter cells produced by each mother cell. Yeast RLS has been proposed as a model for aging of mitotic cells in higher eukaryotes and may be particularly relevant for stem cell aging.

In order to model the mechanisms of aging in postmitotic cells using yeast, an alternative measure of life span, referred to as chronological life span (CLS), monitors the ability of cells to reenter the cell cycle following a period of quiescence-like growth arrest.[25] CLS is typically determined by growing cells to high density in liquid culture, at which point they exit the cell cycle but remain metabolically active. Viability over time is determined by periodically challenging a portion of the aging population for the ability to resume vegetative growth under nutrient rich conditions. The relationship between replicative and chronological aging in yeast remains incompletely defined; however, a link has been suggested by the observation that chronologically aged cells subsequently have a shortened replicative life span.[26]

Data from the deletion collection longevity screens to date have identified many new genetic modulators of aging, some of which can be placed in preexisting pathways, such as the TOR pathway, while others appear to be novel mechanisms of life span control. Although there appears to be a wide variety of genetic modifiers of aging in yeast, many of them fall under distinct function classifications, specifically nutrient response mechanisms, mitochondrial genes and regulators of protein biosynthesis.[27-29]

Multiple large-scale longevity screens in nematodes have been conducted concurrently with the yeast screens using RNA interference (RNAi). Though there is no complete worm deletion set, RNAi has been shown to effectively block gene function in *C. elegans* and two commercially available RNAi libraries cover more than 80% of the predicted worm open reading frames.[30,31] From these studies, more than 300 genes have been identified that are associated with increased life span when their function is diminished in the genome.[32-37] Many of the known longevity genes in *C. elegans* can be grouped into one of four at least partially distinct classes: (1) genes that function in ILS, (2) genes involved in mitochondrial function, (3) genes that promote mRNA translation and (4) genes involved in the response to dietary restriction (DR), which seems to correlate well with the data obtained from yeast. Thus, while there are many genes involved, only a few major pathways appear to regulate longevity.[36]

While many new longevity genes have been identified through these screens, the possibility of false negatives should be considered. For example, in one screen from Hansen et al,[36] only one previously known longevity gene, daf-2, was identified. However, when RNAi clones for previously known longevity genes such as age-1 and akt-1 were tested, they were shown to significantly extend lifespan. This indicated that there was a high level of false negatives in this screen and thus more genes involved in lifespan regulation remain unidentified. One way to further investigate this possibility has been to perform RNAi screens of specific categories of genes, such as developmental arrest genes.[32,33] Studies such as these have strengthened the link between mRNA translation, development and lifespan since the genes identified were enriched for those involved in protein synthesis.

mRNA Translation is a Public Determinant of Longevity

Evidence that mRNA translation is an important modifier of longevity in different species has been provided by data from several independent studies. As mentioned previously, mutations that decrease TOR activity are known to increase life span in yeast, nematodes and flies.[15,23,38,39] Likewise, mutation of the TOR-regulated ribosomal S6 kinase, multiple ribosomal protein genes and multiple translation initiation factors, has also been shown to be sufficient to increase life span in each of these organisms.[14,15,40-47]

The data derived from genome-wide longevity studies in yeast and nematodes have presented a largely unbiased picture of many of the genes that are relevant for aging in these organisms. This, in turn, has allowed for the first quantitative analysis of the degree to which genetic control of aging has been evolutionarily conserved. Smith et al[48] used protein sequence homology algorithms to match reported *C. elegans* longevity genes with corresponding yeast homologs. The RLS of yeast strains carrying deletion alleles in these genes (yeast homologs of worm aging genes) was then determined. In the highest confidence set, defined by a reciprocal BLASTp best-match, a 4-fold enrichment in life span extending deletions, relative to randomly selected deletion strains, was observed. This represents a statistically significant enrichment, suggesting that genetic control of longevity has been evolutionarily conserved between budding yeast and nematodes.

In total, Smith et al[48] identified 25 homolog-pairs that similarly modulate longevity in both yeast and nematodes. The most striking feature of the set of 25 conserved aging genes is the substantial enrichment for genes that code for proteins involved in regulating mRNA translation (Table 1). Among the 25 homolog-pairs, only two were previously known to modulate aging in both yeast and worms: *TOR1/let-363* and *SCH9/rsks-1*. *SCH9* and *rsks-1* are homologs of mammalian ribosomal S6 kinase.[43,49] In addition to TOR and ribosomal S6 kinase homologs, three ribosomal proteins of the large subunit (*RPL19A/rpl-19*, *RPL6B/rpl-6*, *RPL9A/rpl-9*) and three translation initiation factors (*TIF1/inf-1*, *TIF2/inf-1* and *TIF4631/ifg-1*) were identified. Given that TOR and S6K are known to negatively regulate both ribosome biogenesis and translation initiation factor activity, it is reasonable to speculate that all of these factors act in a single conserved longevity pathway (Fig. 1).

Although TOR signaling and regulation of translation initiation are conserved longevity control mechanisms shared between yeast and worms, it is interesting to note that the mechanisms by which reduced ribosome biogenesis influences longevity may be somewhat different in the two species. A functional ribosome is composed of two multi-protein subunits, a 40S "small" subunit and a 60S "large" subunit. In *C. elegans*, RNAi knock-down of ribosomal proteins or rRNA processing factors that mediate production of either ribosomal subunit are associated with increased life span.[14,32,33,43] In yeast, in contrast, decreased abundance of the large ribosomal subunit seems to be much more important for longevity than the small subunit,[47] with potential exceptions.[50] Whether these differences reflect overlapping or entirely different downstream mechanisms of longevity control in yeast and worms is currently unknown.

Is DR Mediated by Reduced mRNA Translation?

As mentioned previously, one appealing hypothesis for DR is that increased life span is mediated at least in part by modulating mRNA translation in response to reduced TOR signaling. Several lines of evidence from different species support this hypothesis, including (1) TOR activity is reduced by DR, (2) reduced TOR activity is sufficient to increase life span under nutrient-replete conditions, (3) reduced TOR activity leads to reduced S6 kinase activity, (4) reduced S6 kinase activity is sufficient to increase life span, (5) reduced TOR/S6 kinase activity leads to reduced mRNA translation via regulation of translation initiation factors and expression of ribosomal proteins and (6) mutations in translation initiation factors and ribosomal proteins have been found to be sufficient to increase life span.

Table 1. Translation-related genes reported to modulate aging in* Saccharomyces cerevisiae *and* Caenorhabditis elegans. *Shading indicates that altered function of the protein has been reported to increase life span in the indicated organism. When applicable, homologs are shown for yeast, nematodes and humans

Type	Yeast	Nematode	Human
Regulatory Kinases	TOR1[48]	let-363[48]	FRAP1
	SCH9[48]	rsks-1[48,81]	RPS6KB1, SGK2
Translation Initiation Factors	TIF1[48]	inf-1[48]	EIF4A2, EIF4A1
	TIF2[48]	inf-1[48]	EIF4A2, EIF4A1
	TIF4631[46,48]	ifg-1[33,48,81]	EIF4G1, EIF4G3
	HYP2	iff-1[35]	EIF5A2, EIF5A
	CDC33	ife-2[14,44]	EIF4E1B, EIF4E
	GCD11	Y39G10AR.8[32]	EIF2S3
	GCD6	D2085.3[32]	EIF2B5
	RPN8	eif-3.F[32]	EIF3F
	SUI3	iftb-1[14,17]	EIF2S2
	RPG1	egl-45[33]	EIF3S10
	PRT1	eif-3.B[33]	EIF3S9
	RRP3	inf-1[33]	EIF4A2
	GPI10	T27F7.3[33]	PIGB
Large Ribosomal Subunit Components	RPL4B	rpl-4[14]	RPL4
	RPL6B[47]	rpl-6[48]	RPL6
	RPL7A[47]	rpl-7	RPL7A
	RPL9A[47]	rpl-9[48]	RPL9
	RPL13A[47]	rpl-13	RPL13A
	RPL10*	rpl-10	RPL10A
	RPL19A[47]	rpl-19[48]	RPL19
	RPL20B[46,47]	rpl-20	-
	RPL21B[47]	rpl-21	RPL18A
	RPL22A[47]	rpl-22	-
	RPL23A[47]	rpl-23	RPL23A
	RPL30	rpl-30[14]	RPL30
	RPL31A[47]	rpl-31	-
	RPL33B[47]	rpl-33	RPL35A
	RPL34B[47]	rpl-34	-
	RPL43B[47]	rpl-43	RPL37A
Small Ribosomal Subunit Components	RPS6B*	rps-6[14]	RPS6
	RPS3	rps-3[33]	RPS3P3

continued on next page

Table 1. Continued

Type	Yeast	Nematode	Human
	RPS5	rps-5[125]	RPS5
	RPS8B	rps-8[33]	RPS8
	RPS10B	rps-10[14]	RPS10
	RPS11A	rps-11[14,33]	RPS11P5
	RPS15	rps-15[14]	RPS15P5
	RPS18A/B*	rps-18	RPS18
	RPS22A	rps-22[14]	RPS15A
	RPS23B	rps-23[125]	RPS23
	RPS26B	rps-26[14]	RPS26P8
Mitochondrial Ribosomal Components	MNP1	W09D10.3[125]	MRPL12
	RSM10	Y37D8A.18[125]	MRPS10
	MRPL4	B0261.4[37]	MRPL47
	MRPS9	F09G8.3[125]	MRPS9
	-	F33D4.5[125]	MRPL1
	-	F59A3.3[32]	MRPL24
	-	tag-264[32]	MRPS30
Nucleolar Proteins	NOP58	nol-5[32]	AC064836.1
	NOP14	Y48G1A.4[32]	NOP14
	NOP12[47]	F11A10.7	-
Others	NPL3	drr-2[36]	WBSCR1
	RRP40	exos-3[32]	EXOSC3
	-	F53F4.11[33]	RSL1D1
	SSF1[47]	lpd-6	-
	LOC1[47]	-	-
	REI1[47]	C16A3.4	ZPR9
	RPP2B[47]	C37A2.7, rla-2	RPLP2
	TMA19[46]	tct-1	TPT1, RP11

Epistasis studies in yeast add additional support for the hypothesis that TOR-regulated mRNA translation control underlies life span extension from DR. In the yeast replicative aging paradigm, deletion of *TOR1*, *SCH9*, or a subset of large subunit ribosomal proteins increases the RLS of cells under nutrient replete conditions, but fails to further increase the RLS of cells under reduced glucose conditions,[15,47] which is used as a model of DR in this organism.[51,52] Similar to the case for DR,[53,54] TOR, Sch9 and ribosomal proteins also modulate RLS in a Sir2-independent manner.[15,47] These data are consistent with the idea that all of these interventions act in a single pathway to modulate RLS in yeast.

The relationship between DR and TOR-regulated translation appears to be more complex in *C. elegans*. Similar to the case in yeast, inhibition of TOR signaling fails to further increase the life span of worms subjected to DR and both TOR signaling and DR are reported to promote longevity by a mechanism independent of the nematode Sir2 ortholog, SIR-2.1.[14,17,55,56] It remains unclear, however, whether S6 kinase and translation initiation factors act downstream of TOR and DR to promote longevity in *C. elegans* or in a parallel pathway. For example, deletion of the gene coding for the nematode S6 kinase homolog *rsks-1* increases life span additively with DR, which differs from the case in yeast.[43] One issue that complicates interpretation of these studies, however, is that multiple methods of DR have been described in *C. elegans* and the relationship between DR, TOR and mRNA translation has only been studied using a subset of these methods. There is also evidence that different methods of DR in *C. elegans* can promote longevity by distinct mechanisms,[57] raising the possibility that decreased mRNA translation could be important for some, but not all, types of DR in this organism.

Based on the complex genetic interactions between DR, TOR and mRNA translation in nematodes, as well as additional data from studies of DR and mTOR in mammals, it seems likely that a simple linear model placing DR upstream of TOR, upstream of S6K and translation initiation factors is too simplistic to accurately represent the underlying biology of DR. For example, TOR signaling interacts with ILS, which is known to modulate longevity in both invertebrates and mammals, through reciprocal cross-talk mediated in part by AKT.[58] TOR also interacts with the hypoxic response, which has recently been shown to modulate longevity in *C. elegans*.[59] It is reasonable to speculate that TOR could be influencing longevity through altered ILS, induction of the hypoxic response, or both. Likewise, as mentioned previously, TOR signaling regulates additional downstream processes, including mitochondrial function and autophagy. It has previously been shown that autophagy, in particular, is required for life span extension in response to reduced ILS,[19] is induced by DR,[60-62] and is required for increased life span.[17-19] Taken together, these observations may suggest that the life span extension associated with DR and reduced TOR signaling is likely to require multiple TOR-regulated outputs, including enhanced autophagy, altered ILS or hypoxic response, as well as reduced mRNA translation.

Possible Mechanisms for How Translation Influences Aging

Regardless of whether mRNA translation plays a primary role in life span extension from DR or reduced TOR signaling, it is clear that mutations to ribosomal S6 kinase, translation initiation factors, or ribosomal proteins is sufficient to increase life span in yeast, nematodes and flies. How might mutations that reduce mRNA translation slow aging? Here we describe experimental evidence supporting three distinct, but not mutually exclusive, hypotheses that may contribute to life span extension in response to altered mRNA translation: (a) differential translation of specific mRNAs, (b) improved protein homeostasis with aging and (c) altered energetics and distribution of resources toward somatic maintenance.

a. **Differential translation.** Perhaps the simplest explanation to account for the observation that some mutations leading to altered mRNA translation also increase life span is differential translation of one (or more) mRNA coding for a protein that modulates aging. For example, consider a hypothetical protein AGE-X that promotes longevity: AGE-X mRNA is poorly translated under normal conditions but is efficiently translated when a specific translation initiation factor is mutated, even though most genes are less efficiently translated. Increased life span is observed when this translation initiation factor is mutated *because* AGE-X protein abundance is increased in both relative and absolute terms.

The yeast transcription factor, Gcn4, is a longevity gene that fits the AGE-X example. In a recent study by Steffen et al[47] it was found that deletion of several genes encoding proteins in the 60S ribosomal subunit, but not the 40S ribosomal subunit, increased yeast RLS. Gcn4 was found to be differentially translated in long-lived mutants and necessary for full life span extension.[47] When functional 60S ribosomal subunits are depleted, global mRNA translation is decreased while translation of Gcn4 is increased. Differential translation of Gcn4 is mediated by post-transcriptional regulation via four inhibitory upstream open reading frames (uORFs) located in the 5′ leader of the *GCN4* mRNA that act to negatively regulate its translation under certain conditions.[63] Under conditions of rapid growth, little Gcn4 protein is produced due to translation of the inhibitory uORFs. It is hypothesized that when 60S subunits are depleted, translation of the inhibitory uORFs is reduced, leading to increased efficiency of translation of the Gcn4 protein.

It is interesting to note that the deletion of 60S ribosomal proteins and the corresponding increase in Gcn4 translation mimics the regulation of Gcn4 in response to stress. Organisms have evolved specific mechanisms to increase their survival during times of stress and one mechanism is genetic reprogramming through increased expression of transcription factors such as Gcn4. It has been shown, for example, that translation of Gcn4 is specifically up-regulated in response to nutritional deprivation.[64] In addition, mammalian ATF proteins such as ATF4 and ATF5 (functional homologs of yeast Gcn4), are also translationally regulated in response to various stresses including food deprivation and amino acid limitation.[65, 66] It is thus logical to hypothesize that the differential translational regulation of transcription factors such as Gcn4 and mammalian ATFs may be a conserved translational response to DR. In support of this, Gcn4 is required for the full lifespan extension by DR as well as that seen in long-lived TOR and Sch9 deletion mutants that epistasis experiments have placed in the same pathway as DR.[47] Thus, DR may act as a nutritional stressor which signals to an organism to increase translation of transcription factors, such as Gcn4 and mammalian ATFs, thereby activating a genetic program that will increase its chance of survival during times of nutrient scarcity. Along this same notion, reducing translation by inhibition of TOR signaling may also work in this manner. For example it was recently shown in a mammalian cellular system that the Tuberous Sclerosis Complex Proteins 1 and 2 (TSC1 and TSC2) negatively regulate protein translation by inhibition of TOR upon serum withdrawal.[67] Interestingly, this study also found that the TOR complex 1 inhibitor, rapamycin, specifically inhibited a subset of mRNAs that included many ribosomal proteins and components of the translational machinery, suggesting that a major function of TOR complex 1 in mammals, as in yeast, is to positively regulate ribosome biogenesis. Thus, when TOR signaling is inhibited by lack of nutrients, global translation decreases, in part due to reduced translational machinery; on the other hand, specific mRNAs such as Gcn4 and mammalian ATFs increase, because their translational regulation is geared to be activated by nutritional stressors in order to increase the survival of the organism. While it appears that mammalian ATFs are regulated by translation in a similar manner as Gcn4, further studies will be needed to determine whether this mechanism of longevity control extends beyond yeast.

b. Protein homeostasis. A second possibility to account for how reduced mRNA translation might promote longevity is that protein homeostasis during aging is better preserved under such conditions. It has been speculated that failure to maintain protein homeostasis may be a primary molecular cause of aging.[4] Age-associated loss of protein homeostasis could occur from accumulation of damaged, aggregated and misfolded proteins.[68] Evidence that such accumulation does occur with aging is abundant and recent studies have suggested that at least some mutations resulting in increased life span also confer enhanced resistance to proteotoxic stress. For example, at least three distinct longevity pathways in *C. elegans* (DR, insulin-like signaling and the hypoxic response) also modulate toxicity of aggregation prone peptides,[59,69-72] and reduced TOR signaling has been found to increase resistance to polyglutamine toxicity in both flies and mice.[73]

If failure to maintain protein homeostasis due to accumulation of damaged proteins is a limiting factor for longevity, then it is reasonable to speculate that a global reduction in protein synthesis, as is observed in long-lived mutants with impaired mRNA translation, could ameliorate this process. In cases such as Huntington's disease, in which the disease-causing proteotoxicity can be causally linked to a single toxic protein, reduced translation might improve outcome simply by reducing production of that protein. Evidence in support of this idea was recently published in which the inhibitor of TOR, rapamycin, was tested in a cellular model of Huntington's disease. In this study, the authors sought to understand whether rapamycin clears toxic polyglutamine aggregates solely through an autophagic mechanism or if it also acts by reduction of protein synthesis. They found that rapamycin could reduce both the total amount of protein and the amount of soluble exogenously expressed polyglutamine protein in both autophagy-competent

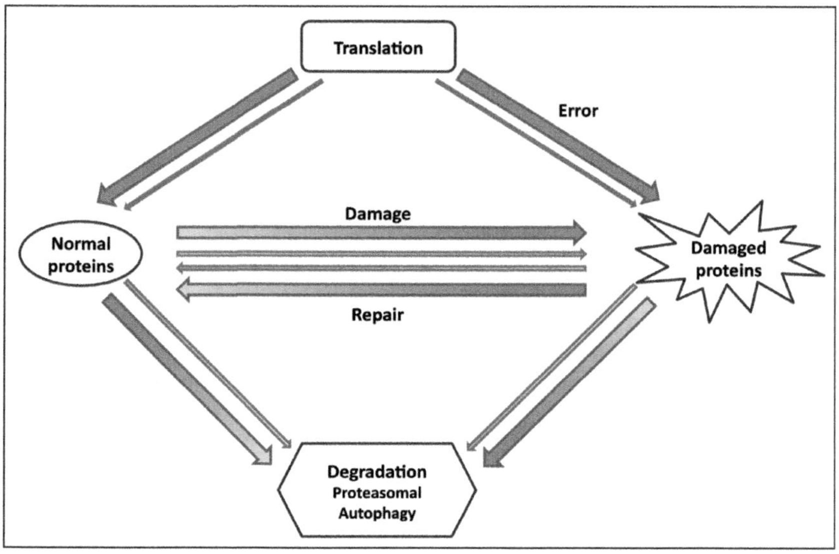

Figure 2. Protein homeostasis and aging. With normal aging (thick red arrows), accumulation of damaged proteins may eventually overburden the repair and removal mechanisms (shown by darker gradient in the thick red arrows). Regulation of normal proteins may then be disrupted, causing an imbalance between normal and damaged proteins which may limit longevity. An overall reduction in the synthesis of new proteins (thin green arrows) could ameliorate the overload of repair and clearance mechanisms and reduce the amount of damaged proteins to below the critical threshold, thereby restoring the regulation of normal proteins and possibly increasing longevity. A color version of this figure is available online at www.landesbioscience.com/curie.

and incompetent cells, suggesting that in addition to autophagy, rapamycin also acts through a global reduction of protein synthesis to prevent polyglutamine-induced toxicity.[74] This finding has important implications for normal aging as well, since it is likely that accumulation of a variety of damaged protein species contributes to loss of protein homeostasis. In this case, a general reduction in protein synthesis could improve protein homeostasis both by reducing the flux of newly synthesized proteins into the system and by allowing the endogenous repair and degradation machinery to maintain a lower steady-state burden of misfolded and damaged peptides (Fig. 2).

c. **Improved energy balance.** A third possible mechanism by which reduced translation could influence aging is by redistributing resources toward somatic maintenance at the expense of fecundity. A large investment of energy is required in order to maintain high levels of protein synthesis. As such, mRNA translation is generally tightly coupled to nutrient availability and environmental conditions related to growth.[75-77] In general, evolutionary pressures are likely to have selected for individuals that are able to respond to growth-favoring conditions by developing rapidly and reproducing abundantly. In contrast, many long-lived mutants develop more slowly and show reduced fecundity.[78,79] This has been taken as evidence that one strategy for slowing aging is to shift the balance of energy expenditure away from reproduction toward somatic maintenance.

Evidence from both yeast and nematodes suggests that long-lived mutants with altered mRNA translation do, indeed, have reduced fecundity. In worms, *rsks-1* and *ifg-1* deletion mutants have been shown to have as much as a 50% decrease in mean brood size.[80,81] *Iftb-1*, *rps-15* and *let-363* (TOR) RNAi treatments have been shown to decrease fecundity by 83%, 59% and 18% respectively.[14] In the yeast replicative aging paradigm, reproductive fitness is most analogous growth rate. For yeast in the wild, under conditions favoring growth, those (clonal) individuals able to grow most rapidly will have a selective advantage. Although the relationship between replicative longevity and growth rate has not been comprehensively examined, at least within the class of longevity genes represented by TOR, SCH9 and translation-related genes, there is a significant inverse correlation between doubling time and RLS.[47]

Does mRNA Translation Modulate Aging in Mammals?

Although it remains unclear whether mRNA translation factors will play a conserved role in determining the longevity of mammals, translation has been indirectly implicated in the mammalian aging process. As covered in detail above, multiple mutants with altered mRNA translation have shown a moderate life span extension in invertebrate animal models. Specifically, TORC1 activity, which promotes mRNA translation through at least two different downstream targets, ribosomal S6 kinase and eukaryotic initiation factor 4E (eIF4E) binding proteins (4E-BPs), has been implicated in the aging process, as well as its down-stream effectors. Activation of ribosomal S6 kinase promotes the activity of other translation initiation factors, such as eukaryotic initiation factor 4B (eIF4B) and stimulates production of ribosomal proteins and ribosome biogenesis (Table 2).[82] Decreased ribosomal S6 kinase activity, translation initiation factor activity, or ribosomal protein expression have all been found to promote longevity in yeast, worms and flies. Analogous mutants, however, have yet to be tested for life span extension in the context of a mammalian system. Rather, the role of mRNA translation in mammalian aging has been implicated through its role in age-related diseases such as cancer, diabetes, cardiovascular defects and neurodegenerative diseases.

mRNA Translation and Cancer

The major hallmark of DR in rodents is a significant reduction in age-associated cancer incidence.[83] It has been speculated that the reduction of cancer occurrence may account for a large fraction of the longevity benefit associated with DR in rodents.[84] Mutations of mTOR pathway

Table 2. Representative mouse models with altered translation

Mouse Model	Phenotype
mTOR knockout[126,127]	Early postimplantation lethality; mTOR (+/–) have reduced S6 kinase phosphorylation but normal S6 phosphorylation.
Raptor knockout[109]	Early embryonic lethality
Rictor knockout[109]	Early embryonic lethality
mLST8 knockout[109]	Early embryonic lethality
S6 kinase 1 (S6k1) knockout[104,128]	Decreased body size from embryonic stage compared to WT. Compensatory up-regulation of S6 kinase 2 (S6k2). Protected against age- and diet-induced obesity. Increased insulin sensitivity.
Adipose-specific raptor knockout[108]	Leanness, protection against diet-induced obesity and hypercholesterolemia, improved insulin sensitivity.
S6k1/S6k2 double knockout[129]	Perinatal lethality; compensatory S6 phosphorylation by MAP kinase in response to mitogens.
eIF6 knockout[130]	Preimplantation lethality; eIF6 (+/–) mice have reduced body weight. 70% of eIF6 (+/–) mice have reduced insulin-stimulated initiation of translation in liver.
4E-BP1/4E-BP2 knockout[131]	Increased s6 kinase activity. Increased sensitivity to diet-induced obesity and insulin resistance.
RPL22 knockout[132]	Impairs $\alpha\beta$ T-cell development.

components (the effectors of which are involved in mRNA translation) and/or elevated mTOR activity have been discovered in many cancers including lymphomas, melanomas, gliomas and CNS malignancies as well as carcinomas of lung, bladder, kidney, ovary, breast, prostate, stomach, pancreas and head and neck.[85] Up-regulation of mTOR signaling by mutation in its upstream negative regulators, TSC1 or TSC2, results in an increased risk of renal cell carcinoma.[86,87] Furthermore, up-regulation of mTOR signaling is correlated with poor tumor prognosis for melanoma,[88] prostate,[89] pancreatic,[90] thyroid,[91] breast,[92] and cervical cancers.[93] These findings have prompted extensive development and clinical testing of rapamycin derivatives (termed rapalogs) as therapeutic treatments toward a wide variety of tumors.[94-98]

mRNA Translation and Diabetes

The mTOR signaling pathway has also been implicated in both Type 1 and Type 2 diabetes. Type 1 diabetes is characterized by reduced insulin production due to destruction of pancreatic β-cells, while Type 2 diabetes is characterized by a form of insulin resistance. Insulin-induced protein translation and proliferation[99] as well as growth induced by glucose and branched amino acids[100] is dependent on mTOR signaling in pancreatic β-cells. Desensitization of cells to insulin has been linked in part to S6 kinase-dependent phosphorylation and inhibition of the insulin-receptor substrate proteins (IRS1 and 2).[101,102] Furthermore, S6K1-deficient mice have smaller islets and exhibit hypersensitivity to insulin and glucose intolerance.[103] These mice also display reduced weight gain and resistance to obesity due to age or challenge with a high calorie diet compared to wild-type mice.[104] Similarly, disruption of 4E-BP1, another downstream component of mTOR signaling, results in mice with reduced adipose tissue.[105] Taken all together, this suggests that chronic TOR activation results in insulin resistance (i.e., hyperglycemia and Type 2 diabetes).

In addition mTOR has also been shown to play a role in lipid metabolism. mTOR expression is increased during adipogenesis and dosing with rapamycin results in the inhibition of adipocyte differentiation and a reduction of adipogenesis and lipogenesis.[106,107] Mice containing an adipose-specific knockout of the TOR complex 1 accessory protein Raptor are lean and are resistant to diet-induced obesity and insulin resistance.[108] These adipose-specific Raptor knockout mice also have enhanced mitochondrial uncoupling in adipose, which leads to increased energy expenditure. Not surprisingly, the consequences of Raptor knockout vary depending on the specific tissue and a complete knockout results in developmental lethality.[109] For example, while an adipose-specific knockout of Raptor seems to have beneficial consequences, skeletal muscle-specific knockout mice have a muscular dystrophy phenotype.[110]

mRNA Translation and Cardiovascular Disease

Cardiac hypertrophy, an enlargement of the heart due to increased cell size and protein synthesis, is directly linked to elevated activity of the PI3K/Akt/mTOR pathway.[111] mTOR inhibition by rapamycin treatment has been shown to regress cardiac hypertrophy.[111-113] These effects are accompanied by suppression of ribosomal S6 protein and eIF4E phosphorylation, which occurs due to pressure overload.[113] These data indicate that the specific activity of mTOR regulating mRNA translation may play a key role in cardiac health. With age there is an increase in the prevalence of cardiovascular diseases, such as congestive heart failure and coronary artery disease. Studies are indicate that regulation of mRNA translation through modulation of mTOR activity may prove a viable cardioprotective strategy.

mRNA Translation and Neurodegenerative Disease

DR has been long been suggested to have a neuroprotective effect and to decrease age-related neuronal loss, though the underlying mechanism remains poorly understood.[114,115] mTOR signaling has been shown to have a protective role against polyglutamine toxicity,[116-118] which is associated with several neurodegenerative diseases in people, including Huntington's disease.[119] Thus such disorders, as well as normal brain function, may benefit from pharmacological modulation of the mTOR signaling pathway. It has been speculated that an enhanced autophagic response induced by DR or rapamycin treatment may help remove toxic protein aggregates; however, it remains possible that altered mRNA translation also plays an important mechanistic role. Further studies will be needed to determine precisely which aspects of TOR signaling are most relevant for enhanced resistance to proteotoxic stress in both invertebrate and mammalian models.

Conclusion

It is clear that mRNA translation control is a central determinant of longevity in invertebrates. Indeed, other than DR, only two proteins have been shown to modulate longevity in yeast, nematodes and flies: TOR and ribosomal S6 kinase. Multiple additional translation factors have been found to modulate aging in both yeast and nematodes and it seems likely that this will be extended to flies in the near future. The importance of this pathway in mammalian aging has yet to be determined. In the absence of mammalian longevity data, however, there is reason to be optimistic that reduced TOR signaling may be beneficial for a variety of age-associated diseases in mammals. For example, mice treated with rapamycin show resistance to cancer, neurodegeneration and cardiac disease.[16,113] Additionally, raptor and S6 kinase knockout mice show phenotypes consistent with a partial genetic mimetic of DR, including improved insulin sensitivity, reduced adiposity and resistance to age-associated obesity.[104] There is also emerging data that inhibition of TOR is likely to have beneficial health effects in humans. Rapamycin (Sirolimus) is used clinically as an immunosuppressant and to prevent coronary stent restenosis,[120] and is also in clinical trials as an anti-cancer therapeutic.[121] It is noteworthy that reduced age-associated cancer incidence is a primary feature of DR in rodents, suggesting that rapamycin mimics at least some DR phenotypes in humans. It will be of great interest to determine whether rapamycin exposure can prevent or treat other age-associated human diseases. The mechanistic explanations underlying these beneficial effects are likely to be complicated and the consequences of reduced TOR or SK6 activity may well be different from tissue to tissue. It is of vital importance to know whether reduced TOR activity enhances longevity in mammals. Fortunately, longevity experiments with mice fed a diet supplemented with rapamycin are well underway as part of the National Institute on Aging Interventions Testing Program[122] and results are likely to be available by the time this chapter is published.

In the last few years, the relationship between protein translation and aging has begun to come into sharper focus.[123,124] The next few years should see increasingly successful efforts to elucidate the mechanisms underlying enhanced longevity resulting from reduced TOR signaling and/or translation initiation, as well as a better understanding of whether the translation control genes identified from longevity studies in model organisms are likely to be useful therapeutic targets for treating age-related diseases in humans.

References

1. Bohr VA, Anson RM. DNA damage, mutation and fine structure DNA repair in aging. Mutat Res 1995; 338:25-34.
2. de Magalhaes JP, Faragher RG. Cell divisions and mammalian aging: integrative biology insights from genes that regulate longevity. Bioessays 2008; 30:567-78.
3. Jeyapalan JC, Sedivy JM. Cellular senescence and organismal aging. Mech Ageing Dev 2008; 129:467-74.
4. Morimoto RI, Cuervo AM. Protein homeostasis and aging: taking care of proteins from the cradle to the grave. J Gerontol A Biol Sci Med Sci 2009; 64:167-70.
5. Ramasamy R, Vannucci SJ, Yan SS et al. Advanced glycation end products and RAGE: a common thread in aging, diabetes, neurodegeneration and inflammation. Glycobiology 2005; 15:16R-28R.
6. Sohal RS, Weindruch R. Oxidative stress, caloric restriction and aging. Science 1996; 273:59-63.
7. Warner HR. Superoxide dismutase, aging and degenerative disease. Free Radic Biol Med 1994; 17:249-58.
8. Kaeberlein M. Genome-wide approaches to understanding human ageing. Hum Genomics 2006; 2:422-8.
9. Masoro EJ. Overview of caloric restriction and ageing. Mech Ageing Dev 2005; 126:913-22.
10. Weindruch RH, Walford RL. The Retardation of Aging and Disease by Dietary Restriction. (Thomas, Springfield, IL., 1988).
11. Colman RJ, Beasley TM, Allison DB et al. Attenuation of sarcopenia by dietary restriction in rhesus monkeys. J Gerontol A Biol Sci Med Sci 2008; 63:556-9.
12. Ingram DK, Young J, Mattison JA. Calorie restriction in nonhuman primates: assessing effects on brain and behavioral aging. Neuroscience 2007; 145:1359-64.

13. Messaoudi I, Warner J, Fischer M et al. Delay of T-cell senescence by caloric restriction in aged long-lived nonhuman primates. Proc Natl Acad Sci USA 2006; 103:19448-53.
13a. Colman RJ, Anderson RM, Johnson SC, et al. Caloric restriction delays disease onset and mortality in rhesus monkeys. Science 2009; 325:201-4.
14. Hansen M, Taubert S, Crawford D et al. Lifespan extension by conditions that inhibit translation in Caenorhabditis elegans. Aging Cell 2007; 6:95-110.
15. Kaeberlein M, Powers RW 3rd, Steffen KK et al. Regulation of yeast replicative life span by TOR and Sch9 in response to nutrients. Science 2005; 310:1193-6.
16. Wullschleger S, Loewith R, Hall MN. TOR signaling in growth and metabolism. Cell 2006; 124:471-84.
17. Hansen M, Chandra A, Mitic LL et al. A role for autophagy in the extension of lifespan by dietary restriction in C. elegans. PLoS Genet 2008; 4:e24.
18. Jia K, Levine B. Autophagy is required for dietary restriction-mediated life span extension in C. elegans. Autophagy 2007; 3:597-9.
19. Meléndez A, Tallóczy Z, Seaman M et al. Autophagy genes are essential for dauer development and life-span extension in C. elegans. Science 2003; 301:1387-91.
20. Kaeberlein M, Kennedy BK. Protein translation, 2007. Aging Cell 2007; 6:731-4.
21. Winzeler EA, Shoemaker DD, Astromoff A et al. Functional characterization of the S. cerevisiae genome by gene deletion and parallel analysis. Science 1999; 285:901-6.
22. Kaeberlein M, Kennedy BK. Large-scale identification in yeast of conserved ageing genes. Mech Ageing Dev 2005; 126:17-21.
23. Powers RW 3rd, Kaeberlein M, Caldwell SD et al. Extension of chronological life span in yeast by decreased TOR pathway signaling. Genes Dev 2006; 20:174-84.
24. Kaeberlein M. in Handbook of models for human aging (ed. Conn, P.M.) 109-120 (Elsevier Press, Boston, 2006).
25. Fabrizio P, Longo VD. The chronological life span of Saccharomyces cerevisiae. Aging Cell 2003; 2:73-81.
26. Ashrafi K, Sinclair D, Gordon JI et al. Passage through stationary phase advances replicative aging in Saccharomyces cerevisiae. Proc Natl Acad Sci USA 1999; 96:9100-5.
27. Jazwinski SM. Molecular mechanisms of yeast longevity. Trends Microbiol 1999; 7:247-52.
28. Jazwinski SM. Yeast longevity and aging—the mitochondrial connection. Mech Ageing Dev 2005; 126:243-8.
29. Steinkraus KA, Kaeberlein M, Kennedy BK. Replicative aging in yeast: the means to the end. Annu Rev Cell Dev Biol 2008; 24:29-54.
30. Kamath RS, Fraser AG, Dong Y et al. Systematic functional analysis of the Caenorhabditis elegans genome using RNAi. Nature 2003; 421:231-7.
31. Rual JF, Ceron J, Koreth J et al. Toward improving Caenorhabditis elegans phenome mapping with an ORFeome-based RNAi library. Genome Res 2004; 14:2162-8.
32. Chen D, Pan KZ, Palter JE et al. Longevity determined by developmental arrest genes in Caenorhabditis elegans. Aging Cell 2007; 6:525-33.
33. Curran SP, Ruvkun G. Lifespan regulation by evolutionarily conserved genes essential for viability. PLoS Genet 2007; 3:e56.
34. Dillin A, Hsu AL, Arantes-Oliveira N et al. Rates of behavior and aging specified by mitochondrial function during development. Science 2002; 298:2398-401.
35. Hamilton B et al. A systematic RNAi screen for longevity genes in C. elegans. Genes Dev 2005; 19:1544-55.
36. Hansen M, Hsu AL, Dillin A et al. New genes tied to endocrine, metabolic and dietary regulation of lifespan from a Caenorhabditis elegans genomic RNAi screen. PLoS Genet 2005; 1:119-28.
37. Lee SS et al. A systematic RNAi screen identifies a critical role for mitochondria in C. elegans longevity. Nat Genet 2003; 33:40-8.
38. Jia K, Chen D, Riddle DL. The TOR pathway interacts with the insulin signaling pathway to regulate C. elegans larval development, metabolism and life span. Development 2004; 131:3897-906.
39. Vellai T, Takacs-Vellai K, Zhang Y et al. Genetics: influence of TOR kinase on lifespan in C. elegans. Nature 2003; 426:620.
40. Fabrizio P, Pletcher SD, Minois N et al. Chronological aging-independent replicative life span regulation by Msn2/Msn4 and Sod2 in Saccharomyces cerevisiae. FEBS Lett 2004; 557:136-42.
41. Fabrizio P, Pozza F, Pletcher SD et al. Regulation of longevity and stress resistance by Sch9 in yeast. Science 2001; 292:288-90.
42. Kapahi P, Zid BM, Harper T et al. Regulation of lifespan in Drosophila by modulation of genes in the TOR signaling pathway. Curr Biol 2004; 14:885-90.
43. Pan KZ, Palter JE, Rogers AN et al. Inhibition of mRNA translation extends lifespan in Caenorhabditis elegans. Aging Cell 2007; 6:111-9.

44. Syntichaki P, Troulinaki K, Tavernarakis N. eIF4E function in somatic cells modulates ageing in Caenorhabditis elegans. Nature 2007; 445:922-6.
45. Chiocchetti A, Zhou J, Zhu H et al. Ribosomal proteins Rpl10 and Rps6 are potent regulators of yeast replicative life span. Exp Gerontol 2006.
46. Managbanag JR, Witten TM, Bonchev D et al. Shortest-path network analysis is a useful approach toward identifying genetic determinants of longevity. PLoS ONE 2008; 3:e3802.
47. Steffen KK, MacKay VL, Kerr EO et al. Yeast life span extension by depletion of 60s ribosomal subunits is mediated by Gcn4. Cell 2008; 133:292-302.
48. Smith ED, Tsuchiya M, Fox LA et al. Quantitative evidence for conserved longevity pathways between divergent eukaryotic species. Genome Res 2008; 18:564-70.
49. Urban J, Soulard A, Huber A et al. Sch9 is a major target of TORC1 in Saccharomyces cerevisiae. Mol Cell 2007; 26:663-74.
50. Chiocchetti A, Zhou J, Zhu H et al. Ribosomal proteins Rpl10 and Rps6 are potent regulators of yeast replicative life span. Exp Gerontol 2007; 42:275-86.
51. Lin SJ, Defossez PA, Guarente L. Requirement of NAD and SIR2 for life-span extension by calorie restriction in Saccharomyces cerevisiae. Science 2000; 289:2126-8.
52. Lin SJ et al. Calorie restriction extends Saccharomyces cerevisiae life span by increasing respiration. Nature 2002; 418:344-48.
53. Jiang JC, Jaruga E, Repnevskaya MV et al. An intervention resembling caloric restriction prolongs life span and retards aging in yeast. FASEB J 2000; 14:2135-7.
54. Kaeberlein M, Kirkland KT, Fields S et al. Sir2-independent life span extension by calorie restriction in yeast. PLoS Biol 2004; 2:E296.
55. Kaeberlein TL, Smith ED, Tsuchiya M et al. Lifespan extension in Caenorhabditis elegans by complete removal of food. Aging Cell 2006.
56. Lee GD, Wilson MA, Zhu M et al. Dietary deprivation extends lifespan in Caenorhabditis elegans. Aging Cell 2006.
57. Greer EL, Brunet A. Different dietary restriction regimens extend lifespan by both independent and overlapping genetic pathways in C. elegans. Aging Cell 2009; 8:113-27.
58. Avruch J, Hara K, Lin Y et al. Insulin and amino-acid regulation of mTOR signaling and kinase activity through the Rheb GTPase. Oncogene 2006; 25:6361-72.
59. Mehta R, Steinkraus KA, Sutphin GL et al. Proteasomal regulation of the hypoxic response modulates aging in C. elegans. Science 2009.
60. Juhasz G, Erdi B, Sass M et al. Atg7-dependent autophagy promotes neuronal health, stress tolerance and longevity but is dispensable for metamorphosis in Drosophila. Genes Dev 2007; 21:3061-6.
61. Morck C, Pilon M. C. elegans feeding defective mutants have shorter body lengths and increased autophagy. BMC Dev Biol 2006; 6:39.
62. Takeshige K, Baba M, Tsuboi S et al. Autophagy in yeast demonstrated with proteinase-deficient mutants and conditions for its induction. J Cell Biol 1992; 119:301-11.
63. Hinnebusch AG. Translational regulation of yeast GCN4. A window on factors that control initiator-trna binding to the ribosome. J Biol Chem 1997; 272:21661-4.
64. Yang R, Wek SA, Wek RC. Glucose limitation induces GCN4 translation by activation of Gcn2 protein kinase. Mol Cell Biol 2000; 20:2706-17.
65. Shimizu YI, Morita M, Ohmi A et al. Fasting induced up-regulation of activating transcription factor 5 in mouse liver. Life Sci 2009.
66. Zhou D, Palam LR, Jiang L et al. Phosphorylation of eIF2 directs ATF5 translational control in response to diverse stress conditions. J Biol Chem 2008; 283:7064-73.
67. Bilanges B, Argonza-Barrett R, Kolesnichenko M et al. Tuberous sclerosis complex proteins 1 and 2 control serum-dependent translation in a TOP-dependent and -independent manner. Mol Cell Biol 2007; 27:5746-64.
68. Balch WE, Morimoto RI, Dillin A et al. Adapting proteostasis for disease intervention. Science 2008; 319:916-9.
69. Cohen E, Bieschke J, Perciavalle RM et al. Opposing activities protect against age-onset proteotoxicity. Science 2006; 313:1604-10.
70. Hsu AL, Murphy CT, Kenyon C. Regulation of aging and age-related disease by DAF-16 and heat-shock factor. Science 2003; 300:1142-5.
71. Morley JF, Morimoto RI. Regulation of longevity in Caenorhabditis elegans by heat shock factor and molecular chaperones. Mol Biol Cell 2004; 15:657-64.
72. Steinkraus KA, Smith ED, Davis C et al. Dietary restriction suppresses proteotoxicity and enhances longevity by an hsf--1-dependent mechanism in Caenorhabditis elegans. Aging Cell 2008.
73. Ravikumar B, Vacher C, Berger Z et al. Inhibition of mTOR induces autophagy and reduces toxicity of polyglutamine expansions in fly and mouse models of Huntington disease. Nat Genet 2004; 36:585-95.

74. King MA, Hands S, Hafiz F et al. Rapamycin inhibits polyglutamine aggregation independently of autophagy by reducing protein synthesis. Mol Pharmacol 2008; 73:1052-63.
75. Barbet NC, Schneider U, Helliwell SB et al. TOR controls translation initiation and early G1 progression in yeast. Mol Biol Cell 1996; 7:25-42.
76. Dennis PB, Fumagalli S, Thomas G. Target of rapamycin (TOR): balancing the opposing forces of protein synthesis and degradation. Curr Opin Genet Dev 1999; 9:49-54.
77. Holcik M, Sonenberg N. Translational control in stress and apoptosis. Nat Rev Mol Cell Biol 2005; 6:318-27.
78. Barnes AI, Partridge L. Costing reproduction. Animal Behaviour 2003; 66:199-204.
79. Kenyon C. The plasticity of aging: Insights from long-lived mutants. Cell 2005; 120:449-460.
80. Friberg J. in Umeå Centre for Molecular Pathogenesis 39 (Umea University, Umea, 2006).
81. Pan KZ, Palter JE, Rogers AN et al. Inhibition of mRNA translation extends lifespan in Caenorhabditis elegans. Aging Cell 2007; 6:111-119.
82. Jastrzebski K, Hannan KM, Tchoubrieva EB et al. Coordinate regulation of ribosome biogenesis and function by the ribosomal protein S6 kinase, a key mediator of mTOR function. Growth Factors 2007; 25:209-26.
83. Weindruch R, Walford RL. The Retardation of Aging and Disease by Dietary Restriction. Springfield: Charles C. Thomas, 1988.
84. Spindler SR. Rapid and reversible induction of the longevity, anticancer and genomic effects of caloric restriction. Mech Ageing Dev 2005; 126:960-6.
85. Guertin DA, Sabatini DM. An expanding role for mTOR in cancer. Trends Mol Med 2005; 11:353-61.
86. Sampson JR, Patel A, Mee AD. Multifocal renal cell carcinoma in sibs from a chromosome 9 linked (TSC1) tuberous sclerosis family. J Med Genet 1995; 32:848-50.
87. Al-Saleem T, Wessner LL, Scheithauer BW et al. Malignant tumors of the kidney, brain and soft tissues in children and young adults with the tuberous sclerosis complex. Cancer 1998; 83:2208-16.
88. Daido S, Yamamoto A, Fujiwara K et al. Inhibition of the DNA-dependent protein kinase catalytic subunit radiosensitizes malignant glioma cells by inducing autophagy. Cancer Res 2005; 65:4368-75.
89. Mita MM, Mita AC, Chu QS et al. Phase I trial of the novel mammalian target of rapamycin inhibitor deforolimus (AP23573; MK-8669) administered intravenously daily for 5 days every 2 weeks to patients with advanced malignancies. J Clin Oncol 2008; 26:361-7.
90. Yamamoto S, Tomita Y, Hoshida Y et al. Prognostic significance of activated Akt expression in pancreatic ductal adenocarcinoma. Clin Cancer Res 2004; 10:2846-50.
91. Vasko V, Saji M, Hardy E et al. Akt activation and localisation correlate with tumour invasion and oncogene expression in thyroid cancer. J Med Genet 2004; 41:161-70.
92. Zhou BP, Liao Y, Xia W et al. Cytoplasmic localization of p21Cipl/WAF1 by Akt-induced phosphorylation in HER-2/neu-overexpressing cells. Nat Cell Biol 2001; 3:245-52.
93. Faried LS, Faried A, Kanuma T et al. Predictive and prognostic role of activated mammalian target of rapamycin in cervical cancer treated with cisplatin-based neoadjuvant chemotherapy. Oncol Rep 2006; 16:57-63.
94. Blando J, Portis M, Benavides F et al. PTEN deficiency is fully penetrant for prostate adenocarcinoma in C57BL/6 mice via mTOR-dependent growth. Am J Pathol 2009; 174:1869-79.
95. Ebi H, Tomida S, Takeuchi T et al. Relationship of deregulated signaling converging onto mTOR with prognosis and classification of lung adenocarcinoma shown by two independent in silico analyses. Cancer Res 2009; 69:4027-35.
96. Kim SH, Zukowski K, Novak RF. Rapamycin effects on mTOR signaling in benign, premalignant and malignant human breast epithelial cells. Anticancer Res 2009; 29:1143-50.
97. Meier F, Guenova E, Clasen S et al. Significant response after treatment with the mTOR inhibitor sirolimus in combination with carboplatin and paclitaxel in metastatic melanoma patients. J Am Acad Dermatol 2009; 60:863-8.
98. Raimondi AR, Molinolo A, Gutkind JS. Rapamycin prevents early onset of tumorigenesis in an oral-specific K-ras and p53 two-hit carcinogenesis model. Cancer Res 2009; 69:4159-66.
99. McDaniel ML, Marshall CA, Pappan KL et al. Metabolic and autocrine regulation of the mammalian target of rapamycin by pancreatic beta-cells. Diabetes 2002; 51:2877-85.
100. Briaud I, Dickson LM, Lingohr MK et al. Insulin receptor substrate-2 proteasomal degradation mediated by a mammalian target of rapamycin (mTOR)-induced negative feedback down-regulates protein kinase B-mediated signaling pathway in beta-cells. J Biol Chem 2005; 280:2282-93.
101. Manning BD. Balancing Akt with S6K: implications for both metabolic diseases and tumorigenesis. J Cell Biol 2004; 167:399-403.
102. Harrington LS, Findlay GM, Gray A et al. The TSC1-2 tumor suppressor controls insulin-PI3K signaling via regulation of IRS proteins. J Cell Biol 2004; 166:213-23.

103. Pende M, Kozma SC, Jaquet M et al. Hypoinsulinaemia, glucose intolerance and diminished beta-cell size in S6K1-deficient mice. Nature 2000; 408:994-7.
104. Um SH, Frigerio F, Watanabe M et al. Absence of S6K1 protects against age- and diet-induced obesity while enhancing insulin sensitivity. Nature 2004; 431:200-5.
105. Tsukiyama-Kohara K, Poulin F, Kohara M et al. Adipose tissue reduction in mice lacking the translational inhibitor 4E-BP1. Nat Med 2001; 7:1128-32.
106. Cho HJ, Park J, Lee HW et al. Regulation of adipocyte differentiation and insulin action with rapamycin. Biochem Biophys Res Commun 2004; 321:942-8.
107. Kim JE, Chen J. regulation of peroxisome proliferator-activated receptor-gamma activity by mammalian target of rapamycin and amino acids in adipogenesis. Diabetes 2004; 53:2748-56.
108. Polak P, Cybulski N, Feige JN et al. Adipose-specific knockout of raptor results in lean mice with enhanced mitochondrial respiration. Cell Metab 2008; 8:399-410.
109. Guertin DA, Stevens DM, Thoreen CC et al. Ablation in mice of the mTORC components raptor, rictor, or mLST8 reveals that mTORC2 is required for signaling to Akt-FOXO and PKCalpha, but not S6K1. Dev Cell 2006; 11:859-71.
110. Bentzinger CF, Romanino K, Cloëtta D et al. Skeletal muscle-specific ablation of raptor, but not of rictor, causes metabolic changes and results in muscle dystrophy. Cell Metab 2008; 8:411-24.
111. Shioi T, McMullen JR, Tarnavski O et al. Rapamycin attenuates load-induced cardiac hypertrophy in mice. Circulation 2003; 107:1664-70.
112. Sanada S, Node K, Asanuma H et al. Opening of the adenosine triphosphate-sensitive potassium channel attenuates cardiac remodeling induced by long-term inhibition of nitric oxide synthesis: role of 70-kDa S6 kinase and extracellular signal-regulated kinase. J Am Coll Cardiol 2002; 40:991-7.
113. Gao XM, Wong G, Wang B et al. Inhibition of mTOR reduces chronic pressure-overload cardiac hypertrophy and fibrosis. J Hypertens 2006; 24:1663-70.
114. Gillette-Guyonnet S, Vellas B. Caloric restriction and brain function. Curr Opin Clin Nutr Metab Care 2008; 11:686-92.
115. Levenson CW, Rich NJ. Eat less, live longer? New insights into the role of caloric restriction in the brain. Nutr Rev 2007; 65:412-5.
116. Malagelada C, Jin ZH, Greene LA. RTP801 is induced in Parkinson's disease and mediates neuron death by inhibiting Akt phosphorylation/activation. J Neurosci 2008; 28:14363-71.
117. Millino C, Fanin M, Vettori A et al. Different atrophy-hypertrophy transcription pathways in muscles affected by severe and mild spinal muscular atrophy. BMC Med 2009; 7:14.
118. Sarkar S, Ravikumar B, Rubinsztein DC. Autophagic clearance of aggregate-prone proteins associated with neurodegeneration. Methods Enzymol 2009; 453:83-110.
119. Bonini NM, La Spada AR. Silencing polyglutamine degeneration with RNAi. Neuron 2005; 48:715-8.
120. Cheng-Lai A, Frishman WH. Sirolimus-eluting coronary stents: novel devices for the management of coronary artery disease. Am J Ther 2004; 11:218-28.
121. Weil RJ. Incorporating molecular tools into early-stage clinical trials. PLoS Med 2008; 5:e21.
122. Miller RA, Harrison DE, Astle CM et al. An aging interventions testing program: study design and interim report. Aging Cell 2007; 6:565-75.
123. Syntichaki P, Troulinaki K, Tavernarakis N. Protein synthesis is a novel determinant of aging in Caenorhabditis elegans. Ann N Y Acad Sci 2007; 1119:289-95.
124. Tavernarakis N. Ageing and the regulation of protein synthesis: a balancing act? Trends Cell Biol 2008; 18:228-35.
125. Kim Y, Sun H. Functional genomic approach to identify novel genes involved in the regulation of oxidative stress resistance and animal lifespan. Aging Cell 2007; 6:489-503.
126. Gangloff YG, Mueller M, Dann SG et al. Disruption of the mouse mTOR gene leads to early postimplantation lethality and prohibits embryonic stem cell development. Mol Cell Biol 2004; 24:9508-16.
127. Murakami M, Ichisaka T, Maeda M et al. mTOR is essential for growth and proliferation in early mouse embryos and embryonic stem cells. Mol Cell Biol 2004; 24:6710-8.
128. Shima H, Pende M, Chen Y et al. Disruption of the p70(s6k)/p85(s6k) gene reveals a small mouse phenotype and a new functional S6 kinase. EMBO J 1998; 17:6649-59.
129. Pende M, Um SH, Mieulet V et al. S6K1(−/−)/S6K2(−/−) mice exhibit perinatal lethality and rapamycin-sensitive 5′-terminal oligopyrimidine mRNA translation and reveal a mitogen-activated protein kinase-dependent S6 kinase pathway. Mol Cell Biol 2004; 24:3112-24.
130. Gandin V, Miluzio A, Barbieri AM et al. Eukaryotic initiation factor 6 is rate-limiting in translation, growth and transformation. Nature 2008; 455:684-8.
131. Le Bacquer O, Petroulakis E, Paglialunga S et al. Elevated sensitivity to diet-induced obesity and insulin resistance in mice lacking 4E-BP1 and 4E-BP2. J Clin Invest 2007; 117:387-96.
132. Anderson SJ, Lauritsen JP, Hartman MG et al. Ablation of ribosomal protein L22 selectively impairs alphabeta T-cell development by activation of a p53-dependent checkpoint. Immunity 2007; 26:759-72.

CHAPTER 3

Protein Synthesis and the Antagonistic Pleiotropy Hypothesis of Aging

Pankaj Kapahi*

Abstract

Growth and somatic maintenance are thought to be antagonistic pleiotropic traits, but the molecular basis for this tradeoff is poorly understood. Here it is proposed that changes in protein synthesis mediate the tradeoffs that take place upon genetic and environmental manipulation in various model systems including yeast, worms, flies and mice. This hypothesis is supported by evidence that inhibition of the TOR (target of rapamycin) pathway and various translation factors that inhibit protein synthesis lead to slowing of growth and development but extend lifespan. Furthermore, dietary restriction (DR) that leads to antagonistic changes in growth and lifespan, also mediates this change by inhibiting protein synthesis. Direct screens to identify genes that extend lifespan from a subset of genes that are essential for growth and development have also uncovered a number of genes involved in protein synthesis. Given the conserved mechanisms of protein synthesis across species, I discuss potential mechanisms that mediate the lifespan extension by inhibition of protein synthesis that are likely to be important for aging and age-related disorders in humans.

Evolution of Aging

'Nothing in Biology makes Sense except in Light of Evolution' Dobzhansky.

Though a number of single gene mutations have yielded important insights into the cellular processes involved in determining lifespan, a better understanding of the aging process is likely to be achieved by understanding their function in terms of evolution. The underlying principle of evolutionary theories of aging is that the force of natural selection declines with age.[1] Medawar observed that the inevitable loss of individuals in a population over time due to accidents or diseases leaves fewer animals alive even in the absence of senescence or any age related decline. Therefore, the younger animals contribute the most to the succeeding generations and the gene pool. Two possible theories have been proposed to explain the existence of aging based on this principle. The first proposed by Medawar himself, is known as the mutation accumulation theory. According to this, deleterious mutations that manifest their effects late in life would accumulate in a population and lead to senescence as they are not removed due to lack of selective pressure. The second, proposed by Williams[2] is termed the *antagonistic pleiotropy theory*. In his classic paper, Williams postulated the existence of pleiotropic genes that endow benefits early in life at the cost of deleterious effects later to explain the evolution of senescence.[2] Williams also predicted "*It would be expected that if development could be completely arrested there would be no senescence*". He gave calcification of bones as an example of a pleiotropic process that strengthens skeletal structures early in life but causes deleterious effects later in life due to

*Pankaj Kapahi—Buck Institute for Age Research, 8001 Redwood Blvd., Novato, California 94945. USA. Email: pkapahi@buckinstitute.org

Protein Metabolism and Homeostasis in Aging, edited by Nektarios Tavernarakis.

calcification of arteries. These alleles confer an early advantage and are favored in a population. However, these alleles may have deleterious effects later on (post-reproductive period) but are not removed due to reduced selective pressure. The *disposable soma theory* is based on resource allocation and is compatible with the antagonistic pleiotropy theory. According to this theory aging takes place due to tradeoffs resulting from partitioning of finite energy resources spent either on somatic maintenance or growth and reproduction. Antagonistic pleiotropy has been suggested as a 'public' mechanism of aging which would be common to various species, compared to mutation accumulation which may be a 'private' mechanism.[3] Strong support for antagonistic pleiotropic mechanisms comes from various methods of lifespan extension that are conserved across species. However the molecular mechanisms of how antagonistic pleiotropic phenotypes are mediated remain unclear and hold keys for important insight into the aging process. Genetic mutations in various signaling pathways offer hope to probe into understanding the mechanisms of antagonistic pleiotropy and are described below.

Insulin-Like Signaling (ILS)

Antagonistic pleiotropic mechanisms that modulate lifespan are at play in single gene mutants that extend lifespan. Modulation of genes in the ILS pathway, known to alter nutrient sensing,[4] has been shown to extend lifespan in various species.[3,5-7] Inhibition of ILS pathway using genetic mutants, mirrors the cellular and organismal effects of starvation and leads to a reduction in body size in *Drosophila*.[4,8] On the other hand hypomorphic mutations in *Inr* (insulin-like receptor) and *chico* (insulin receptor substrate), both components of the insulin signaling pathway, extend lifespan.[9,10] In *C. elegans* crowding and starvation conditions lead to formation of dauer larvae, an alternative developmental state that is nonreproducing, stress resistant and long lived.[11,12] Mutants in the ILS pathway, such as *daf-2* (insulin receptor) and *age-1* (PI3 kinase), lead to significant lifespan extensions, but the animals are also more prone to dauer formation.[13,14] Both the dauer formation and lifespan extension phenotypes are suppressed by mutations in *daf-16*, a forkhead family transcription factor[13,15,16] that regulates stress responsive genes.[17,18] Competition experiments in worms show that insulin pathway mutants have greatly reduced fitness which is exacerbated under conditions of starvation.[19] These results together demonstrate that modulation of the ILS pathway mediates tradeoffs between somatic maintenance, by modulating stress responses and fitness early on life.

TOR Pathway

Similar to mutants in the ILS pathway mutants in the TOR pathway, a nutrient sensing pathway conserved from yeast to humans show antagonistically pleiotropic phenotypes. In *D. melanogaster*, larvae lacking TOR show similarities to amino acid-deprived animals, such as reduced nucleolar size, developmental arrest and lipid vesicle aggregation in the larval fat body.[20] However, inhibition of the TOR signaling pathway, including the products of the tuberous sclerosis complex genes (Tsc1 and Tsc2), TOR and S6K, extend lifespan in *D. melanogaster*.[21] In *C. elegans*, deletion of *CeTOR* leads to developmental arrest at the L3 larval stage and intestinal atrophy.[22] A similar phenotype was observed for homozygous *daf-15* mutants.[23] DAF-15, the worm ortholog of the mammalian protein raptor (regulatory associated protein of mammalian TOR), forms a stoichiometric complex with TOR. Interestingly, neither the developmental arrest nor the fat accumulation of homozygous *daf-15* mutants is suppressible by *daf-16* mutations.[23] Inhibition of genes in the TOR pathway in worms can extend lifespan.[24-28] In *C. elegans* inhibition of S6 Kinase reduces body size and growth but extends lifespan.[26] Thus, although mutations that affect the TOR pathway can result in phenotypes similar to those caused by mutations that affect the ILS pathway, the TOR pathway is either acting downstream or in parallel to the ILS pathway in *C. elegans*. ILS is believed to have evolved to allow coordination of growth signals in multicellular organisms, while the TOR pathway plays a cell autonomous role in growth.[29]

Protein Synthesis

Inhibition of protein synthesis is one of the key outputs of reducing the flux through the ILS and TOR pathways and is likely to mediate the phenotypic effects of modulating these pathways at least in mammals.[30,31] Protein synthesis is a regulated cellular process that plays a key role in keeping organismal growth and development in tune with environmental conditions.[32] Direct inhibition of protein synthesis by attenuating expression of various mRNA translation factors or reducing protein in the diet has been shown to antagonistically slow growth and development on one hand and enhance somatic maintenance and lifespan on the other in multiple species.[24-26,28,33-36] A well-studied translation factor in the context of aging is *ifg-1*, the worm ortholog of the eIF4G translation factor. A reduction in overall protein translation upon inhibition of *ifg-1* has been previously observed.[26] Inhibition of *ifg-1* during development causes larval arrest while its reduced expression during adulthood increases lifespan and stress resistance in *C. elegans*.[22,25,26,28,37] eIF-4G acts as a scaffold for the highly conserved eIF-4F cap-binding complex that bridges eIF-4E, which binds the 5' methylated cap, to the poly-A binding protein (PABP), which associates with the poly-A tail, aiding in the circularization of mRNAs and increasing translation initiation.[26,38] Inhibition of a number of components in the eIF4F complex leads to lifespan extension.[26] Over-expression of eIF4E in cultured mammalian cells causes transformation of fibroblasts and increased cell size, which can be reversed by increasing the abundance of 4EBP.[30,39] eIF4e is also critical for growth in *Drosophila*.[40] Overexpression of eIF4E has been shown to increase cellular senescence in mammalian cells as determined by β-galactosidase staining.[41] Inhibition of eIF4E in somatic cells has been shown to be sufficient to extend lifespan in *C. elegans* in a DAF-16 independent manner.[24] Following mRNA translation initiation, the ribosome plays a key role in catalyzing the assembly of amino acids into polypeptide chains. Ribosomal proteins have an established role in protein synthesis and growth in various species. Studies from both *S. cerevisiae* and *C. elegans* demonstrate a conserved role for inhibition of ribosomal proteins in lifespan extension.[25,33] Together, these observations support the idea that inhibition of protein synthesis regulates growth and somatic maintenance antagonistically, a phenomenon that is likely to be conserved across species.

Direct Screens to Identify Genes That Antagonistically Regulate Growth and Longevity

Genome-wide RNAi screens for longevity genes have been performed by several groups.[42,43] Surprisingly, only 3 genes were common in the two screens, which identified a total 112 genes, suggesting the screens for extended longevity have not been saturated. In these screens, animals were exposed to RNAi either from an embryonic or the first larval stage.[42,43] Thus, genes that cause lethality and/or developmental arrest upon inhibition, were excluded from lifespan analysis. However, the antagonistic pleiotropy theory of aging suggests that genes playing essential roles during development may influence lifespan in adulthood. Two recent studies tested this idea and revealed that a significant proportion of genes that cause developmental arrest or lethality in *C. elegans* can extend lifespan when inhibited after development.[37,44]

To get a better understanding of the molecular mechanisms of antagonistic pleiotropy, our group examined the effect on adult lifespan of genes that cause larval arrest based on the previous genome-wide screen performed by the Ahringer lab.[45] We specifically chose genes that show larval arrest, as mutants in the TOR pathway display a similar phenotype. Inhibition of *ifg-1* results in developmental arrest accompanied by the appearance of large, refractile intestinal vesicles and severe intestinal atrophy resembling that of CeTOR deficiency.[22] From a pool of 57 genes that cause developmental arrest after inhibition using RNA interference, 24 genes that extend lifespan were identified when inactivated during adulthood. Many of these genes are involved in regulation of mRNA translation and mitochondrial functions. Genetic epistasis experiments indicate that the mechanisms of lifespan extension by inactivating the identified genes may be different from those of the insulin/insulin-like growth factor 1 (IGF-1) and dietary restriction pathways. Inhibition of many of these genes also results in increased stress resistance and decreased fecundity, supporting the notion that they may mediate the trade-offs between somatic maintenance and reproduction. In the second study 2,700 genes essential for *C. elegans* development were screened and identified 64 genes that extend lifespan when inactivated during adulthood. Genes that are part of the insulin and metabolic pathways, as expected, but also genes involved in protein synthesis and chromatin factors came out of this study.

The two studies identified some common and significant features relevant to the genetics of aging. One important finding was the identification of highly conserved genes with orthologs in flies, mammals and humans, which are likely to extend lifespan by conserved mechanisms across species. Secondly, a number of genes found by both studies showed longer lifespan extension than previously identified in the field by RNAi of other genes. Remarkably, the number of positives from these screens was much higher than those achieved so far, with the first study identifying 23 out 57 genes to be long lived while the second identified 64 out of 2700 genes tested. This was much higher than the 1-2% hit rate obtained from unbiased genome-wide screens in *C. elegans*.[43,46] These studies suggest that the key determinants of aging are likely to have antagonistic pleiotropic phenotypes. There is evidence that protein synthesis may play a role in contributing to evolution of differences in lifespan between species. An inverse correlation across species between rate of protein synthesis and longevity has been previously described.[47,48] Future studies in this area will reveal insights into how protein synthesis mediates the intrinsic link between organism development and adult lifespan.

Dietary Restriction (DR), Protein Synthesis and Antagonistic Pleiotropy

DR has also been hypothesized to influence lifespan by antagonistically pleiotropic mechanisms. It has been proposed that a shift in metabolic investment away from reproduction and growth toward somatic maintenance ensures extended survival under DR.[49] Trade-offs due to genetic changes that influence evolution of species-specific lifespan are important mechanisms invoked for understanding the effects of particular genotypes on life history comparisons in metazoans. Trade-offs can also result from fluctuating environments that require a shift in an organism's physiology to adapt to the environmental change. I postulate that inhibition of protein synthesis plays an important role in both genetic and environmental trade-offs. Given the high cost of maintaining mRNA translation in a cell, with as high as 35-45%[32] of genes involved in this process, it is likely that under periods of dietary restriction there is a shift of investment from continuing protein synthesis for growth to somatic maintenance.

There are a number of observations that support the idea that a reduction in protein synthesis plays an important role in extending lifespan upon DR. It was previously believed that total caloric intake was the most important aspect of extending lifespan by manipulating diet. However a number of studies over the past two decades have demonstrated that restricting protein is more important than restricting the total number of calories. Results comparable to restricting *ad libitum* feeding have been demonstrated by restricting intake of a particular amino acid, such as methionine or tryptophan[34-36,50] in rodent studies. Although total body weight was reduced in methionine-restricted animals, they actually ate more per unit of body mass than the controls.[36] These results suggest that a decrease in energy intake or expenditure on a body weight basis is not necessary for lifespan extension. In *D. melanogaster*, restriction of total food yields extended lifespan.[51-53] Fly media primarily consists of yeast, the major source of protein and sugar. It has been shown by a number of groups that reducing yeast alone is also sufficient to extend lifespan in flies.[21,54-57] The importance of individual macronutrients was underscored by a recent study validating that reduction of yeast and not total calories in the diet extends lifespan in *D. melanogaster*.[57]

The TOR pathway, which responds to the presence of amino acids, has been shown to play important roles in DR-dependent lifespan extension in *D. melanogaster*,[21] *S. cerevisiae*[58] and *C. elegans*.[59,60] The effects of TOR in mediating its lifespan extension effects have been suggested to be through inhibiting protein synthesis in *D. melanogaster*.[61] 4EBP (eukaryotic initiation factor eIF4E binding protein), a translational repressor that acts as a downstream effector of the TOR pathway, is a key player in mediating the lifespan extension effects of DR in flies. 4EBP is phosphorylated upon insulin stimulation and is also targeted by TOR, to regulate mRNA translation and growth.[30,31] When inactive, the hypophosphorylated form acts as a translational repressor by binding the protein synthesis initiation factor eIF4E, thus blocking the activity of the eIF4F complex. The eukaryotic initiation factor 4F (eIF4F) initiation complex mediates growth-dependent protein synthesis.[30-32] Together the observations that reduction of protein in the diet and the role of the amino acid sensing TOR pathway in mediating lifespan extension by TOR support the idea that modulation of protein synthesis plays a role in mediating the lifespan extension by DR.

Mechanism of Lifespan Extension by Inhibition of Protein Synthesis

What are the mechanisms by which inhibiting protein synthesis mediates antagonistic pleiotropic phenotypes? Given that amino acids are building blocks of proteins, it is easy to conceive how reduction in protein synthesis slows growth and development but why this would extend lifespan is not intuitive. This is also apparently paradoxical if one takes into account that there is a global decline in protein synthesis with age. A model depicting how inhibition of protein synthesis leads to antagonistic pleiotropic effects on growth and somatic maintenance is shown in Figure 1. One possible mechanism for the lifespan extension is that the inhibition in protein synthesis enhances lifespan by reducing the burden on the protein homeostatic apparatus.[62,63] The advent of new technologies that can assess the half-life of individual proteins and quantify the damage inflicted to individual proteins will be of great use in examining the hypothesis. It would also be of great interest to understand if the autophagy machinery plays a role in altering protein homeostasis under conditions of low protein synthesis.

Another important question with regards to the mechanisms of lifespan extension is, whether there is simply a global decrease in protein synthesis, or differential translation of particular mRNAs despite an overall decrease in mRNA translation. Previous studies have pointed out the importance of regulation of mRNA translation especially under environmental stresses and the apparent lack of correlation between the transcriptome and proteome.[32,64,65] A recent study shows that RAS and Akt contribute to growth and tumor formation only modestly by transcriptional changes but largely by altered recruitment of mRNAs to large polysomes.[66] This study found that genes that regulate growth, transcriptional regulation, cell-to-cell interactions and morphology were mostly altered in their mRNA translation state. Selective increase in translation of particular mRNAs

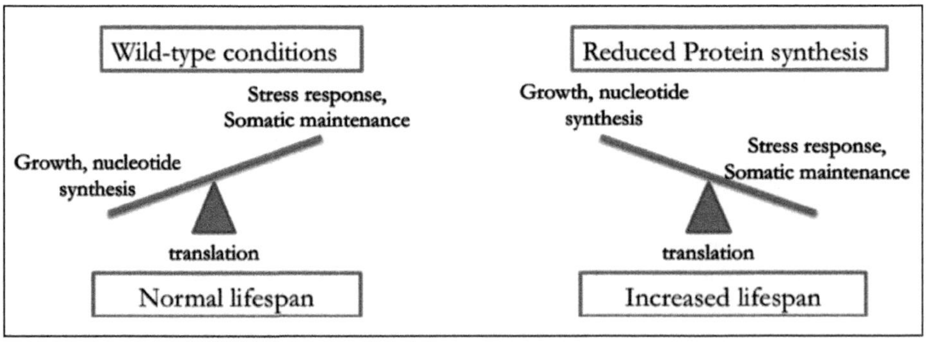

Figure 1. A model for how inhibition of protein synthesis leads to antagonistic pleiotropic effects on growth and somatic maintenance. Under normal conditions there is higher protein synthesis, which leads to increased resources for growth and reproduction. Under conditions of reduced protein synthesis there is a shift in the balance leading to an increased investment in somatic maintenance and a reduction in growth.

under conditions that result in most mRNAs being translationally repressed has been described in other systems. Inhibition of eIF4E in mammalian cells has been shown to decrease cell growth and protein synthesis but increase the protein levels of HSP 90, HSP 70, HSP 65 and HSP 27. This increase in protein was found to be due to an increased association of the respective mRNAs with larger polysomes.[67] Furthermore, under well-fed conditions, yeast transcriptional regulator GCN4 mRNAs are attached to only a single ribosome. Upon starvation, its mRNAs are found associated with polysomes due to an increased rate of translation initiation.[68] While the global rate of protein synthesis decreases under these conditions, production of GCN4 protein increases as suggested by the increase in ribosomal loading for GCN4-specific mRNA, driven by short upstream open reading frames in the 5'UTR of the gene.[69] A recent study has shown that GCN4 is important for mediating changes in lifespan in *Saccharomyces cerevisiae*. The transcriptional regulator GCN4 has been shown to be translationally upregulated and contributes to increased lifespan when overall mRNA translation is reduced via inhibition of 60S ribosomal subunit expression.[33] In order to investigate regulation of mRNA translation, a method to study genome-wide mRNA translation state has been previously established.[65,70] This translational profiling method assesses the mRNA translation state based on the separation of mRNAs bound to varying number of polysomes via density gradient centrifugation. This method is likely to prove very useful in identifying translational changes in gene expression which will help reveal new insights into how mRNA translation regulation plays a role in mediating the antagonistic pleiotropic effects of inhibiting protein synthesis.

Conclusion

In summary, the effects of inhibition of protein synthesis in different model systems, provides a novel paradigm to explore the mechanisms of antagonistic pleiotropy. The advent of new technologies to examine mRNA translation state, protein turnover and de novo protein synthesis are likely to play a critical role in understanding the mechanisms of phenotypic changes upon inhibition of protein synthesis. Given the high degree of genetic homology between cellular processes, especially mRNA translation regulation, in humans and model systems like worms, flies and yeast and given the importance of protein translation in human disease, this area is likely to be of extreme significance for understanding various age-related human diseases.

Acknowledgements

We are grateful to Di Chen, Aric Rogers, Keith Lizotte and Gordon Lithgow for sharing their ideas and helpful discussions. This work was funded by grants from the Ellison Medical Foundation, American Foundation for aging research, Hillblom foundation, a Nathan Shock Startup award, a gift from the Harold J. and Reta Haynes Family Foundation and the NIH (RL1AAG032113, 1R21AG028241-01).

References

1. Medawar PB. An Unsolved Problem of Biology. London: H.K. Lewis. 1952.
2. Williams GC. Pleiotropy, natural selection and evolution of senescence. Evolution 1957; 11:398-411.
3. Martin GM, Austad SN, Johnson TE. Genetic analysis of ageing: role of oxidative damage and environmental stresses. Nat Genet 1996; 13(1):25-34.
4. Britton JS, Lockwood WK, Li L et al. Drosophila's insulin/PI3-kinase pathway coordinates cellular metabolism with nutritional conditions. Dev Cell 2002; 2(2):239-49.
5. Tatar M, Bartke A, Antebi A. The endocrine regulation of aging by insulin-like signals. Science 2003; 299(5611):1346-51.
6. Partridge L, Gems D. Mechanisms of ageing: public or private? Nat Rev Genet 2002; 3(3):165-75.
7. Guarente L, Kenyon C. Genetic pathways that regulate ageing in model organisms. Nature 2000; 408(6809):255-62.
8. Bohni R, Riesgo-Escovar J, Oldham S et al. Autonomous control of cell and organ size by CHICO, a Drosophila homolog of vertebrate IRS1-4. Cell 1999; 97(7):865-75.
9. Tatar M, Kopelman A, Epstein D et al. A mutant Drosophila insulin receptor homolog that extends life-span and impairs neuroendocrine function. Science 2001; 292(5514):107-10.

10. Clancy DJ, Gems D, Harshman LG et al. Extension of life-span by loss of CHICO, a Drosophila insulin receptor substrate protein. Science 2001; 292(5514):104-6.
11. Riddle DL, Swanson MM, Albert PS. Interacting genes in nematode dauer larva formation. Nature 1981; 290(5808):668-71.
12. Klass M, Hirsh D. Non-ageing developmental variant of Caenorhabditis elegans. Nature 1976; 260(5551):523-5.
13. Kenyon C, Chang J, Gensch E et al. A C. elegans mutant that lives twice as long as wild type. Nature 1993; 366(6454):461-4.
14. Johnson TE. Increased life-span of age-1 mutants in Caenorhabditis elegans and lower Gompertz rate of aging. Science 1990; 249(4971):908-12.
15. Vowels JJ, Thomas JH. Genetic analysis of chemosensory control of dauer formation in Caenorhabditis elegans. Genetics 1992; 130(1):105-23.
16. Ogg S, Paradis S, Gottlieb S et al. The Fork head transcription factor DAF-16 transduces insulin-like metabolic and longevity signals in C. elegans. Nature 1997; 389(6654):994-9.
17. Murphy CT, McCarroll SA, Bargmann CI et al. Genes that act downstream of DAF-16 to influence the lifespan of Caenorhabditis elegans. Nature 2003; 424(6946):277-83.
18. McElwee J, Bubb K, Thomas JH. Transcriptional outputs of the Caenorhabditis elegans forkhead protein DAF-16. Aging Cell 2003; 2(2):111-21.
19. Walker DW, McColl G, Jenkins NL et al. Evolution of lifespan in C. elegans. Nature 2000; 405(6784):296-7.
20. Zhang H, Stallock JP, Ng JC et al. Regulation of cellular growth by the Drosophila target of rapamycin dTOR. Genes Dev 2000; 14(21):2712-24.
21. Kapahi P, Zid BM, Harper T et al. Regulation of lifespan in Drosophila by modulation of genes in the TOR signaling pathway. Curr Biol 2004; 14(10):885-90.
22. Long X, Spycher C, Han ZS et al. TOR deficiency in C. elegans causes developmental arrest and intestinal atrophy by inhibition of mRNA translation. Curr Biol 2002; 12(17):1448-61.
23. Jia K, Chen D, Riddle DL. The TOR pathway interacts with the insulin signaling pathway to regulate C. elegans larval development, metabolism and life span. Development 2004; 131(16):3897-906.
24. Syntichaki P, Troulinaki K, Tavernarakis N. eIF4E function in somatic cells modulates ageing in Caenorhabditis elegans. Nature 2007; 445(7130):922-6.
25. Hansen M, Taubert S, Crawford D et al. Lifespan extension by conditions that inhibit translation in Caenorhabditis elegans. Aging Cell 2007; 6(1):95-110.
26. Pan KZ, Palter JE, Rogers AN et al. Inhibition of mRNA translation extends lifespan in Caenorhabditis elegans. Aging Cell 2007; 6(1):111-9.
27. Vellai T, Takacs-Vellai K, Zhang Y et al. Genetics: influence of TOR kinase on lifespan in C. elegans. Nature 2003; 426(6967):620.
28. Henderson ST, Bonafe M, Johnson TE. daf-16 protects the nematode Caenorhabditis elegans during food deprivation. J Gerontol A Biol Sci Med Sci 2006; 61(5):444-60.
29. Kapahi P, Zid B. TOR pathway: linking nutrient sensing to life span. Sci Aging Knowledge Environ 2004; 2004(36):PE34.
30. Shamji AF, Nghiem P, Schreiber SL. Integration of growth factor and nutrient signaling: implications for cancer biology. Mol Cell 2003; 12(2):271-80.
31. Harris TE, Lawrence JC Jr. TOR signaling. Sci STKE 2003; 2003(212):re15.
32. Sonenberg N, Hershey JWB, Mathews BM. Translational Control of Gene Expression. Cold Spring Harbor: Cold Spring Harbor Laboratory Press, 2000.
33. Steffen KK, MacKay VL, Kerr EO et al. Yeast life span extension by depletion of 60s ribosomal subunits is mediated by Gcn4. Cell 2008; 133(2):292-302.
34. Miller RA, Buehner G, Chang Y et al. Methionine-deficient diet extends mouse lifespan, slows immune and lens aging, alters glucose, T4, IGF-I and insulin levels and increases hepatocyte MIF levels and stress resistance. Aging Cell 2005; 4(3):119-25.
35. Richie JP Jr, Leutzinger Y, Parthasarathy S et al. Methionine restriction increases blood glutathione and longevity in F344 rats. FASEB J 1994; 8(15):1302-7.
36. Orentreich N, Matias JR, DeFelice A et al. Low methionine ingestion by rats extends life span. J Nutr 1993; 123(2):269-74.
37. Curran SP, Ruvkun G. Lifespan regulation by evolutionarily conserved genes essential for viability. PLoS Genet 2007; 3(4):e56.
38. Hershey JWB, Merrick WC. The pathway and mechanism of initiation of protein synthesis. In Translational Control of Gene Expression (eds Sonenberg et al). 2000:33-8.
39. Lazaris-Karatzas A, Montine KS, Sonenberg N. Malignant transformation by a eukaryotic initiation factor subunit that binds to mRNA 5' cap. Nature 1990; 345(6275):544-7.

40. Lachance PE, Miron M, Raught B et al. Phosphorylation of eukaryotic translation initiation factor 4E is critical for growth. Mol Cell Biol 2002; 22(6):1656-63.
41. Ruggero D, Montanaro L, Ma L et al. The translation factor eIF-4E promotes tumor formation and cooperates with c-Myc in lymphomagenesis. Nat Med 2004; 10(5):484-6.
42. Hamilton B, Dong Y, Shindo M et al. A systematic RNAi screen for longevity genes in C. elegans. Genes Dev 2005; 19(13):1544-55.
43. Hansen M, Hsu AL, Dillin A. New genes tied to endocrine, metabolic and dietary regulation of lifespan from a Caenorhabditis elegans genomic RNAi screen. PLoS Genet 2005; 1(1):119-28.
44. Chen D, Pan KZ, Palter JE et al. Longevity determined by developmental arrest genes in Caenorhabditis elegans. Aging Cell 2007.
45. Kamath RS, Fraser AG, Dong Y et al. Systematic functional analysis of the Caenorhabditis elegans genome using RNAi. Nature 2003; 421(6920):231-7.
46. Lee SS, Lee RY, Fraser AG et al. A systematic RNAi screen identifies a critical role for mitochondria in C. elegans longevity. Nat Genet 2002.
47. Finch CE. Longevity, Senescence and the Genome. Chicago: University of Chicago Press; 1990.
48. Munro HN. Evolution of protein metabolism in mammals. In: Munro HN, Allison JB, eds. Mammalian Protein Metabolism, vol. 3. Academic Press Inc, 1969:133-82.
49. Holliday R. Food, reproduction and longevity: is the extended lifespan of calorie-restricted animals an evolutionary adaptation? Bioessays 1989; 10(4):125-7.
50. Zimmerman JA, Malloy V, Krajcik R et al. Nutritional control of aging. Exp Gerontol 2003; 38(1-2):47-52.
51. Clancy DJ, Gems D, Hafen E et al. Dietary restriction in long-lived dwarf flies. Science 2002; 296(5566):319.
52. Rogina B, Helfand SL, Frankel S. Longevity regulation by Drosophila Rpd3 deacetylase and caloric restriction. Science 2002; 298(5599):1745.
53. Mair W, Goymer P, Pletcher SD et al. Demography of dietary restriction and death in Drosophila. Science 2003; 301(5640):1731-3.
54. Chippindale AK, Leroi AM, Kim SB et al. Phenotypic plasticity and selection in Drosophila life-history evolution. I. Nutrition and the cost of reproduction. J Evol Biology 1993; 6:171-93.
55. Nusbaum J, Rose MR. The effects of nutritional manipulation and laboratory selection on lifespan in Drosophila melanogaster. Journal of Gerontology: Biological Sciences 1999; 54A:B192-B8.
56. Good TP, Tatar M. Age-specific mortality and reproduction respond to adult dietary restriction in Drosophila melanogaster. J Insect Physiol 2001; 47(12):1467-73.
57. Mair W, Piper MD, Partridge L. Calories do not explain extension of life span by dietary restriction in Drosophila. PLoS Biol 2005; 3(7):e223.
58. Kaeberlein M, Powers RW 3rd, Steffen KK et al. Regulation of yeast replicative life span by TOR and Sch9 in response to nutrients. Science 2005; 310(5751):1193-6.
59. Chen D, Thomas EL, Kapahi P. HIF-1 modulates dietary restriction-mediated lifespan extension via IRE-1 in Caenorhabditis elegans. PLoS Genet 2009; 5(5):e1000486.
60. Hansen M, Chandra A, Mitic LL et al. A role for autophagy in the extension of lifespan by dietary restriction in C. elegans. PLoS Genet 2008; 4(2):e24.
61. Zid BM, Rogers A, Katewa SD et al. 4E-BP modulates lifespan and mitochondrial translation upon dietary restriction in Drosophila. Cell 2009; Manuscript accepted.
62. Hipkiss AR. On why decreasing protein synthesis can increase lifespan. Mech Ageing Dev 2007.
63. Kaeberlein M, Kennedy BK. Protein translation, 2007. Aging Cell 2007; 6(6):731-4.
64. Sonenberg N, Hinnebusch AG. New modes of translational control in development, behavior and disease. Mol Cell 2007; 28(5):721-9.
65. Zong Q, Schummer M, Hood L et al. Messenger RNA translation state: the second dimension of high-throughput expression screening. Proc Natl Acad Sci USA 1999; 96(19):10632-6.
66. Rajasekhar VK, Viale A, Socci ND et al. Oncogenic Ras and Akt signaling contribute to glioblastoma formation by differential recruitment of existing mRNAs to polysomes. Mol Cell 2003; 12(4):889-901.
67. Joshi-Barve S, De Benedetti A, Rhoads RE. Preferential translation of heat shock mRNAs in HeLa cells deficient in protein synthesis initiation factors eIF-4E and eIF-4 gamma. J Biol Chem 1992; 267(29):21038-43.
68. Tzamarias D, Roussou I, Thireos G. Coupling of GCN4 mRNA translational activation with decreased rates of polypeptide chain initiation. Cell 1989; 57(6):947-54.
69. Hinnebusch AG. Translational regulation of GCN4 and the general amino acid control of yeast. Annu Rev Microbiol 2005; 59:407-50.
70. Serikawa KA, Xu XL, MacKay VL et al. The Transcriptome and Its Translation during Recovery from Cell Cycle Arrest in Saccharomyces cerevisiae. Mol Cell Proteomics 2003; 2(3):191-204.

Chapter 4

Proteasome Function Determines Cellular Homeostasis and the Rate of Aging

Niki Chondrogianni and Efstathios S. Gonos*

Abstract

Homeostasis is a key feature of cellular lifespan. Maintenance of cellular homeostasis influences the rate of aging and its efficiency is determined by the cooperation between protein stability and resistance to stress, protein refolding, protein repair and proteolysis of damaged proteins. Protein degradation is predominately catalyzed by the proteasome which is responsible for cell clearance of abnormal, denatured or in general damaged proteins as well as for the regulated degradation of short-lived proteins. Impaired proteasome function has been tightly correlated to aging both in vivo and in vitro and thus, emphasis has been given recently in identifying ways of its activation. A number of studies have shown that the proteasome can be activated by genetic manipulations as well as by factors that affect either its conformation and stability or the expression of its subunits and the rate of proteasome assembly. This "readjustment" has been shown to have a great impact on retention of cellular homeostasis since it promotes lifespan extension. This chapter focuses on protein homeostasis and its direct link to proteasome function, dysfunction and manipulation and provides insights regarding the activation of proteasome-mediated protein degradation that, in turn, ensures health maintenance.

Protein Homeostasis and Aging: Which Are the Key Players?

Maintenance of cellular homeostasis is a basic feature that determines organismal lifespan. Cellular homeostasis is affected by several intracellular functions since there are different pathways that need to be properly protected and maintained in order to assure organismal homeostasis.[1] Protein homeostasis is one of the major nodal points that need to be preserved in order to retain homeostatic balance. Efficient protein maintenance that is determined by protein stability and resistance to stress per se is accompanied by potential protein repair but also by efficient protein degradation, processes that in turn influence organismal homeostasis and lifespan.

Proteome homeostasis is characterized by "preventing" as well as by "repairing" steps. Preventing functions that contribute to protein homeostasis include resistance of proteins per se to oxidative and other cytotoxic damage(s). Several studies have revealed the central role of protein stability and resistance to oxidative stress as determinants of longevity. These studies have revealed the central role of the cellular antioxidant mechanisms to the ameliorated survival and extended lifespan, depending on the model used[2] and their positive correlation to maximum lifespan potential.[3]

*Corresponding Author: Efstathios S. Gonos—National Hellenic Research Foundation, Institute of Biological Research and Biotechnology, 48 Vas. Constantinou Ave., Athens 11635, Greece. Email: sgonos@eie.gr

Protein Metabolism and Homeostasis in Aging, edited by Nektarios Tavernarakis.

More recently, studies in long-lived species have further revealed the role of the "preventing" step of protein homeostasis in extended lifespan. More precisely, Chaudhuri and colleagues have examined the possible causes of extended lifespan in the longest-living rodent, the naked mole-rat[4] as well as in long-lived bats.[5] Specifically, two species of long-lived bats exhibit enhanced resistance to protein oxidation and increased resistance of their proteins to unfolding per se. Moreover, these levels of protein self-protection seem to be adequate for the recorded extended lifespan of these species since, surprisingly, they do not possess any other enhanced protective systems, like for instance increased protein removal.[5] In contrast, the naked mole-rats display enhanced resistance to oxidative stress with concominant ability of the native proteins to resist against in vitro unfolding. Moreover these animals are characterized with attenuated age-related changes in cystein oxidation and in general a tightly modulated protein-redox state, systems that may also contribute to their exceptional longevity.[4]

Given that the proteome of most species has not yet shown to be equipped with the above "preventing" characteristics, the "repairing" steps against oxidative stress are of equal or even higher importance. Therefore, when protein damage occurs, the homeostatic mechanisms are activated to refold the protein, to repair the protein damage (when it is possible) or to degrade the damaged molecules in order to prevent their accumulation and aggregation inside the cell. Three "cellular players" are involved: the chaperones, the protein repair enzymes that reverse some of the protein oxidative damage and the cell clearance arsenal that includes the systems that are involved in the removal of damaged proteins. Regarding the chaperones, their role in aging is well established as they constitute the intermediate point of protein rescue if, at all, possible.[6] Depending on the amino acids that are subjected to oxidation/damage, there are also few protein repair enzymes; the glutaredoxin/glutathione/glutathione reductase, the thioredoxin/thioredoxin reductase and the methionine sulfoxide reductases are the best characterized systems. These enzymes can reduce catalytically the oxidation of sulfur-containing amino acids, cysteine and methionine that have been shown to be the most susceptible to oxidation.[7] Among these few protein repair enzymes, MsrA and MsrB enzymes have been shown to play a major role in protein homeostasis during aging. Their role has been largely addressed once they were overexpressed, leading to extended lifespan and better coping against oxidative challenges.[8,9] In support, Friguet and colleagues have also shown that replicative senescence of human fibroblasts is accompanied with their reduced expression and function.[10] However, despite the high importance of the above mentioned repairing enzymes, the major "repairing" system against protein damage is protein degradation. Protein degradation can be mediated by lysosomes,[11] proteasomes[12] and the Lon protease in mitochondria.[13] Although all three proteolytic machineries are highly important, the proteasome represent the basic cellular machinery governing regulated proteolysis of both normal and abnormal proteins.

An Introduction to the Proteasome Biology

Controlled protein degradation is predominately catalyzed by the proteasome. The proteasome is responsible for cell clearance of abnormal, denatured or in general damaged proteins as well as for the regulated degradation of short-lived proteins.[12,14] The proteolytic core, the 20S proteasome is a 700 kDa multisubunit enzyme complex composed by 7 different α- and 7 different β-subunits arranged as a barrel-shape stack of four heptameric rings localized in both cytoplasm and nucleus. The two outer α-subunits rings ($\alpha_{1\text{-}7}$) embrace two central head-to-head oriented rings containing β-subunits ($\beta_{1\text{-}7}$). The proteolytic active sites are hosted in the internal chamber that is composed by β-subunits. Three of the β-subunits, β_1, β_2 and β_5, are responsible for the proteasome hydrolyzing activities that cleave peptide bonds on the carboxyl site of acidic (peptidylglutamylpeptide hydrolyzing activity, PGPH), basic (trypsin-like activity, T-L) and hydrophobic (chymotrypsin-like activity, CT-L) amino acids respectively (Fig. 1).[12]

Once the 20S particle is capped by 19S regulatory complexes, the 26S proteasome is assembled which is central for the ATP/ubiquitin-dependent protein degradation. The 19S particle is composed by two subcomplexes, namely the lid and the base (Fig. 1). The lid covers

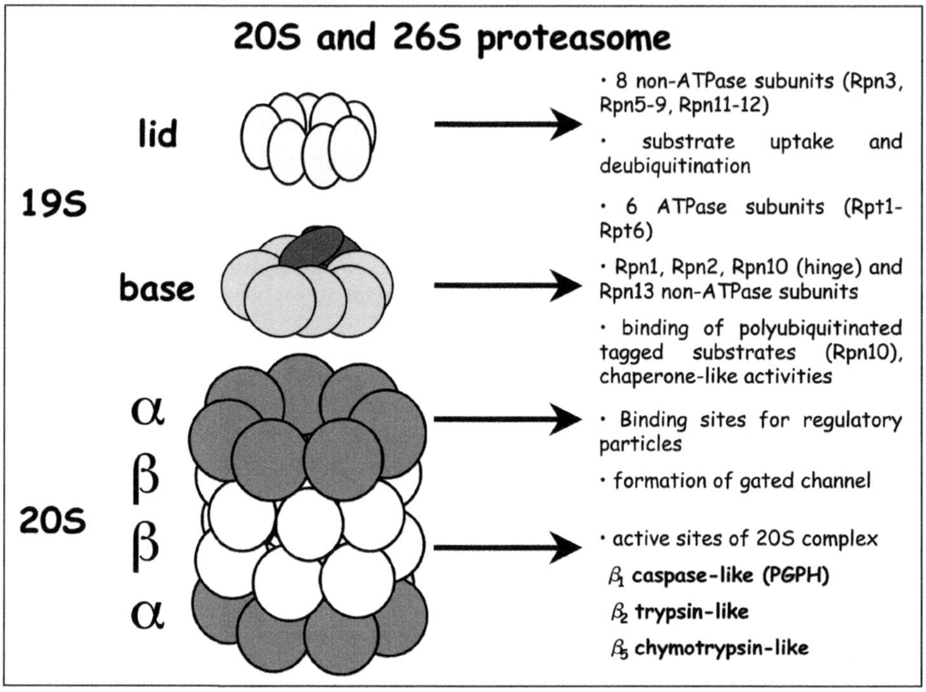

Figure 1. Schematic representation of the 20S proteasome and the 19S regulatory particle.

the base and it is involved in the substrate uptake and deubiquitination. The base is consisted of six ATPase subunits as well as four non-ATPase subunits involved in the recognition, unfolding and further translocation of the substrate.[15] The target protein is firstly labeled via covalent attachment of multiple ubiquitin molecules before the degradation by the 26S complex. Three steps are involved in the conjugation of ubiquitin to the substrate. Ubiquitin is activated by E1, the ubiquitin-activating enzyme, then transferred by an E2 enzyme, the ubiquitin-carrier protein or ubiquitin-conjugating enzyme, to a member of the ubiquitin-protein ligase family, E3, to which the target protein is specifically bound. This enzyme catalyzes the repeated conjugation of ubiquitin to the substrate prior of the 26S complex-mediated degradation.[14]

Proteasome has also been implicated in antigen presentation process and immune response, mainly through immunoproteasomes. In this proteasome type, the constitutively expressed β_1, β_2 and β_5 subunits are substituted during de novo proteasome biosynthesis by β_{1i}, β_{2i} and β_{5i} subunits respectively. As a result, although these particles digest proteins at rates similar to those for constitutive proteasomes, they generate a higher fraction of peptides with the appropriate C-termini and length to serve in antigen presentation. In immunoproteasomes, PA28/11S activator can replace the 19S complex.[16] Finally additional tissues may also have their own type of proteasomes as it was shown for cortical thymic epithelial cells where the β_{5i} subunit is substituted by β_{5t}.[17]

Proteasome biogenesis is an accurately ordered multistep event. It is initiated with the biosynthesis of α- and β-subunits, followed by the organization of α-subunits in seven-member rings under the supervision and assistance of proteasome assembling chaperones, 1, 2 and 3 (PAC1-3).[18] Then the precursor β-subunits assemble[19] and associate with UMP1/POMP accessory protein that, in conjunction to the α-subunits rings, give rise to half proteasomes. Upon dimerization of two half proteasome precursors, the preholeoproteasome is produced that is then activated through the cleavage of the precursor and still inactive β-subunits, thus giving rise to the mature 20S proteasomes.[20]

Proteasome during Aging

Aging is a natural biological process that involves the gradual decline of a variety of physiological functions and the eventual failure of homeostasis. Mammalian aging can be studied in vitro as mitotic cells loose their proliferative capacity following serial passaging, a process termed as replicative senescence.[21] Senescent cells are characterized by several morphological and biochemical alterations when compared with their young counterparts, including changes in proteasome function. Specifically, a number of studies have shown an ~50% decline of proteasomal activities in several aged tissues in humans (muscle, lens, lymphocytes and epidermis) as well as in other mammals such as mice, rats and bovines (liver, spinal cord, lens, heart and retina).[22] A similar decline of proteasome activities has also been revealed in human primary cultures undergoing replicative senescence,[23] whereas, in support, proteasome inhibition in young cells induces premature senescence.[24] Earlier work has attributed the observed decrease of proteasome function to the accumulation of damaged proteins, such as lipofuscin, during aging and senescence.[25] To understand the molecular basis of these findings, work in our laboratory has determined that the decreased function of proteasome in senescence is, in addition to the accumulation of damaged proteins, primarily due to the reduced rates of proteasome biosynthesis and assembly. Moreover, we have determined that replicative senescence is accompanied with reduced levels of the β-type subunits, thus acting as the "rate-limiting" subunits for efficient proteasome assembly.[23] Immunoproteasomes have also been investigated during aging. Ponnappan and coworkers[26] have recently reported lower expression of α- and β-type proteasome subunits in T-cells derived from old donors, while we have reported the inability of senescent human fibroblasts to induce proteasome immunosubunits following interferon-γ treatment as opposed to their young counterparts.[27] Interestingly, in tissues irrelevant to the immune system, like the muscle, an age-related up-regulation of immunoproteasomes in parallel to the decreased amount of constitutive proteasomes has been demonstrated, thereby implying for a potential compensatory mechanism which may result in an overall proteasome function fine-tuning.[28] Additional work in 19S complex has identified several subunits to be down-regulated during replicative senescence,[23,29, our unpublished data] whereas a lower ability of proteasome activators to assemble with 20S complex in aged rat muscle has also been reported.[28,30] In support, studies in lower eukaryotes like *Drosophila melanogaster* have shown that aging also perturbs 26S proteasome assembly,[31] whereas more recently Miura and colleagues have provided genetic evidence in *Drosophila melanogaster* that links age-dependent attenuation of the 26S proteasome function and assembly with the aging process.[32] Finally, aging has been also shown to have profound effects in the whole axis of ubiquitin/proteasome. Some of the compounds of this machinery, e.g., E1 and several E2 enzymes, like $E2_{20k}$ and $E2_{25k}$, have been correlated to aging and oxidative stress in lens. Specifically, Shang et al[33] have found increased ubiquitination activities following oxidative stress without any change, however, of the mentioned E2 protein expression levels. The same work highlights that lens have the ability to increase ubiquitin conjugation activity in response to oxidative stress and this ability is attenuated upon aging. Recently, components of the Skp1-Cul1-F-Box E3 ligase complex have been identified to regulate the lifespan of *Caenorhabditis. elegans*.[34] In complementary studies, worms lacking AIP-1, the homologue of mammalian AIRAP (arsenic-inducible proteasomal 19S regulatory particle-associated protein), were not only impaired in their ability to resist exposure to arsenite but also exhibited shortened lifespan and hypersensitivity to misfolding-prone proteins under normal laboratory conditions.[35] In contrast, the overexpression of this regulator protected animals against Aβ toxicity.[36] Apart from the above mentioned studies that show mainly alterations in the proteasome content or in the assembly of the 26S complex during aging, several studies have also demonstrated proteasome inactivation through modification of proteasome subunits and inhibition of its function through endogenously heavily modified and cross-linked proteins that practically block the proteasome.[22] In summary, these data strongly suggest that aging has profound effects on proteasome biosynthesis, assembly and function and moreover, that the progression of the process is highly affected or even accelerated by proteasome dysfunction.

Proteasome Activation: Is There a Way to Restore Proteasome Function?

Proteasome can be activated by several regulators, such as the 19S and 11S complexes, for performing its ordinary functions as described previously.[12,14] However, over the last years it became evident that proteasome, in addition to its known regulators, can also be activated by genetic or other means. This activation is mediated either through conformational changes of the structure of the proteasome or through increased expression of the proteasome subunits and regulators and the downstream enhanced assembly of the complex.

Genetic Activation of the Proteasome

Goldberg and coworkers[37] were the first to demonstrate the enhancement of CT-L and T-L proteasome activities following β5i transfection in lymphoblasts and HeLa cells. The same group in a follow-up study has shown a similar stimulation of T-L and PGPH activities following β1i and β1 subunit overexpression in HeLa cells.[38] We have also overexpressed the β5 subunit stably in WI38/T and HL60 cells.[39] We found that transfection of the β5 subunit resulted to upregulation of other β-type subunits, thus implying for a common regulatory loop that is still under investigation and to the recruitment of "free" α-type subunits that are normally found in excess to produce new and functional proteasomes. This was evident as the developed "proteasome activated cell lines" exhibited increased rates of proteolysis as well as enhanced resistance against various oxidative challenges. Importantly, a similar β5 subunit overexpression in primary human embryonic fibroblasts (IMR90 cells) resulted in the extension of lifespan by ~15-20% with the parallel maintenance of the young phenotype for longer periods.[39] The observed coregulation of β-type subunits has also been confirmed in other cell types, such as the lens epithelial cells[40] or the murine neuroblastoma cells,[41] following overexpression of the β5 subunit. Moreover, it was recently shown that restoration of the normal level of proteasome subunits in aged human fibroblasts reduces the levels of various aging biomarkers,[42] thereby confirming our assumption regarding the vital association between optimal proteasome function and retention of cellular homeostasis. Additionally, we have achieved proteasomal upregulation via overexpression of hUMP1/POMP protein.[43] hUMP1/POMP stable clones exhibited increased rates of proteasome assembly and function as assayed by their enhanced recovery ability following administration of oxidative stressors. In support, Keller and colleagues have shown that UMP1 overexpression in yeast enhances viability during oxidative stress mainly through the enhanced preservation of proteasome-mediated protein degradation which, in turn, extends lifespan.[44] Moreover, silencing of UMP1 by RNA interference has been shown to induce proteasome inhibition in yeast that eventually promotes a robust increase of protein oxidation that results to autophagic death,[45] whereas, deletion of UMP1 has a neutral effect on cellular viability under normal growth conditions, but impairs the ability of yeast cells to survive under stationary phase conditions.[46] Finally, more recently, Miura and colleagues suggested that improving the amount and/or activity of the 26S proteasome by overexpressing a lid subunit such as Rpn11, could provide an extension to the mean lifespan and suppress the age-related progression of neurodegenerative diseases.[32] In total, these data reveal that the proteasome system (20S and 26S) can be genetically modulated and have great impact on the progression of aging and longevity.

Proteasome Activation by Natural or Chemical Compounds

Structural alterations of proteasome subunits have been shown to affect proteasome activity through changes of the 20S barrel conformation. Initially, SDS and some fatty acids have been shown to stimulate proteasome activities in the test tube,[47] whereas potassium chloride has a negative effect by favoring the open or the closed conformation of the proteasome,[48] respectively. Proteasome-activating hydrophobic peptides have been shown to be bound as modifiers at noncatalytic sites, thus mimicking the effect of 11S complex by opening the α-gated pore.[49] On the basis of these studies, we have recently isolated oleuropein, the most abundant of the phenolic compounds in Olea europaea leaf extract, olive oil and olives[50] and we have demonstrated that it has a stimulatory impact on proteasome activities in vitro.[51] The increased activities promote cellular resistance to oxidants and confer extension of human fibroblasts lifespan.[51] Oleuropein most likely acts through structural changes of the 20S α-gated channels conformation in a similar manner to

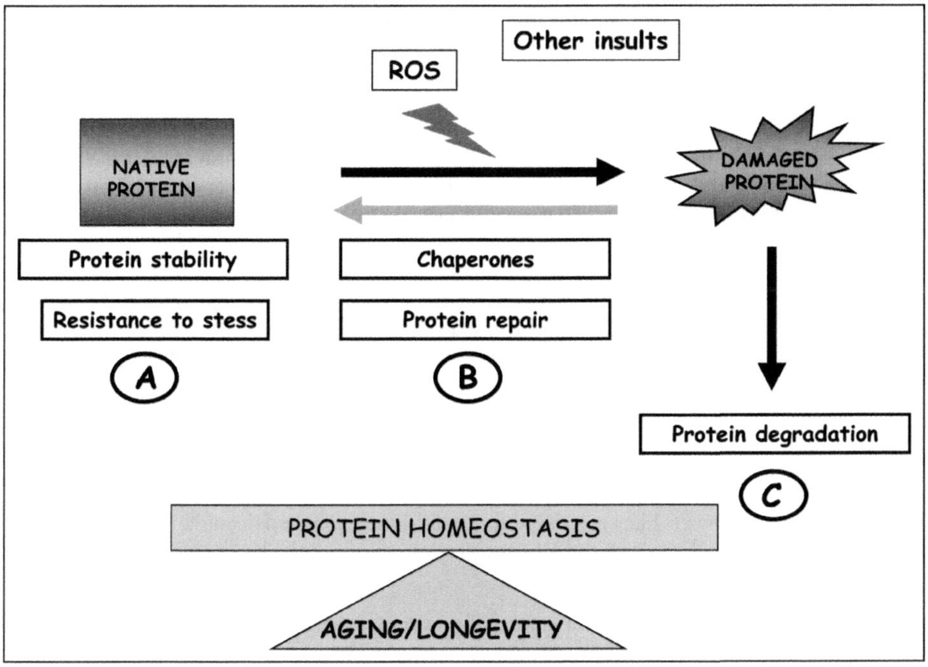

Figure 2. Protein maintenance determines cellular homeostasis. Protein homeostasis plays a central role in order to retain cellular homeostatic balance and can be preserved due to protein stability per se as well as to resistance to stress (A). Upon protein damage through the action of ROS or other cytotoxic insults, efficient protein maintenance is determined by chaperones that may refold the unfolded proteins or by protein repair enzymes that restore damage through specific enzymatic reactions (B). Finally, if the above mentioned mechanisms are not efficient in converting a damaged protein into its native form, protein degradation occurs (C). The equilibrium between these mechanisms may promote longevity and reduction of the rate of aging; in contrast dysfunction of this balance confers the opposite results.

SDS, albeit its effects are considerably stronger. A similar stimulatory effect on proteasomes has been observed by an algae extract of human keratinocytes, as it provides protection from UVA and UVB irradiation.[52]

Proteasome activation has also been achieved following treatment with various antioxidants. Specifically, Kensler and coworkers have shown natural antioxidants, such as dithiolethione and sulforafane, to enhance mammalian proteasome expression through the Keap1-Nrf2 signaling pathway, resulting to increased protection against various oxidants[53,54] and lately against Aβ cytotoxicity.[55] Nrf2 is a key molecule involved in cellular protection against chemically induced oxidative stress, since it is a central transcription factor of Phase 2 detoxifying enzymes and antioxidative genes.[56] Nevertheless, the correlation between aging and Nrf2 pathway is still under investigation. For example, Nrf2 knock out mice were not identified to exhibit further alterations of lifespan following caloric restriction.[57] In contrast we have shown that treatment with a Nrf2 inducer results to enhance cell survival following oxidative stress while continuous treatment leads to lifespan extension in human fibroblasts. These positive results have been shown to be linked with proteasome activation.[58]

In Vivo Evidence of Proteasome Activation

The previously reported studies establish the beneficial effects of proteasome activation in various in vitro systems. To test the in vivo relevance of these findings, we have determined proteasome biosynthesis and function in cultures derived from skin fibroblasts biopsies of several healthy donors of various ages including healthy centenarians. Although we observed a proteasome function decline with the age of the donors, importantly, we found that healthy centenarians exhibit a more functional proteasome when compared with the elderly donors.[59] We hypothesize that their functional proteasome provides enhanced rates of proteolysis of damaged proteins, thus indirectly contributing to their longevity. In support various studies have shown that in caloric restricted animals the proteasome function is maintained or even enhanced, thus possibly contributing to the observed extended lifespan of these animals.[60,61,62]

Conclusion

It is clear that cellular and organismal homeostasis is highly related to protein homeostasis which is affected by multiple molecular and biochemical pathways as mentioned above (Fig. 2). Although the direct interactions and cross-talks between all those pathways are not yet fully elucidated, it is obvious that "recalibration" of one of them triggers a wave of readjustments of the whole system that eventually lead to better survival, maintenance and longevity. This intra-molecular connection gives important value to the system of protein homeostasis and opens new dimensions for future therapeutic interventions, not only regarding aging and age-related diseases but also in many pathological conditions that are related to proteotoxicity. The elucidation of the unknown links between different pathways like protein stability, refolding, repair and degradation will constitute the new "holy grail" in designing strategies for promotion of cellular health maintenance.

References

1. Petropoulou C, Chondrogianni N, Simões D et al. Aging and longevity: A paradigm of complementation between homeostatic mechanisms and genetic control? Ann NY Acad Sci 2000; 908:133-142.
2. Trougakos IP, Chondrogianni N, Pimenidou A et al. Slowing down cellular aging in vitro. In: S.I.S. Rattan, ed. Modulating Aging and Longevity Kluwer Academic Publishers, 2003:65-83.
3. Sohal RS, Sohal BH, Brunk UT. Relationship between antioxidant defenses and longevity in different mammalian species. Mech Aging Dev 1990; 53:217-227.
4. Pérez VI, Buffenstein R, Masamsetti V et al. Protein stability and resisatnce to oxidative stress are determinants of longevity in the longest-living rodent, the naked mole-rat. Proc Natl Acad Sci USA 2009; 106:3059-3064.
5. Salmon AB, Leonard S, Masamsetti V et al. The long lifespan of two bat species is correlated with resistance to protein oxidation and enhanced protein homeostasis. FASEB J 2009; 23:in press.
6. Söti C, Csermely P. Protein stress and stress proteins: implications in aging and disease. J Biosci 2007; 32:511-515.
7. Petropoulos I, Friguet B. Maintenance of proteins and aging: the role of oxidized protein repair. Free Radic Res 2006; 40:1269-1276.
8. Picot CR, Petropoulos I, Perichon M et al. Overexpression of MsrA protects WI-38 SV40 human fibroblasts against H_2O_2-mediated oxidative stress. Free Radic Biol Med 2005; 39:1332-1341.
9. Cabreiro F, Picot CR, Perichon M et al. Overexpression of mitochondrial methionine sulfoxide reductase B2 protects leukemia cells from oxidative stress-induced cell death and protein damage. J Biol Chem 2008; 283:16673-16681.
10. Picot CR, Perichon M, Cintrat JC et al. The peptide methionine sulfoxide reductases, MsrA and MsrB (hCBS-1), are downregulated during replicative senescence of human WI-38 fibroblasts. FEBS Lett 2004; 558:74-78.
11. Cuervo AM, Bergamini E, Brunk UT et al. Autophagy and aging: the importance of maintaining "clean" cells. Autophagy 2007; 1:131-140.
12. Goldberg AL. Functions of the proteasome: from protein degradation and immune surveillance to cancer therapy. Biochem Soc Trans 2007; 35:12-17.
13. Friguet B, Bulteau AL, Petropoulos I. Mitochondrial protein quality control: implications in ageing. Biotechnol J 2008; 3:757-764.
14. Ciechanover A. Intracellular protein degradation: from a vague idea through the lysosome and the ubiquitin-proteasome system and onto human diseases and drug targeting. Cell Death Differ 2005; 12:1178-1190.
15. Hanna J, Finley D. A proteasome for all occasions. FEBS Lett 2007; 581:2854-2861.
16. Borissenko L, Groll M. Diversity of proteasomal missions: fine tuning of the immune response. Biol Chem 2007; 388:947-955.

17. Murata S, Sasaki K, Kishimoto T et al. Regulation of CD8+ T-cell development by thymus-specific proteasomes. Science 2007; 316:1349-1353.
18. Hirano Y, Hayashi H, Iemura S et al. Cooperation of multiple chaperones required for the assembly of mammalian 20S proteasomes. Mol Cell 2006; 24:977-984.
19. Hirano Y, Kaneko T, Okamoto K et al. Dissecting β-ring assembly pathway of the mammalian 20S proteasome. EMBO J 2008; 27:2204-2213.
20. Heinemeyer W, Ramos PC, Dohmen RJ. The ultimate nanoscale mincer: assembly, structure and active sites of the 20S proteasome core. Cell Mol Life Sci 2004; 61:1562-1578.
21. Herbig U, Sedivy JM. Regulation of growth arrest in senescence: Telomere damage is not the end of the story. Mech Aging Dev 2006; 127:16-24.
22. Chondrogianni N, Gonos ES. Proteasome dysfunction in mammalian aging: steps and factors involved. Exp Gerontol 2005; 40:931-938.
23. Chondrogianni N, Stratford FLL, Trougakos IP et al. Central role of the proteasome in senescence and survival of human fibroblasts: induction of a senescence-like phenotype upon its inhibition and resistance to stress upon its activation. J Biol Chem 2003; 278:28026-28037.
24. Chondrogianni N, Gonos ES. Proteasome inhibition induces a senescence-like phenotype in primary human fibroblasts cultures. Biogerontology 2004; 5:55-61.
25. Grune T, Jung T, Merker K et al. Decreased proteolysis caused by protein aggregates, inclusion bodies, plaques, lipofuscin, ceroid and 'aggresomes' during oxidative stress, aging and disease. Int J Biochem Cell Biol 2004; 36:2519-2530.
26. Ponnappan S, Ovaa H, Ponnappan U. Lower expression of catalytic and structural subunits of the proteasome contributes to decreased proteolysis in peripheral blood T-lymphocytes during aging. Int J Biochem Cell Biol 2007; 39:799-809.
27. Stratford FLL, Chondrogianni N, Trougakos IP et al. Proteasome response to interferon-gamma is altered in senescent human fibroblasts. FEBS Lett 2006; 580:3989-3994.
28. Ferrington DA, Husom AD, Thompson LV. Altered proteasome structure, function and oxidation in aged muscle. FASEB J. 2005; 19:644-646.
29. Ly DH, Lockhart DJ, Lerner RA et al. Mitotic misregulation and human aging, Science 2000; 287:2486-2492.
30. Husom AD, Peters EA, Kolling EA et al. Altered proteasome function and subunit composition in aged muscle. Arch Biochem Biophys 2004; 421:67-76.
31. Vernace VA, Arnaud L, Schmidt-Glenewinkel T et al. Aging perturbs 26S proteasome assembly in Drosophila melanogaster. FASEB J 2007; 21:2672-2682.
32. Tonoki A, Kuranaga E, Tomioka T et al. Genetic evidence linking age-dependent attenuation of the 26S proteasome with the aging process. Mol Cell Biol 2009; 29:1095-1106.
33. Shang F, Gong X, Palmer HJ et al. Age-related decline in ubiquitin conjugation in response to oxidative stress in the lens. Exp Eye Res 1997; 64:21-30.
34. Ghazi A, Henis-Korenblit S, Kenyon C. Regulation of Caenorhabditis elegans lifespan by a proteasomal E3 ligase complex. Proc Natl Acad Sci USA 2007; 104:5947-5952.
35. Yun C, Stanhill A, Yang Y et al. Proteasomal adaptation to environmental stress links resistance to proteotoxicity with longevity in Caenorhabditis elegans. Proc Natl Acad Sci USA 2008; 105:7094-7099.
36. Hassan WM, Merin DA, Fonte V et al. AIP-1 ameliorates {beta}-amyloid peptide toxicity in a Caenorhabditis elegans Alzheimer's disease model. Hum Mol Genet 2009; in press.
37. Gaczynska M, Rock K, Spies T et al. Peptidase activities of proteasomes are differentially regulated by the major histocompatibility complex-encoded genes for LMP2 and LMP7. Proc Natl Acad Sci USA 1994; 91:9213-9217.
38. Gaczynska M, Goldberg AL, Tanaka K et al. Proteasome subunits X and Y alter peptidase activities in opposite ways to the interferon-gamma-induced subunits LMP2 and LMP7. J Biol Chem 1996; 271:17275-17280.
39. Chondrogianni N, Tzavelas C, Pemberton AJ et al. Overexpression of proteasome beta 5 subunit increases the amount of assembled proteasome and confers ameliorated response to oxidative stress and higher survival rates. J Biol Chem 2005; 280:11840-11850.
40. Liu Y, Liu X, Zhang T et al. Cytoprotective effects of proteasome bet5 subunit overexpression in lens epithelial cells. Mol Vis 2007; 13:31-38.
41. Kwak MK, Cho JM, Huang B et al. Role of increased expression of the proteasome in the protective effects of sulphoraphane against hydrogen peroxide-mediated cytotoxicity in murine neuroblastoma cells. Free Radic Biol Med 2007; 43:809-817.
42. Hwang JS, Hwang JS, Chang I et al. Age-associated decrease in proteasome content and activities in human dermal fibroblasts: restoration of normal level of proteasome subunits reduces aging markers in fibroblasts from elderly persons. J Gerontol A Biol Sci Med Sci 2007; 62:490-499.

43. Chondrogianni N, Gonos ES. Overexpression of hUMP1/POMP proteasome accessory protein enhances proteasome-mediated antioxidant defence. Exp Gerontol 2007; 42:899-903.
44. Chen Q, Thorpe J, Dohmen JR et al. Ump1 extends yeast lifespan and enhances viability during oxidative stress: central role for the proteasome? Free Radic Biol Med 2006; 40:120-126.
45. Chen Q, Ding Q, Thorpe J et al. RNA interference toward UMP1 induces proteasome inhibition in Saccharomyces cerevisiae: evidence for protein oxidation and autophagic cell death. Free Radic Biol Med 2005; 38:226-234.
46. Chen Q, Thorpe J, Ding Q et al. Proteasome synthesis and assembly are required for survival during stationary phase. Free Radic Biol Med 2004; 37:859-868.
47. Dahlmann B, Rutschmann M, Kuehn L et al. Activation of the multicatalytic proteinase from rat skeletal muscle by fatty acids or sodium dodecyl sulphate. Biochem J 1985; 228:171-177.
48. Kohler A, Cascio P, Leggett DS et al. The axial channel of the proteasome core particle is gated by the Rpt2 ATPase and controls both substrate entry and product release. Mol Cell 2001; 7:1143-1152.
49. Kisselev AF, Kaganovich D, Goldberg AL. Binding of hydrophobic peptides to several noncatalytic sites promotes peptide hydrolysis by all active sites of 20S proteasomes. Evidence for peptide-induced channel opening in the alpha-rings. J Biol Chem 2002; 277:22260-22270.
50. Bonoli M, Bendini A. Qualitative and semiquantitative analysis of phenoloic compounds in extra virgin olive oils as a function of the ripening degree of olive fruits by different analytical techniques. J Agric Food Chem 2004; 52:7026-7032.
51. Katsiki M, Chondrogianni N, Chinou I et al. The olive constituent oleuropein exhibits proteasome stimulatory properties in vitro and confers life span extension of human embryonic fibroblasts. Rejuvenation Res 2007; 10:157-172.
52. Bulteau AL, Moreau M, Saunois A et al. Algae extract-mediated stimulation and protection of proteasome activity within human keratinocytes exposed to UVA and UVB irradiation. Antioxid Redox Signal 2006; 8:136-143.
53. Kwak MK, Wakabayashi N, Greenlaw JL et al. Antioxidants enhance mammalian proteasome expression through the Keap1-Nrf2 signaling pathway. Mol Cell Biol 2003; 23:8786-8794.
54. Kwak MK, Cho JM, Huang B et al. Role of increased expression of the proteasome in the protective effects of sulforaphane against hydrogen peroxide-mediated cytotoxicity in murine neuroblastoma cells. Free Radic Biol Med 2007; 43:809-817.
55. Park HM, Kim JA, Kwak MK. Protection against amyloid beta cytotoxicity by sulforaphane: role of the proteasome. Arch Pharm Res 2009; 32:109-115.
56. Kwak MK, Wakabayashi N, Kensler TW. Chemoprevention through the Keap1-Nrf2 signaling pathway by phase 2 enzyme inducers. Mutat Res 2004; 555:133-148.
57. Pearson KJ, Lewis KN, Price NL et al. Nrf2 mediates cancer protection but not prolongevity induced by caloric restriction. Proc Natl Acad Sci USA 2008; 105:2325-2330.
58. Kapeta S, Chondrogianni N, Gonos ES. Nuclear erythroid factor 2 (Nrf2) mediated proteasome activation delays senescence in human fibroblasts. J Biol Chem 2010; in press.
59. Chondrogianni N, Petropoulos I, Franceschi C et al. Fibroblast cultures from healthy centenarians have an active proteasome. Exp Gerontol 2000; 35:721-728.
60. Bonelli MA, Desenzani S, Cavallini G et al. Low-level caloric restriction rescues proteasome activity and Hsc70 level in liver of aged rats. Biogerontology 2008; 9:1-10.
61. Goto S, Takahashi R, Radak Z et al. Beneficial biochemical outcomes of late-onset dietary restriction in rodents. Ann NY Acad Sci 2007; 1100:431-441.
62. Salesby JT, Judge AR, Yimlamai T et al. Life long calorie restriction increases heat shock proteins and proteasome activity in soleus muscles of Fisher 344 rats. Exp Gerontol 2005; 40:37-42.

Chapter 5

Autophagy and Longevity:
Lessons from *C. elegans*

Kailiang Jia* and Beth Levine*

Abstract

Aging is a process in which individuals undergo an exponential decline in vitality, leading to death. In the last two decades, the study of the molecular regulation of aging in model organisms, particularly in *C. elegans*, has greatly expanded our knowledge of aging. Multiple longevity pathways, such as insulin-like growth factor signaling, TOR signaling, dietary restriction and mitochondrial activity, control aging in *C. elegans*. Recent genetic studies indicate that autophagy, an evolutionary conserved lysosomal degradation pathway, interacts with various longevity signals in the regulation of *C. elegans* life span. Here, we review the current progress in understanding the role of autophagy in the regulation of *C. elegans* life span.

Introduction

Biogerontologists have long argued that the aging process is under genetic control. A major aim of aging research is to identify longevity-determining genes and their functions. Studies in model organisms such as *C. elegans* have provided significant insight into the molecular mechanisms underlying aging.

C. elegans is a free-living soil nematode. In abundant food, the postembryonic development of *C. elegans* consists of four larval stages (L1-L4) and the adult (Fig. 1). The mean adult life span is about two weeks. However, if food is scarce and the animals are overcrowded (the population density is measured by the constitutively secreted pheromone), the animals will enter a specialized third larval stage called the dauer larva (Fig. 1). The dauer larva is a developmentally arrested dispersal stage that can live up to six months. Aging seems to be suspended in dauer larvae.[1] Indeed, four *daf* (dauer formation) genes, *daf-2, age-1, daf-16* and *daf-18,* have been shown to regulate adult life span.[2-4] *daf-2* and *age-1* mutants form dauer larva constitutively. However, if these mutants are grown at low temperature, they do not form dauer larvae, but grow to adults with extended life spans.[2-4] It has been proposed that a specific life span extension program is activated in dauer larvae and this program is inappropriately activated in *daf-2* and *age-1* adults.[5]

C. elegans is well suited to aging studies; it has a rapid life cycle and a short life span and is amenable to genetic analysis. Like higher eukaryotic organisms, *C. elegans* goes through a visible aging process.[5] In addition, a variety of cellular and developmental processes in higher organisms, including mammals, have been elucidated through studies of similar processes in *C. elegans*. Thus, findings in *C. elegans* are often applicable to more complex animals. Furthermore, the technique of RNAi, coupled with the availability of the complete genomic sequence of *C. elegans* has made

*Corresponding Authors: Kailiang Jia—Florida Atlantic University, 777 Glades Rd, Boca Raton, Florida 33431, USA. Email: kjia@fau.edu; Beth Levine—UT Southwestern Medical Center, HHMI, 5323 Harry Hines Blvd, Dallas, Texas 75390-9113, USA.
Email: beth.levine@utsouthwestern.edu

Protein Metabolism and Homeostasis in Aging, edited by Nektarios Tavernarakis.

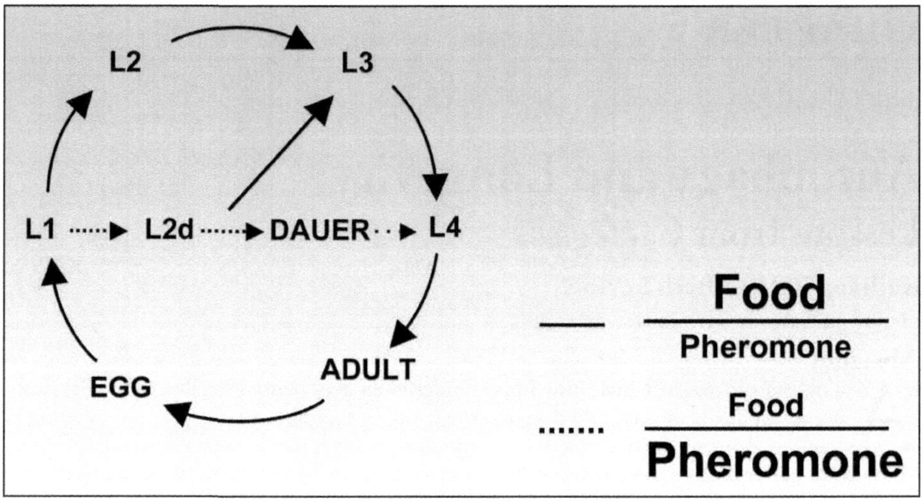

Figure 1. The life cycle of *C. elegans*. *C. elegans* has two alternative developmental pathways that are determined by at least three environmental conditions: food supply, temperature and the constitutively secreted dauer-inducing pheromone. The pheromone level is an indicator of the population density. Under favorable growth conditions (abundant food and/or low pheromone level), development proceeds continuously through four larval stages (L1, L2, L3 and L4) to the adult. However, if food is scarce and/or the pheromone level is high prior to the L1 stage, development will lead to the predauer (L2d) stage. The L2d will enter the dauer stage if environmental conditions remain adverse; otherwise it will switch to continuous development to the adult. When environmental conditions improve, the dauer larva can resume development, molt to the L4 and grow to the adult.

it easy to study gene function at a global level. In the last two decades, *C. elegans* model systems have led the way in the elucidation of molecular mechanisms in life span regulation.

In this chapter, we will first review genetic pathways regulating *C. elegans* life span including the DAF-2 insulin-like signaling pathway, dietary restriction and mitochondrial factors (Fig. 2). Then we will briefly review the process of autophagy (Fig. 3) and discuss its interaction with longevity pathways in *C. elegans* (Fig. 2, Table 1). Quite strikingly, this protein degradation pathway is emerging as a central mediator of life span extension.

DAF-2 Insulin/IGF-1-Like Signaling

Mutations in single genes, *daf-2*[2] and *age-1*,[3,4] significantly extend the life span of nematodes. *daf-2* encodes an insulin receptor-like tyrosine kinase[6] and *age-1* encodes a homologue of the catalytic subunit of phosphatidylinositol 3-kinase (PI3K).[7] PI3K acts downstream of dimeric growth factor receptor tyrosine kinases, including the insulin receptor family.[8] DAF-16 is a major target of DAF-2/AGE-1 signaling and *daf-16* activity is required for the enhanced longevity of *daf-2* and *age-1* mutants. *daf-16* encodes a FOXO (forkhead box O) transcription factor.[9,10] Upon stimulation by insulin-like peptide, the DAF-2 receptor kinase activates PI3K that in turn activates the serine-threonine kinase Akt. The activated Akt kinase phosphorylates. The DAF-16/FOXO transcription factor, which results in the export of DAF-16 from the nucleus to the cytoplasm.[11] However, in the absence of DAF-2 ligand, dephosphorylated DAF-16 enters the nucleus and turns on the transcription of a spectrum of genes required for longevity such as antioxidant genes and heat shock resistance genes.[12] Thus, DAF-2 and AGE-1 are major signaling molecules and DAF-16 is a major transcription factor that function in the control of aging (Fig. 2).

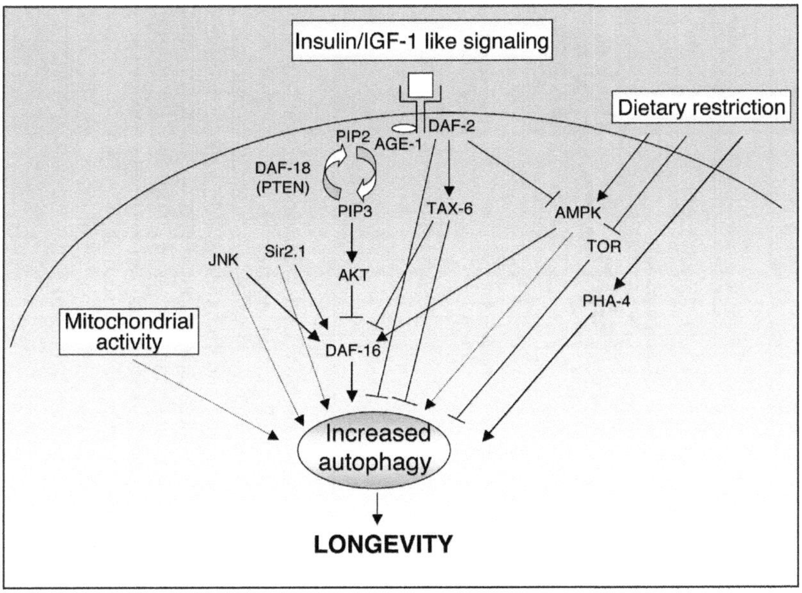

Figure 2. Model depicting potential interrelationships between autophagy and longevity pathways in *C. elegans* based on studies in *C. elegans* and other organisms. Proposed wild-type functions are shown, with arrows indicating stimulation of activities and T-bars indicating inhibition, but the steps do not indicate direct protein interactions. Dashed arrows signify that the proposed interaction has not been determined in *C. elegans* but has been found in other organisms. This model focuses on three longevity pathways: the *daf-2* insulin-like signaling pathway, dietary restriction and mitochondrial activity. *Daf-2* signaling negatively regulates the DAF-16 Foxo transcription factor, a key player in life span regulation. The phosphorylation by AKT kinase of DAF-16 prevents its nuclear localization and transcriptional activation of genes required for longevity. In contrast, the phosphorylation of DAF-16 by JNK kinase promotes its nuclear localization. Sir2.1 can enter the nucleus and form a complex with DAF-16 to turn on gene transcription. TAX-6 acts downstream of DAF-2 and is partially dependent on DAF-16 to control life span. Dietary restriction is another conserved life span extension mechanism and functions through at least three pathways, AMPK, TOR and PHA-4, depending on the dietary restriction method applied to worms. AMPK activity is also influenced by DAF-2 signaling. When AMPK functions downstream of DAF-2, it controls life span independently of DAF-16 whereas DAF-16 is required for AMPK to regulate life span under certain dietary restriction protocols. The third group of longevity mutants increases *C. elegans* life span in a DAF-16 independent manner by reducing mitochondrial activity. Notably, all these longevity pathways regulate autophagy, directly or indirectly. Downstream of *daf-2* signalling, DAF-16 stimulates autophagy whereas TAX-6 inhibits autophagy. DAF-2 signalling may also inhibit autophagy through an unknown protein(s) independently of DAF-16. Besides interacting with DAF-16, Sir2.1 directly interacts with the autophagy proteins and deacetylates them to induce autophagy in mammalian cells. JNK kinase may stimulate autophagy through DAF-16. In addition, JNK induces autophagy in starvation conditions in mammalian cells by regulating the interaction of Beclin 1 and Bcl-2 (mammalian orthologs of BEC-1 and CED-9, respectively). Under dietary restriction, TOR is inhibited and autophagy is upregulated. Dietary restriction also induces autophagy through the PHA-4 transcription factor and AMPK. Free reactive oxygen species produced by mitochondria can induce autophagy and presumably promote the autophagic clearance of damaged mitochondria. Taken together, current studies of various longevity pathways place autophagy at a central position in the regulation of *C. elegans* life span. It has been shown that autophagy is required for life span extension conferred by reduction of *daf-2* signaling and TOR signaling, dietary restriction, decreased TAX-6 activity and reduced mitochondrial activity. Further studies are required to evaluate the role of autophagy induction in JNK, AMPK, Sir2.1 and PHA-4 regulation of *C. elegans* life span.

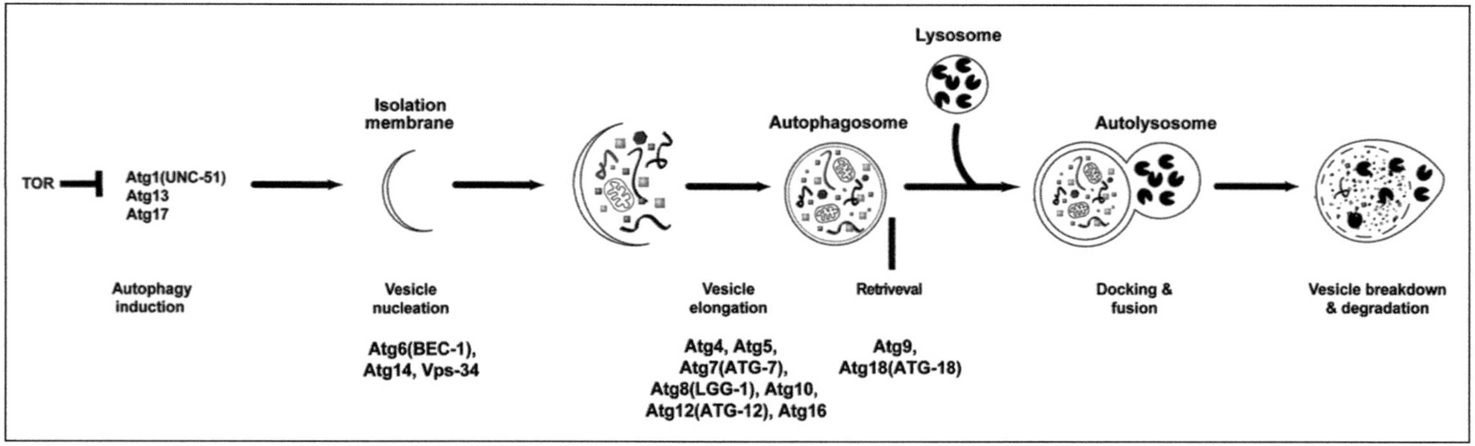

Figure 3. Schematic illustration of the autophagy process in yeast and mammalian cells. 1) Autophagy induction. Under starvation conditions, TOR is inhibited. A serine-threonine kinase complex including Atg1, Atg13 and Atg17 (that is repressed by TOR kinase) initiates the autophagy process. 2) Vesicle nucleation. A Class III PI3K complex including Vps34, Atg6 and Atg14 is required for the initiation of the vesicle nucleation process. 3) Vesicle expansion. Two ubiquitin-like (Ubl) protein conjugation systems are essential for vesicle expansion and completion of the autophagosome. Two Ub1 proteins include Atg8 and Atg12. Atg7 is a ubiquitin-activating enzyme. Atg3 and Atg10 are homologous to ubiquitin-conjugating enzymes. Atg8 associates with the autophagosomal membrane as a phosphatidylethanolamine-conjugated form during the process of autophagy. 4) Retrieval. After the formation of the autophagosome, autophagy proteins except for Atg8 are retrieved; Atg9 and Atg18 are required for this retrieval step. 5) Fusion and vesicle breakdown. The outer membrane of the autophagosome fuses with a lysosome to form an autolysosome. The inner membrane and sequestered cytosolic components are degraded by lysosomal enzymes inside the autolysosome. Shown are the names of yeast protein names; the names of *C. elegans* ortholog gene products that have been targeted in RNAi studies in longevity mutants are listed in parentheses after the corresponding yeast autophagy protein.

***Table 1. Role of autophagy genes in life span extension in long-lived mutant* C. elegans**

C. elegans Longevity Mutation	Life Span Effect	Autophagy Gene(s)	References
daf-2	Increase (loss-of-function)	Required: *bec-1, atg-7, lgg-1, atg-12, atg-18, vps-34*	Melendez et al, 2003 Hars et al, 2007 Hansen, et al, 2008 Toth et al, 2008
eat-2	Increase (loss-of-function)	Required: *unc-51, bec-1, atg-7, vps-34*	Jia and Levine, 2007 Hansen et al, 2008 Toth et al, 2008
TOR	Increase (loss-of-function)	Required: *unc-51, bec-1, atg-18*	Hansen et al, 2008 Toth et al, 2008
atp-3	Increase (loss-of-function)	Required: *unc-51, atg-18*	Toth et al, 2008
clk-1	Increase (loss-of-function)	Required: *unc-51, atg-18* Not required: *bec-1*	Toth et al, 2008 Hansen et al, 2008
isp-1	Increase (loss-of-function)	Not required: *bec-1*	Hansen et al, 2008
tax-6	Increase (loss-of-function)	Required *bec-1, atg-7*	Dwivedi et al, 2009
sir2.1	Increase (overexpression)	Not determined	
JNK	Increase (overexpression)	Not determined	
AMPK	Increase (overexpression)	Not determined	

The activated PI3K converts the plasma membrane lipid phosphatidylinositol-4,5-bisphosphate (PIP2) to phosphatidylinositol-3,4,5-trisphosphate (PIP3). PIP3 can bind to various signaling proteins including Akt and activate them[8] and PI3K signaling is terminated by the degradation of PIP3. The phosphatase PTEN (Phosphatase and Tensin homolog deleted on chromosome TEN) dephosphorylates the 3 position of PIP3 to produce PIP2.[8] *daf-18* encodes a *C. elegans* homolog of PTEN that opposes the activity of AGE-1 (Fig. 2).[13-15] In fact, the pathway from the insulin-like receptor tyrosine kinase to the FOXO transcription factor is conserved from *C. elegans* to mammals.[12,15] Interestingly, longevity resulting from reduced insulin-like signaling is observed in a variety of species, ranging from *C. elegans*, to *Drosophila* and mice.[12,17]

The nuclear localization of DAF-16 is also regulated by other kinases. For example, JNK-1 (c-Jun N-terminal kinase), a member of the MAPK superfamily, phosphorylates DAF-16 and promotes the translocation of DAF-16 into the nucleus.[18] As expected, overexpression of JNK-1 results in DAF-16-dependent life span extension.[18] In mammals, the FOXO transcription factor is deacetylated by sirtuin, an NAD-dependent protein deacetylase.[19,20] DAF-16 may be also deacetylated by sirtuin although biochemical evidence for this is lacking. Interestingly, overexpression of the *C. elegans* sirtuin, SIR-2.1, increases the life span of wild type worms in a DAF-16-dependent manner, suggesting that sirtuin may modify DAF-16 activity.[21] In addition to being post-translationally modified, DAF-16 interacts with several different transcription factors to control the expression of genes that extend life span by promoting stress resistance.[17] As a central regulator of life span downstream of DAF-2 signaling, the identification of additional DAF-16 target genes and regulators of DAF-16 function will further our understanding of how insulin-like signaling controls life span in *C. elegans* and other organisms.

Dietary Restriction

Of all the known longevity pathways, dietary restriction (DR) is the only intervention conclusively and reproducibly shown to slow aging and maintain health and vitality in metazoans. DR has been shown to extend life span in *C. elegans* and several other organisms, including yeasts, *Drosophila* and mammals (including primates).[12,22-24] Two methods are widely used to apply dietary restriction to *C. elegans*.[23,25] One method is to feed worms with diluted bacteria[26] (different dilution protocols have been developed.[27]). Worms show a parabolic response to different dilutions of bacteria and maximum life span extension is reached at specific dilutions.[23] Dietary restriction is also achieved by using a genetic mutant called *eat-2*.[28] *Eat-2* mutants have defects in pharyngeal muscle function leading to insufficient food intake and a starvation phenotype. The *eat-2* mutation prolongs life span, presumably by restricting the food intake of the worm.[28]

The molecular mechanisms mediating dietary restriction have been studied in yeast,[29] *C. elegans*[23] and *Drosophila*.[30] In *C. elegans*, they involve the DAF-2/insulin-like growth factor pathway, the TOR (Target of Rapamycin) pathway and other proteins such as sirtuins and the transcription factor PHA-4.[23] However, none of these mechanisms is the master regulator of dietary restriction. They are required to extend life span by dietary restriction only under very specific conditions. The complexity of these effects may be partly due to lack of a consistent methodology to apply dietary restriction to *C. elegans*.[27]

Mitochondrial Activity

All organisms live in an environment that contains free radicals (ROS: reactive oxygen species) because mitochondrial respiration constantly generates ROS by leaking intermediates from the electron transport chain. The free radical theory postulates that free radicals cause the accumulation of oxidant-damaged molecules and organelles that result in senescence.[31] Consistent with this theory, *C. elegans* mitochondrial mutations either extend or decrease life span depending on their effects on mitochondrial activity

isp-1 encodes the *C. elegans* iron sulfur protein of mitochondrial complex III. A mutation in this gene results in a decrease of oxygen consumption, insensitivity to ROS and extension of life span.[32] Mutations in the *C. elegans* coenzyme Q biosynthetic gene *clk-1* also cause an increase in life span.[33] Coenzyme Q is an essential component of the electron transport chain of mitochondria and is required to biosynthesize ubiquinone Q. *clk-1* mutants accumulate high levels of the precursor, demethoxy ubiquinone (DMQ).[34] The increased life span of *clk-1* mutants may result from the decrease of ROS due to the presence of DMQ.[35] By contrast, mutations in a subunit of the enzyme succinate dehydrogenase cytochrome *b* (*mev-1*) cause oxidative stress and short life spans.[36] Interestingly, the treatment of *C. elegans* with free radical detoxifying mimetics results in a 44% increase in life spans of wild type animals.[37] In addition, these compounds rescue the life span defect of the *mev-1* mutants.[37]

Consistent with studies of these mitochondrial mutants, a group of mitochondrial genes has been identified in systematic RNA interference (RNAi) screens of *C. elegans* genes involved in life span regulation.[38,39] RNAi-induced mitochondrial defects result in increased life span and decreased production of ATP.[38,39] The long-lived worms become more resistant to hydrogen peroxide and heat shock treatment.[39] However, the extended life span of worms with defective mitochrondria cannot simply be assigned to lower free radical production because some of them show a heightened sensitivity to the superoxide-inducing agent paraquat.[39] This conclusion is supported by a recent study in which overexpression of the antioxidant enzyme superoxide dismutase (SOD-1) extends the life span of worms but increases their sensitivity to paraquat.[40] Thus, the superoxide radical appears not to be a major determinant of aging in *C. elegans*. Therefore, although mitochondria clearly play an important role in regulation of *C. elegans* life span, the exact mechanisms by which mitochondrial mutations extend life span warrant further investigation.

Autophagy

In recent years, increasing evidence has emerged that the cellular pathway of autophagy may be a central regulator of life span that is required for the effects of DAF-2 insulin/IGF-1-like signaling, dietary restriction and some mitochondrial mutations (e.g., *ATP-synthase-3* RNAi inactivation) on nematode longevity. The autophagy pathway has been extensively reviewed in numerous excellent recent reviews.[41-43] In this section, we will provide a brief overview of autophagy as background for understanding the potential role(s) of autophagy in anti-aging pathways (Fig. 3).

Autophagy (derived from the Greek meaning "to eat oneself") is an evolutionary conserved lysosomal degradation pathway. It is present in all eukaryotic cells and conserved from yeast to humans.[44] There are three types of autophagy, macroautophagy, microautophagy and chaperone-mediated autophagy (CMA). Macroautophagy is a dynamic process involving the rearrangement of subcellular membranes to sequester cytoplasm and organelles for delivery to the lysosome or vacuole, where the sequestered cargo is degraded and recycled.[44] By contrast, microphagy is a direct engulfment of cellular components by invagination of the lysosomal membrane. Unlike macroautophagy and microautophagy that involve the rearrangement of subcellular membranes, CMA is used specifically to degrade cytosolic proteins that bear a particular pentapeptide consensus motif and are directly targeted to the lysosome.[45] A cytosolic chaperone protein recognizes and binds these proteins to form a substrate—chaperone complex that is delivered to the lysosome through the lysosome-associated membrane protein (LAMP) Type 2A, a lysosomal membrane receptor for the CMA pathway.[46] Of note, macroautophagy is the only known cellular pathway for removing mitochondria and other organelles and is a major cellular pathway for the degradation of long-lived and aggregate-prone proteins. Furthermore, genetic studies of tissue-specific deletion of macroautophagy genes in mice reveals a critical role for this pathway in vivo in protein and organelle quality control, cellular and tissue homeostasis and the prevention of the accumulation of ROS, ubiquitinated cellular aggregates and abnormal organelles.[47]

In this chapter, we will focus exclusively on macroautophagy (herein referred to as autophagy). Autophagy is involved in many different biological functions such as starvation adaptation, tumor suppression, development and differentiation, cell death, innate immunity and anti-aging.[44] Basal levels of autophagy occur constitutively at low levels even under normal growth conditions, which is critical for "cellular housekeeping" functions.[48] Autophagy activity can also be induced by nutrient starvation, deprivation of growth factors and various intracellular and extracellular stress conditions such as hypoxia, high temperature, or overcrowding.[47] Nutrient starvation and repression of TOR strongly induce autophagy from yeast to mammals, including in *C. elegans*.[49] In multicellular eukaryotic organisms, various signal transduction pathways including insulin, TOR, stress-activated kinases (e.g., JNK) and eukaryotic initiation factor 2α kinase signaling pathways are orchestrated to respond to nutrient status in the regulation of autophagy activity.[47]

The autophagy process involves the formation of a double-membraned structure called the autophagosome, which involves sequential steps including induction, vesicle nucleation, vesicle expansion and completion and retrieval, that are then followed by fusion with the lysosome and degradation of the sequestered contents (Fig. 3). In yeast, more than thirty autophagy-related (*ATG*) genes have been identified by genetic screens and they all function downstream of TOR kinase. Under normal nutrient conditions in yeast, TOR inhibits the interaction between Atg13 and Atg1 by causing hyperphosphorylation of Atg13 through an unknown mechanism. However, under starvation conditions, this inhibition is released and the activated Atg13-Atg1 complex results in the induction of autophagy.[44] In the first step of autophagosome formation, an isolated membrane called a phagophore forms at the phagophore assembly site and sequesters cytosolic proteins and organelles.[50] A Class III PI3K complex involving Vps34 and Atg6 in yeast (Beclin 1 in mammalian cells) is required for the initiation of this vesicle nucleation process.[44,50] The elongation of the phagophore membrane results in the formation of autophagosome. Two ubiquitin-like protein system are required for this step.[50] Atg8, one of the ubiquitin-like proteins in these systems, associates with the autophagosomal membrane as a phosphatidylehanolamine-conjugated form during the process of autophagy.[50] Therefore, GFP-labeled ATG8 is widely used as a marker to visualize autophagosomes that appear as green puncta. After the formation of the autophagosome, autophagy proteins except for Atg8 are retrieved; Atg9 and Atg18 are required for this retrieval step. Then the outer membrane of the autophagosome fuses with the lysosome and the inner membrane along with sequestered cytosolic components are delivered into the lysosome. The fused vesicles are called autolysosomes, in which cytosolic components are degraded by lysosomal enzymes. The resulting breakdown products such as amino acids and free fatty acids are released into cytosol and recycled to synthesize new macromolecules and to maintain cellular energy homeostasis.[44]

Orthologs of many yeast *ATG* genes have been identified in different species including *C. elegans*, *Drosophila*, mice and humans.[44] Sixteen worm orthologs of *ATG* genes are known to exist in the *C. elegans* genome and their putative functions in the autophagy process have been reviewed.[51] Moreoever, recent studies have shown that autophagy plays a role in survival during starvation, dauer development, aging, cell death and neurodegenerative diseases in *C. elegans*.[51] In the remainder of this chapter, we will discuss the role of autophagy in aging, as revealed by genetic studies in *C. elegans*.

Autophagy and *C. elegans* Longevity Pathways

Many longevity genes have been identified in *C. elegans* and recent studies indicate that autophagy is involved in several highly conserved longevity pathways (Table 1), including insulin-like signaling, dietary restriction and mitochondrial respiratory function.

One of the best-studied longevity pathways in the nematode is the DAF-2 insulin/IGF-1-like signaling pathway. In 2003, the first study of autophagy in *C. elegans* showed that the autophagy gene *bec-1* (ortholog of yeast *ATG6*) is essential for the longevity phenotype of *daf-2* mutant animals.[52] Subsequent studies showed that five additional autophagy genes, *atg-7*, *lgg-1* (ortholog of yeast *atg8*), *atg-12*, *atg-18* and *vps-34* are also required for life span extension in *daf-2* mutants.[53-55] Consistent with the life span data, several studies have shown that autophagy is induced in *daf-2* dauer larvae and adults, using worms that transgenically express a fluorescent autophagy marker, GFP fused to LGG-1 (the worm ortholog of Atg8).[52,54,56] Together, these studies suggest that the upregulation of autophagy in the setting of decreased insulin-like signaling mechanistically underlies the longevity phenotype of nematodes with a loss-of-function mutation in the insulin-like receptor tyrosine kinase, *daf-2*. To date, no studies have been published that examine whether autophagy genes are also, as one would predict, required for *age-1*-mediated life span extension.

An open question is whether autophagy regulation by DAF-16 contributes to *daf-2*-mediated life span extension. A recent report indicates that overexpression of the DAF-16/Foxo transcription factor induces autophagy in *C. elegans*.[56] Furthermore, FOXO overexpression induces autophagy in *Drosophila*[57] and Foxo3 regulates autophagy in mouse muscle cells,[58,59] suggesting that Foxo transcription factors play an evolutionarily conserved role in autophagy induction. However, Hansen et al found that a *daf-16* null mutation does not block autophagy induction in *daf-2* mutants.[54] One potential explanation for this apparent discrepancy is that another unidentified protein may function redundantly with DAF-16 in autophagy regulation. In such a case, the requirement for both DAF-16 and for autophagy genes in *daf-2*-mediated life span extension may reflect a dual necessity for autophagy-independent DAF-16-regulated functions and for autophagy (that can be activated in a DAF-16-independent manner) (Fig. 2).

Dietary restriction is another mechanism regulating *C. elegans* life span, which functions, at least in part, by reducing the activities of the TOR signaling pathway. Mutations in *C. elegans* that result in TOR inhibition extend *C. elegans* life span.[60,61] Dietary restriction cannot further extend the increased life span produced by TOR RNAi, suggesting that TOR mediates the effect of dietary restriction in regulating *C. elegans* life span.[62] As in other organisms, TOR negatively regulates autophagy in *C. elegans*.[54] Importantly, autophagy gene RNAi blocks life span extension that results from reduced TOR activity.[54,55] Thus, it is predicted that dietary restriction inhibits TOR, which in turn induces autophagy. Indeed, increased autophagy activity is observed not only in TOR RNAi animals, but also in *eat-2* and *eat-3* dietary restriction mutants[54,55,63] and in dietary-restricted wild type worms.[54] Accordingly, it not surprising that RNAi knockdown of autophagy genes suppresses the extended life span of *eat-2* mutants.[54,55,64] Thus, autophagy is a central cellular target regulated by both dietary restriction and *daf-2*/insulin-like signaling pathway that is important in the control of *C. elegans* life span (Fig. 2). One caveat is that all studies performed to date to demonstrate that autophagy is required for dietary restriction-mediated *C. elegans* life span extension utilized the *eat-2* genetic mutant;[54,55,64] it will be critical to confirm this conclusion by using the traditional bacteria dilution method which is a more physiological model for dietary restriction.

It is important to note that dietary restriction is considered to represent a pathway that is parallel to the *daf-2*/insulin-like signaling pathway in the control of *C. elegans* life span, based on the differential requirements of downstream transcription factors in each pathway. As noted above, the longevitiy phenotypes of *daf-2* mutants requires DAF-16, whereas the longevity phenotype of *eat-2* mutants is independent of DAF-16.[28] Instead, the FOXA transcription factor PHA-4 is required for the life span extension of animals subjected to dietary restriction.[65] Interestingly, PHA-4 is also required for dietary restriction to induce autophagy.[54] Therefore, it will be important to determine whether PHA-4-dependent autophagy is an essential mechanism by which dietary restriction extends life span in *C. elegans* and more generally, whether mammalian FOXA transcription factors regulate dietary restriction-induced autophagy and/or life span extension.

Similar to dietary restriction, mitochondrial mutants extend *C. elegans* life span in a DAF-16-independent manner. Mutations in *atp-3* and *clk-1* decrease mitochondrial respiration and extend *C. elegans* life span.[33,38] RNAi of *atp-3* and *clk-1* extend life spans of wild type worms but not of autophagy mutants, suggesting that autophagy may be required for life span extension in these two mitochondrial mutants.[55] However, RNAi of *bec-1* during adulthood cannot suppress the longevity phenotype of the mitochondrial mutants *isp-1* and *clk-1*.[54] Therefore, further experiments are required to confirm the involvement of autophagy in the mitochondrial pathway of life span regulation. Since one postulated mechanism by which autophagy may function in longevity is through the removal of damaged mitochondria and through decreasing intracellular levels of ROS,[47] it is possible that the autophagic machinery may be superfluous in certain longevity phenotypes that result from decreased mitochondrial activity.

In addition to the *daf-2*/insulin-like signaling pathway, dietary restriction and mitochondrial mutants, there are many other genes that influence *C. elegans* life span. For example, overexpression of Sir2.1, a worm ortholog of sirtuin, extends *C. elegans* life span, which depends on DAF-16 activity.[21] Under stress conditions, Sir2.1 enters the nucleus and forms a complex with nuclear DAF-16 to turn on the transcription of longevity genes.[66] It has not been tested whether autophagy is required for Sir2.1 to extend life span. However, interestingly, one of the mammalian sirtuins, Sirt1, forms a molecular complex with autophagy proteins including Atg5, Atg7 and Atg8 and directly deacetylates these proteins.[67] Furthermore, transient overexpression of Sirt1 stimulates basal level of autophagy.[67] Therefore, it is tempting to speculate that autophagy may be regulated by Sir2.1 in *C. elegans* and required for life extension conferred by Sir2.1 overexpression.

In contrast to Sir2.1, TAX-6 or CNA-1, the *C. elegans* calcineurin A protein that is the catalytic subunit of a serine or threonine protein phosphatase, controls *C. elegans* life span in a DAF-16-partially-dependent manner.[68] *Tax-6* loss-of-function mutants display a weak *daf-2* mutant phenotype including life span extension and dauer formation.[68] A loss-of-function mutation of *daf-16* incompletely suppresses these phenotypes.[68] Interestingly, a recent study reported that autophagy is induced in *tax-6* mutants and RNAi of autophagy genes suppresses life span extension of *tax-6* mutants.[69] Therefore, autophagy genes are required for the life span regulation by calcineurin in *C. elegans*.

In addition to these well-characterized pathways that require autophagy in life span regulation, there is reason to speculate that autophagy may be also involved in other longevity pathways, based on known interactions between certain longevity signals in *C. elegans* and their established roles in autophagy induction in other organisms. The two major examples of this are AMP-activated protein kinase (AMPK) and JNK. AMPK is a sensor of low energy levels and becomes active when the AMP:ATP ratio is high.[70] The *C. elegans* AMPKα subunit AAK-2 is activated by AMP and influences life span independently of DAF-16; both *daf-16* and *aak-2* are required for the life span extension of *daf-2* mutants and they act in parallel.[71] However, dietary restriction-induced longevity requires AMPK and this involves multisite phosphorylation of the transcription factor DAF-16.[72]

Consistent with the involvement of AMPK in the *daf-2*/insulin-like signaling pathway, metformin, an oral biguanidine that is widely used in humans for the treatment of Type 2 diabetes, reduces plasma glucose and lipids and improves insulin sensitivity by activating AMPK.[73-75] Furthermore, chronic treatment of female outbred SHR (spontaneously hypertensive rat) mice with metformin increases mean life span by 37.8%.[76] Interestingly, various stressful conditions such as hypoxia and ischemia induce autophagy in an AMPK-dependent manner in mammals,[77,78] suggesting that autophagy is regulated by AMPK. In flies, the γ subunit of AMPK SNF4Aγ regulates developmental and stress-induced autophagy[79] and the yeast AMPK, SNF1p, also regulates autophagy.[80] Therefore, autophagy may also be controlled by AMPK in *C. elegans* and it is possible that such control may play a role in AMPK-mediated life span regulation.

JNK is an evolutionarily conserved stress-activated protein kinase. Overexpression of JNK results in increased stress tolerance and extended life span in flies and worms, by a mechanism that is believed to involve promoting the nuclear localization of Foxo transcription factor.[18,81,82] Interestingly, in flies, JNK signaling is thought to protect animals from oxidative stress by transcriptionally activating the autophagy pathway.[83] In mammals, JNK-1 is required for starvation-induced autophagy through a mechanism that involves Bcl-2 multisite phosphorylation, which leads to release of Bcl-2 (an inhibitor of autophagy) from the Bcl-2/Beclin 1 complex.[84] Thus, JNK may regulate autophagy by both transcriptional and post-translational mechanisms and autophagy regulation by JNK, which is conserved in flies and mammals, may also exist in *C. elegans*. Accordingly, it will be interesting to determine whether autophagy is required for life extension conferred by overexpression of JNK in *C. elegans*. Furthermore, since JNK activity is essential for starvation-induced autophagy, at least in mammals,[84] another important question is whether JNK-mediated autophagy may be involved in life span extension that results from dietary restriction.

One critical dilemma in the field of autophagy and longevity control in *C. elegans* is how autophagy function in aging can be truly separated from other phenotypes that result from autophagy gene inactivation. For example, a null mutation in *bec-1* is embryonically lethal[52] and a null mutation in *atg18* causes decreased brood size (Jia and Levine, unpublished data). The dominant phenotype of *unc-51* (worm ortholog of *atg1*) is a neuronal defect resulting in abnormal movement. Thus, it is unknown how the pleiotropic effects of autophagy gene mutations affect aging. Knockdown of autophagy genes by RNAi has eliminated some of these "non-aging" phenotypes and still shows a requirement for autophagy in life span extension in different longevity mutants. However, some of these RNAi studies were performed during the development stage of the nematode life cycle, raising the question of whether these "aging" phenotypes are merely secondary to the effect of autophagy on development. Indeed, it has been reported that RNAi of *bec-1* in young adults, i.e., after development, does not shorten the life span of long-lived translation mutant and mitochondrial mutants *isp-1* and *clk-1*[54] whereas *bec-1*, *unc-51* and *atg-18* mutations (present during development) do suppress the longevity phenotypes of *atp-3* and *clk-1* mutants.[55] Surprisingly, in one study, RNAi of *unc-51*, *bec-1* or *atg-9* after development was reported to extend life span in wild-type animals and *daf-2* mutants.[85] Therefore, further studies are needed to clarify the conflicting results regarding the role of autophagy in the control of *C. elegans* life span.

Despite certain unresolved questions, taken together, the genetic studies described above provide compelling evidence that many longevity mutants require autophagy to extend life span. Furthermore, it is predicted that other longevity signals, such as AMPK and JNK, may also require autophagy. The central question arises—why is autophagy involved in so many different pathways? Perhaps, these genetic studies are pointing to a central downstream essential function of autophagy as an anti-aging process, upon which different longevity signals converge. In theory, the anti-aging effects of autophagy could either relate to its functions in promoting cell survival during adverse conditions (presumably through nutrient recycling and energy homeostasis), or perhaps more importantly, to its functions in specific aspects of cellular housekeeping. As the genetic link between autophagy and life span extension has been discovered relatively recently, few studies, if any, have addressed the underlying mechanisms by which autophagy is required for life span extension. Nonetheless, an obvious hypothesis is that autophagy is a key cellular defense mechanism against what has historically been referred to as the "free radical theory of aging".

According to this theory, mitochondria are the major source of damaging oxygen radicals and the first compartment in the cell to be damaged by these radicals. The accumulation of mitochondrial DNA damage and decline of mitochondrial function is a common feature in aging cells[86] and thought to be mechanistically important in the aging process. Indeed, as discussed above, studies of aging in *C. elegans* suggest an important role of mitochondria in regulating life span. Of note, mitochondrial damage and cellular oxidative stress induce autophagy[87] and in turn, autophagy may selectively remove damaged mitochondria and decrease intracellular levels of ROS and thereby, protect cells from oxidative damage.[88-100] Thus, one common theory is that cells utilize autophagy to fight against oxidative damage and retard aging. In addition, autophagy plays a role in protecting against cellular damage by removing toxins and harmful aggregate-prone proteins[47] as well as a role in cellular lipid metabolism.[101] It seems likely that all of these, as well as yet-to-be-discovered, functions of autophagy may contribute to its anti-aging effects.

Conclusion

In this chapter, we review our current understanding of the role of autophagy in the regulation of *C. elegans* life span. In the past six years, since the initial discovery of a role of *bec-1* in *daf-2*-mediated life span extension,[52] there is accumulating evidence that multiple different autophagy genes are required for life span extension in multiple different *C. elegans* longevity mutants. Despite these advances, many important questions remain. We still know relatively little about how autophagy is regulated in vivo, whether autophagy impairment contributes directly or indirectly to aging and how autophagy functions at the cellular and molecular level in life span extension. We anticipate that continued studies in *C. elegans*, as well as in other model organisms, will likely shed light on these questions and open new frontiers in our understanding of aging. Meanwhile, perhaps the major lesson we have learned from *C. elegans*—that autophagy is a downstream point of convergence in evolutionarily conserved longevity pathways—can be exploited to extend life span in other species by using lifestyle and/or pharmacological approaches to upregulate autophagy.

Acknowledgements

We apologize to the many authors whose work could not be cited due to space limitations. The work in the authors' own laboratory was supported by an Ellison Medical Foundation New Scholars Award in Aging (K.J.), an Ellison Medical Foundation Senior Scholars Award in Global Infectious Diseases (B.L.) and NIH grants, ROI CA84254 and ROI CA109618, to B.L.

References

1. Albert PS, Riddle DL. In: Riddle DL, Blumenthal T, Meyer BJ, Priess JR, eds. C elegans II. Cold Spring Harbor, New York: Cold Spring Harbor Laboratory Press, 1997:739-68.
2. Kenyon C, Chang J, Gensch E et al. A C. elegans mutant that lives twice as long as wild type. Nature 1993; 366:461-4.
3. Friedman DB, Johnson TE. A mutation in the age-1 gene in Caenorhabditis elegans lengthens life and reduces hermaphrodite fertility. Genetics 1988; 118:75-86.
4. Malone EA, Inoue T, Thomas JH. Genetic analysis of the roles of daf-28 and age-1 in regulating Caenorhabditis elegans dauer formation. Genetics 1996; 143:1193-205.
5. Kenyon C. In: Riddle DL, Blumenthal T, Meyer BJ, Priess JR, eds. C elegans II. Cold Spring Harbor, New York: Cold Spring Harbor Laboratory Press, 1997:791-813.
6. Kimura KD, Tissenbaum HA, Liu Y et al. daf-2, an insulin receptor-like gene that regulates longevity and diapause in Caenorhabditis elegans. Science 1997; 277:942-6.
7. Morris JZ, Tissenbaum HA, Ruvkun G. A phosphatidylinositol-3-OH kinase family member regulating longevity and diapause in Caenorhabditis elegans. Nature 1996; 382:536-9.
8. Cantley LC. The phosphoinositide 3-kinase pathway. Science 2002; 296:1655-7.
9. Lin K, Dorman JB, Rodan A et al. daf-16: An HNF-3/forkhead family member that can function to double the life-span of Caenorhabditis elegans. Science 1997; 278:1319-22.
10. Ogg S, Paradis S, Gottlieb S et al. The Fork head transcription factor DAF-16 transduces insulin-like metabolic and longevity signals in C. elegans. Nature 1997; 389:994-9.
11. Hertweck M, Göbel C, Baumeister R. C. elegans SGK-1 is the critical compnent in the Akt/PKB kinase complex to control stress respnse and life span. Dev Cell 2004; 6:577-88.

12. Kenyon C. The plasticity of aging: insights from long-lived mutants. Cell 2005; 120:449-60.
13. Ogg S, Ruvkun G. The C. elegans PTEN homolog, DAF-18, acts in the insulin receptor-like metabolic signaling pathway. Mol Cell 1998; 2:887-93.
14. Mihaylova VT, Borland CZ, Manjarrez L et al. The PTEN tumor suppressor homolog in Caenorhabditis elegans regulates longevity and dauer formation in an insulin receptor-like signaling pathway. Proc Natl Acad Sci USA 1999; 96:7427-32.
15. Rouault JP, Kuwabara PE, Sinilnikova OM et al. Regulation of dauer larva development in Caenorhabditis elegans by daf-18, a homologue of the tumour suppressor PTEN. Curr Biol 1999; 9:329-32.
16. Salin DA, Brunet A. Fox O transcription factors in the maintenance of cellular homeostasis during aging. Curr Opin Cell Biol 2008; 20:126-36.
17. Antebi A. Genetics of aging in Caenorhabditis elegans. PLoS Genet 2007; 3:1565-71.
18. Oh SW, Mukhopadhyay A, Svrzikapa N et al. JNK regulates lifespan in Caenorhabditis elegans by modulating nuclear translocation of forkhead transcription factor/DAF-16. Proc Natl Acad Sci USA 2005; 102:4494-9.
19. Motta MC, Divecha N, Lemieux M et al. Mammalian SIRT1 represses forkhead transcription factors. Cell 2004; 116:551-63.
20. Brunet A, Sweeney LB, Sturgill JF et al. Stress-dependent regulation of FOXO transcription factors by the SIRT1 deacetylase. Science 2004; 303:2011-5.
21. Tissenbaum HA, Guarente L. Increased dosage of a sir-2 gene extends lifespan in Caenorhabditis elegans. Nature 2001; 410:227-30.
22. Cuervo AM, Bergamini E, Brunk UT et al. Autophagy and aging: the importance of maintaining "clean" cells. Autophagy 2005; 1:131-40.
23. Mair W, Dillin A. Aging and survival: the genetics of life span extension by dietary restriction. Annu Rev Biochem 2008; 77:727-54.
24. Roth GS, Ingram DK, Lane MA. Caloric restriction in primates and relevance to humans. Ann N Y Acad Sci 2001; 928:305-15.
25. Houthoofd K, Vanfleteren JR. The longevity effect of dietary restriction in Caenorhabditis elegans. Exp Gerontol 2006; 41:1026-31.
26. Klass MR. Aging in the nematode Caenorhabditis elegans: major biological and environmental factors influencing life span. Mech Ageing Dev 1977; 6:413-29.
27. Mair W, Panowski SH, Shaw RJ et al. Optimizing dietary restriction for genetic epistasis analysis and gene discovery in C. elegans. PLoS One 2009; 4:e4535.
28. Lakowski B, Hekimi S. The genetics of caloric restriction in Caenorhabditis elegans. Proc Natl Acad Sci USA 1998; 95:13091-6.
29. Bishop NA, Guarente L. Genetic links between diet and lifespan: shared mechanisms from yeast to humans. Nat Rev Genet 2007; 8:835-44.
30. Partridge L, Piper MD, Mair W. Dietary restriction in Drosophila. Mech Ageing Dev 2005; 126:938-50.
31. Finkel T, Holbrook NJ. Oxidants, oxidative stress and the biology of ageing. Nature 2000; 408:239-47.
32. Feng J, Bussiere F, Hekimi S. Mitochondrial electron transport is a key determinant of life span in Caenorhabditis elegans. Dev Cell 2001; 1:633-44.
33. Ewbank JJ, Barnes TM, Lakowski B et al. Structural and functional conservation of the Caenorhabditis elegans timing gene clk-1. Science 1997; 275:980-3.
34. Miyadera H, Amino H, Hiraishi A et al. Altered quinone biosynthesis in the long-lived clk-1 mutants of Caenorhabditis elegans. J Biol Chem 2001; 276:7713-6.
35. Hekimi S, Guarente L. Genetics and the specificity of the aging process. Science 2003; 299:1351-4.
36. Ishii N, Fujii M, Hartman PS et al. A mutation in succinate dehydrogenase cytochrome b causes oxidative stress and ageing in nematodes. Nature 1998; 394:694-7.
37. Melov S, Ravenscroft J, Malik S et al. Extension of life-span with superoxide dismutase/catalase mimetics. Science 2000; 289:1567-9.
38. Dillin A, Hsu AL, Arantes-Oliveira N et al. Rates of behavior and aging specified by mitochondrial function during development. Science 2002; 298:2398-401.
39. Lee SS, Lee RY, Fraser AG et al. A systematic RNAi screen identifies a critical role for mitochondria in C. elegans longevity. Nat Genet 2003; 33:40-8.
40. Doonan R, McElwee JJ, Matthijssens F et al. Against the oxidative damage theory of aging: superoxide dismutases protect against oxidative stress but have little or no effect on life span in Caenorhabditis elegans. Genes Dev 2008; 22:3236-41.
41. Klionsky DJ. Autophagy: from phenomenology to molecular understanding in less than a decade. Nat Rev Mol Cell Biol 2007; 8:931-7.

42. Maiuri MC, Zalckvar E, Kimchi A et al. Self-eating and self-killing: crosstalk between autophagy and apoptosis. Nat Rev Mol Cell Biol 2007; 8:741-52.
43. Mizushima N, Klionsky DJ. Protein turnover via autophagy: implications for metabolism. Annu Rev Nutr 2007; 27:19-40.
44. Levine B, Klionsky DJ. Development by self-digestion: molecular mechanisms and biological functions of autophagy. Dev Cell 2004; 6:463-77.
45. Dice JF. Peptide sequences that target cytosolic proteins for lysosomal proteolysis. Trends Biochem Sci 1990; 15:305-9.
46. Cuervo AM, Dice JF. A receptor for the selective uptake and degradation of proteins by lysosomes. Science 1996; 273:501-3.
47. Levine B, Kroemer G. Autophagy in the pathogenesis of disease. Cell 2008; 132:27-42.
48. Mizushima N. The pleiotropic role of autophagy: from protein metabolism to bactericide. Cell Death Differ 2005; 12(Suppl 2):1535-41.
49. Schmelzle T, Hall MN. TOR, a central controller of cell growth. Cell 2000; 103:253-62.
50. Xie Z, Klionsky DJ. Autophagosome formation: core machinery and adaptations. Nat Cell Biol 2007; 9:1102-9.
51. Kramer JM, Moerman DC. Autophagy in C. elegans. Wormbook 2009.
52. Melendez A, Talloczy Z, Seaman M et al. Autophagy genes are essential for dauer development and life-span extension in C. elegans. Science 2003; 301:1387-91.
53. Hars ES, Qi H, Ryazanov AG et al. Autophagy regulates ageing in C. elegans. Autophagy 2007; 3:93-5.
54. Hansen M, Chandra A, Mitic LL et al. A role for autophagy in the extension of lifespan by dietary restriction in C. elegans. PLoS Genet 2008; 4:e24.
55. Toth ML, Sigmond T, Borsos E et al. Longevity pathways converge on autophagy genes to regulate life span in Caenorhabditis elegans. Autophagy 2008; 4:330-8.
56. Jia K, Thoma C, Akbar M et al. Autophagy genes protect against Salmonella typhimuruim infection and mediate insulin signaling-regulated pathogen resistance. Proc Natl Acad Sci USA 2009; 106; 145:64-9.
57. Juhasz G, Puskas LG, Komonyi O et al. Gene expression profiling identifies FKBP39 as an inhibitor of autophagy in larval Drosophila fat body. Cell Death Differ 2007; 14:1181-90.
58. Zhao J, Brault JJ, Schild A et al. FoxO3 coordinately activates protein degradation by the autophagic/lysosomal and proteasomal pathways in atrophying muscle cells. Cell Metab 2007; 6:472-83.
59. Mammucari C, Milan G, Romanello V et al. FoxO3 controls autophagy in skeletal muscle in vivo. Cell Metab 2007; 6:458-71.
60. Vellai T, Takacs-Vellai K, Zhang Y et al. Genetics: influence of TOR kinase on lifespan in C. elegans. Nature 2003; 426:620.
61. Jia K, Chen D, Riddle DL. The TOR pathway interacts with the insulin signaling pathway to regulate C. elegans larval development, metabolism and life span. Development 2004; 131:3897-906.
62. Hansen M, Taubert S, Crawford D et al. Lifespan extension by conditions that inhibit translation in Caenorhabditis elegans. Aging Cell 2007; 6:95-110.
63. Morck C, Pilon M. C. elegans feeding defective mutants have shorter body lengths and increased autophagy. BMC Dev Biol 2006; 6:39.
64. Jia K, Levine B. Autophagy is required for dietary restriction-mediated life span extension in C. elegans. Autophagy 2007; 3:597-9.
65. Panowski SH, Wolff S, Aguilaniu H et al. PHA-4/Foxa mediates diet-restriction-induced longevity of C. elegans. Nature 2007; 447:550-5.
66. Berdichevsky A, Viswanathan M, Horvitz HR et al. C. elegans SIR-2.1 interacts with 14-3-3 proteins to activate DAF-16 and extend life span. Cell 2006; 125:1165-77.
67. Lee IH, Cao L, Mostoslavsky R et al. A role for the NAD-dependent deacetylase Sirt1 in the regulation of autophagy. Proc Natl Acad Sci USA 2008; 105:3374-9.
68. Dong MQ, Venable JD, Au N et al. Quantitative mass spectrometry identifies insulin signaling targets in C. elegans. Science 2007; 317:660-3.
69. Dwivedi M, Song HO, Ahnn J. Autophagy genes mediate the effect of calcineurin on life span in C. elegans. Autophagy 2009; 5:604-7.
70. Hardie DG, Hawley SA. AMP-activated protein kinase: the energy charge hypothesis revisited. Bioessays 2001; 23:1112-9.
71. Apfeld J, O'Connor G, McDonagh T et al. The AMP-activated protein kinase AAK-2 links energy levels and insulin-like signals to lifespan in C. elegans. Genes Dev 2004; 18:3004-9.
72. Greer EL, Dowlatshahi D, Banko MR et al. An AMPK-FOXO pathway mediates longevity induced by a novel method of dietary restriction in C. elegans. Curr Biol 2007; 17:1646-56.

73. Knowler WC, Barrett-Connor E, Fowler SE et al. Reduction in the incidence of type 2 diabetes with lifestyle intervention or metformin. N Engl J Med 2002; 346:393-403.
74. Musi N, Hirshman MF, Nygren J et al. Metformin increases AMP-activated protein kinase activity in skeletal muscle of subjects with type 2 diabetes. Diabetes 2002; 51:2074-81.
75. Zhou G, Myers R, Li Y et al. Role of AMP-activated protein kinase in mechanism of metformin action. J Clin Invest 2001; 108:1167-74.
76. Anisimov VN, Berstein LM, Egormin PA et al. Metformin slows down aging and extends life span of female SHR mice. Cell Cycle 2008; 7:2769-73.
77. Papandreou I, Lim AL, Laderoute K et al. Hypoxia signals autophagy in tumor cells via AMPK activity, independent of HIF-1, BNIP3 and BNIP3L. Cell Death Differ 2008; 15:1572-81.
78. Takagi H, Matsui Y, Hirotani S et al. AMPK mediates autophagy during myocardial ischemia in vivo. Autophagy 2007; 3:405-7.
79. Lippai M, Csikos G, Maroy P et al. SNF4Aγ, the Drosophila AMPK γ subunit is required for regulation of developmental and stress-induced autophagy. Autophagy 2008; 4:476-86.
80. Wang Z, Wilson WA, Fujino MA et al. Antagonistic controls of autophagy and glycogen accumulation by Snf1p, the yeast homolog of AMP-activated protein kinase and the cyclin-dependent kinase Pho85p. Mol Cell Biol 2001; 21:5742-52.
81. Wang MC, Bohmann D, Jasper H. JNK signaling confers tolerance to oxidative stress and extends lifespan in Drosophila. Dev Cell 2003; 5:811-6.
82. Wang MC, Bohmann D, Jasper H. JNK extends life span and limits growth by antagonizing cellular and organism-wide responses to insulin signaling. Cell 2005; 121:115-25.
83. Wu H, Wang MC, Bohmann D. JNK protects Drosophila from oxidative stress by trancriptionally activating autophagy. Mech Dev 2009.
84. Wei Y, Pattingre S, Sinha S et al. JNK1-mediated phosphorylation of Bcl-2 regulates starvation-induced autophagy. Mol Cell 2008; 30:678-88.
85. Hashimoto Y OS, Nishida E. Lifespan extension by suppression of autophagy genes in Caenorhabditis elegans. Genes Cells 2009; 14:717-720.
86. Wallace DC. A mitochondrial paradigm of metabolic and degenerative diseases, aging and cancer: a dawn for evolutionary medicine. Annu Rev Genet 2005; 39:359-407.
87. Chen Y, Azad MB, Gibson SB. Superoxide is the major reactive oxygen species regulating autophagy. Cell Death Differ 2009; 16:1040-52.
88. Elmore SP, Qian T, Grissom SF et al. The mitochondrial permeability transition initiates autophagy in rat hepatocytes. FASEB J 2001; 15:2286-7.
89. Lam M, Oleinick NL, Nieminen AL. Photodynamic therapy-induced apoptosis in epidermoid carcinoma cells. Reactive oxygen species and mitochondrial inner membrane permeabilization. J Biol Chem 2001; 276:47379-86.
90. Rodriguez-Enriquez S, He L, Lemasters JJ. Role of mitochondrial permeability transition pores in mitochondrial autophagy. Int J Biochem Cell Biol 2004; 36:2463-72.
91. Gu Y, Wang C, Cohen A. Effect of IGF-1 on the balance between autophagy of dysfunctional mitochondria and apoptosis. FEBS Lett 2004; 577:357-60.
92. Donati A, Taddei M, Cavallini G et al. Stimulation of macroautophagy can rescue older cells from 8-OHdG mtDNA accumulation: a safe and easy way to meet goals in the SENS agenda. Rejuvenation Res 2006; 9:408-12.
93. Kim I, Rodriguez-Enriquez S, Lemasters JJ. Selective degradation of mitochondria by mitophagy. Arch Biochem Biophys 2007; 462:245-53.
94. Nowikovsky K, Reipert S, Devenish RJ et al. Mdm38 protein depletion causes loss of mitochondrial K+/H+ exchange activity, osmotic swelling and mitophagy. Cell Death Differ 2007; 14:1647-56.
95. Cavallini G, Donati A, Taddei M et al. Evidence for selective mitochondrial autophagy and failure in aging. Autophagy 2007; 3:26-7.
96. Twig G, Elorza A, Molina AJ et al. Fission and selective fusion govern mitochondrial segregation and elimination by autophagy. EMBO J 2008; 27:433-46.
97. Donati A, Ventruti A, Cavallini G et al. In vivo effect of an antilipolytic drug (3,5′-dimethylpyrazole) on autophagic proteolysis and autophagy-related gene expression in rat liver. Biochem Biophys Res Commun 2008; 366:786-92.
98. Chen YF, Kao CH, Chen YT et al. Cisd2 deficiency drives premature aging and causes mitochondria-mediated defects in mice. Genes Dev 2009; 23:1183-94.
99. Okamoto K, Kondo-Okamoto N, Ohsumi Y. Mitochondria-anchored receptor Atg32 mediates degradation of mitochondria via selective autophagy. Dev Cell 2009; 17:87-97.
100. Mathew R, Karp CM, Beaudoin B et al. Autophagy suppresses tumorigenesis through elimination of p62. Cell 2009; 137:1062-75.
101. Singh R, Kaushik S, Wang Y et al. Autophagy regulates lipid metabolism. Nature 2009; 458:1131-5.

Chapter 6

Autophagy and Aging:
Lessons from Progeria Models

Guillermo Mariño, Alvaro F. Fernández and Carlos López-Otín*

Abstract

Autophagy is an evolutionarily conserved process essential for cellular homeostasis and organismal viability. In fact, this pathway is one of the major protein degradation mechanisms in eukaryotic cells. It has been repeatedly reported that the autophagic activity of living cells decreases with age, probably contributing to the accumulation of damaged macromolecules and organelles during aging. Moreover, autophagy modulation in different model organisms has yielded very promising results suggesting that the maintenance of a proper autophagic activity contributes to extend longevity. On the other hand, recent findings have shown that distinct premature-aging murine models exhibit an extensive basal activation of autophagy instead of the characteristic decline in this process occurring during normal aging. This unexpected autophagic increase in progeroid models is usually associated with a series of metabolic alterations resembling those occurring under calorie restriction or in other situations reported to prolong life-span. In this chapter, we will discuss the current knowledge on the relationship between the autophagy pathway and aging with a special emphasis on the unexpected and novel link between premature aging and autophagy up-regulation.

Introduction

Aging is a natural process which affects most biological functions and appears to be a consequence of the accumulative action of different types of stressors. Thus, oxidative damage, telomere attrition and the decline of DNA repair and protein turnover systems are thought to be major causes of aging.[1] Among the diverse events associated with this age-dependent functional decline, the accumulation of damaged proteins and organelles has been proposed to contribute in a decisive manner to cellular malfunction.[2] In eukaryotic cells, there are two major pathways involved in the quality control and turnover of cellular components: the ubiquitin/proteasome system and the autophagic/lysosomal system. The ubiquitin/proteasome system plays a major role in the maintenance of cellular homeostasis and in protein quality control, but also regulates essential processes such as cell cycle progression or signal transduction.[3,4] Nevertheless, the constitutive degradation of proteins by this ubiquitin/proteasome system is mostly limited to short-lived proteins. By contrast, the autophagic/lysosomal system is able to degrade long-lived proteins and even entire organelles. The autophagic degradation routes can be classified into at least three different pathways: macroautophagy, microautophagy and chaperone-mediated autophagy.[5,6] Macroautophagy is the major lysosomal pathway for the turnover of cytoplasmic components and will hereafter be referred to as autophagy. This process begins with the engulfment of cytoplasmic constituents by

*Corresponding Author: Carlos López-Otín—Departamento de Bioquímica y Biología Molecular, Facultad de Medicina, Instituto Universitario de Oncología, Universidad de Oviedo, Oviedo, 33006 Spain. Email: clo@uniovi.es

Protein Metabolism and Homeostasis in Aging, edited by Nektarios Tavernarakis.

a membrane sac: the isolation membrane. Then, this structure forms a double-membrane vesicle called the autophagosome which contains bulk portions of cytoplasm and eventually fuses with the lysosome. Finally, the inner membrane of the autophagosome and its protein and organelle contents are degraded by lysosomal hydrolases and recycled.

It has been proposed that accumulation of cellular garbage associated with aging is mainly caused by an age-related decline of autophagic and lysosomal activity.[7,8] This hypothesis is based on reports showing an age-related decrease of autophagic degradation rate, although direct evidences supporting a causal relation between age-related autophagy dysfunction and aging—especially in mammals—need to be supported by additional experimental data. In this chapter, we will discuss the current knowledge on the impact and relevance of autophagy in aging and life-span extension, paying special attention on recent and unexpected findings suggesting a detrimental role for abnormally increased autophagy in the development of premature aging syndromes.

Autophagy and Physiological Aging

A characteristic feature of aging is the accumulation of damaged proteins and organelles, which is in fact one of the most persistent alterations widely observed in aged cells. This phenomenon probably derives from the decay of folding and protein turnover control mechanisms. In this regard, many authors have postulated that a gradual age-related decrease of autophagic activity could play an essential role in the functional deterioration of aging organisms.[2,7] Given that the recently generated mouse models with impaired autophagy present dramatic phenotypes showing accumulation of misfolded protein aggregates and defective organelles, this hypothesis seems very reasonable.[9,10] In fact, long before the molecular basis of the autophagic pathway began to be understood, some authors had observed that lysosomal activity decreases with age.[11,12] The observed decrease in this activity is accompanied by a series of alterations, such as lysosome expansion and/or accumulation of undigested material as an autofluorescent pigment called lipofuscin.[13] These alterations ultimately lead to lysosomal dysfunction and to the accumulation of immature autophagic vesicles, which remained undegraded due to deficient lysosomal activity. In fact, lipofuscin accumulation, together with the lysosomal damage caused by toxic protein products, affects the ability of lysosomes to fuse with autophagic vesicles and reduce hydrolases activity, thus delaying or even stopping autophagosome maturation.[14]

In addition to the mentioned lysosomal dysfunction, some authors have proposed a link between the age-related autophagic decline and alterations in the levels of circulating hormones or in the response to them. In fact, an autophagy de-regulation has been observed in aged rodent livers, mainly characterized by a progressive resistance to glucagon signalling in terms of autophagy induction.[12,15] In young organisms, a rise in glucagon levels, occurring in starvation periods, induces autophagosome formation, whereas an increment in circulating insulin, mainly observed after feeding, largely inhibits autophagy.[2,16] This effect is mediated by hormonal regulation of mTOR kinase activity, which is linked to insulin receptor activity, in turn regulated by the insulin/glucagon balance and by other activators as growth factors or cytokines.[17,18] In aged organisms, basal insulin receptor activity is usually increased,[19] which may lead to severe metabolic disorders, such as insulin resistance. The observed age-related autophagy de-regulation may be also related to this event, finally resulting in a constitutive increase in mTOR activity and thus in a stronger basal autophagy inhibition. All these observations showing the existence of an age-related decline of autophagic activity have provided very valuable information, but are somewhat limited by the relative lack of knowledge on the molecular mechanisms involved in autophagy regulation.

Knowledge on the molecular basis of autophagy has considerably improved in the last years, mainly due to the isolation and characterization of autophagy-defective mutants in the yeast *Saccharomyces cerevisiae*.[20-22] The subsequent identification of the orthologues of autophagy-related yeast genes in higher eukaryotes has made possible the dissection and study of the relevance of autophagy in diverse physiological and pathological processes, including aging. In this regard, the selective targeting of autophagy genes in diverse model organisms has shown a tight connection between autophagic activity and longevity (Fig. 1). In the nematode

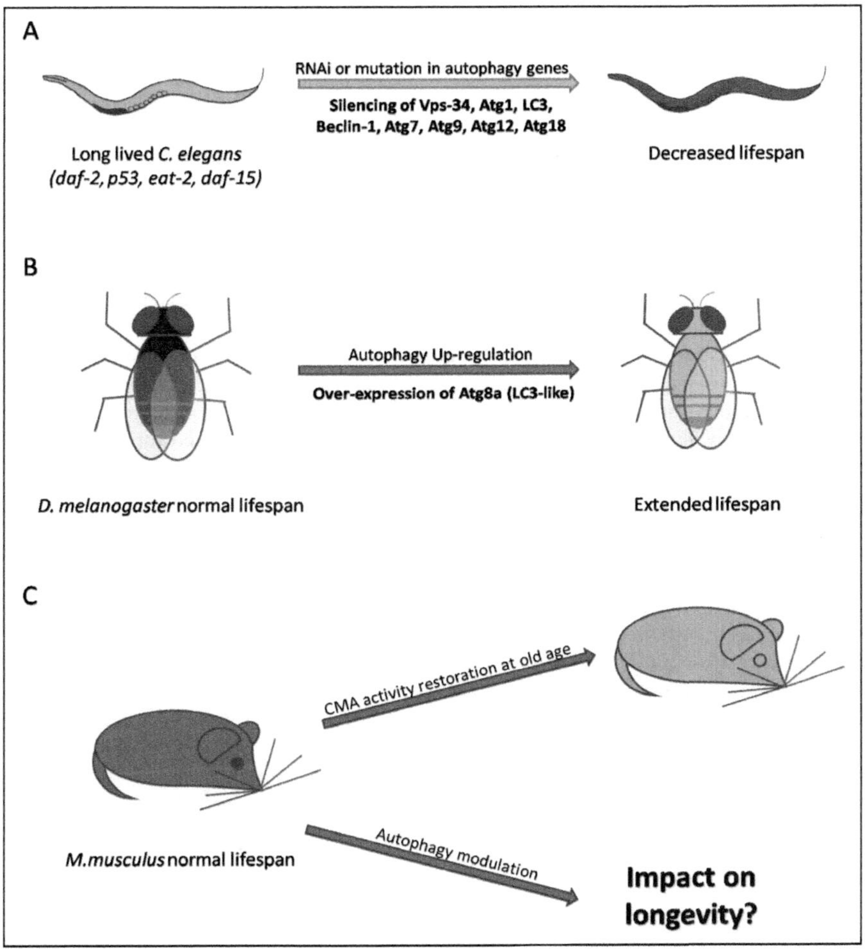

Figure 1. Effects of autophagy modulation in longevity. A) Silencing of a variety of autophagy-related genes results in a decrease in the longevity of long-lived *C. elegans* mutants, which suggests that a proper autophagic activity is required to attain the maximal life-span extension in worms. B) In *D. melanogaster*, overexpression of autophagy-related genes leading to increased autophagic activity results in life-span extension and prevents age-related neuronal damage. C) Although restoration of age-dependent loss of CMA improves hepatic function in mice, additional studied are required to determine the real impact of autophagy modulation in terms of life-span extension in mammals.

Caenorhabditis elegans, disruption of *daf-2*, the functional equivalent for the insulin receptor in mammals, substantially extends life-span.[23] The knock-down of essential proteins for autophagy execution significantly reduces the life-span extension caused by *daf-2* disruption (Fig. 1A).[24] Similarly, a proper autophagic activity seems to be required to achieve the maximum life-span extension obtained by the deletion of mTOR or p53.[25,26] All these results suggest an important role for autophagy in longevity, at least in nematodes. In flies, recent findings also suggest that increased basal autophagy in the nervous system enhances longevity and oxidant resistance (Fig. 1B).[27] By contrast, no life-span studies in mammalian models with reduced autophagy have been reported so far and although several murine models totally lacking autophagic activity have been generated, their phenotypes are so dramatic that make impossible to study the putative implications of this process in aging.[28,29] However, very interesting results have been obtained in mice by manipulating chaperone-mediated autophagy (CMA), a specific form of autophagy in which proteins with a determined amino-acid sequence are directly delivered to the lysosome lumen without the participation of autophagosomes. In fact, restoration of chaperone-mediated autophagy in aging liver improves cellular maintenance and hepatic function (Fig. 1C).[30] Although these results are not a *bona fide* evidence that autophagy up-regulation promotes longevity in mammals, they suggest that maintaining a proper autophagic activity throughout life span prevents or at least slows down some of the functional failures associated with aging. In this regard, a very recent work focused on the use of the immunosuppressant rapamycin, which inhibits mTOR activity and induces autophagy, has shown that this drug extends both median and maximal life-span even when fed late in life.[31] However, as mTOR activity affects multiple and essential processes apart from autophagy, the real implication of this process in life-span extension mediated by rapamycin treatment has to be evaluated in more detail. Finally, it is worthwhile mentioning that caloric restriction (CR), the only 'natural' intervention known to extend longevity, seems to help maintain an adequate basal and induced autophagic activity, possibly due to changes in circulating hormonal and biochemical parameters, as low levels of glucose, insulin or growth factors, observed after CR.[32,33] In this regard, recent studies have demonstrated that CR increases the expression of sirtuin-1 and the life-span extension associated with CR is dependent on sirtuin-1 expression.[34] The fact that sirtuin-1 is able to activate autophagy in cells even in the presence of nutrients also supports the idea of an essential role for autophagy CR-mediated life-span extension in mammals. Consistently, the very recent findings showing that CR in primates not only extends substantially life-span, but also delays the onset of age-associated pathologies[35] highlights the importance of understanding the precise role played by autophagy in CR-mediated life-span extension. In fact, pathologies as cancer, cardiovascular diseases, or brain atrophy have been previously associated with autophagy impairment.[29,36-38] The finding that all these age-related alterations are substantially reduced in rhesus monkeys subjected to CR[35] also supports the idea of an important role of autophagy in mediating the anti-aging effects of CR.

Taken together, all these observations show that autophagic activity decreases with age, mainly due to a decline in lysosomal activity and also to a de-regulation of the autophagic response to circulating hormones. It seems clear that maintaining an adequate autophagic activity is essential for achieving the maximum life-span extension in invertebrates, but additional experimental evidences are necessary to confirm these correlative observations in mammals. However, although the idea of manipulating autophagy to delay human aging—alone or in combination with other treatments—is very tempting, further efforts to analyze the impact of CR, CR pharmacological mimetics, or rapamycin treatment in models with reduced autophagic activity will clarify the real role for this process in aging and/or in life-span extension.

Autophagy and Premature Aging

Although the precise molecular determinants of aging are still very far from being completely understood, our knowledge of the molecular basis of this complex process has improved considerably, in part due to the study of segmental progeroid syndromes.[39,40] These syndromes are dramatic diseases in which certain features of human aging are prematurely developed and mainly derive from deficiencies in DNA-repair systems or from defects in components of the nuclear envelope, which ultimately lead to the activation of DNA-damage and chromosome instability responses.[41-43] Apparently, one would expect that given that premature aging models manifest precociously many features of normal aging, they would show a decrease in autophagic activity. Surprisingly, it has been recently reported that progeroid mice fed *ad libitum* present a clear increase in their tissue autophagic activity. It is remarkable that this unexpected increase in autophagic activity has also been observed in premature aging murine models caused by defects in either nuclear envelope components or DNA repair machinery,[44,45]

Table 1. Mouse models of premature aging also showing autophagy alterations

Model	Phenotype	Autophagy Status	Refs
Zmpste24$^{-/-}$	Growth retardation and premature aging phenotype caused by nuclear envelope alterations	Increased systemic basal autophagy accompanied by metabolic alterations similar to those found in long-lived mice	44,45
XPF$^{-/-}$	Growth retardation and premature aging phenotype caused by deficiencies in DNA repair mechanisms	Increased systemic basal autophagy accompanied by metabolic and somatotroph axis alterations similar to those found in long-lived mice	49,50
CSB$^{-/-}$/*XPA*$^{-/-}$	Growth retardation and premature aging phenotype caused by DNA repair mechanisms deficiencies	Very similar to XPF$^{-/-}$	49,50
Cisd2$^{-/-}$	Premature aging phenotype caused by mitochondrial defects	Increased basal autophagy accompanied by mitochondrial breakdown and dysfunction	46
PTHrP Knock-In	Growth retardation and early lethality showing features of premature aging caused by deletion of the nuclear localization sequence and C-terminus of PTHrp	Increased basal autophagy accompanied by hypoglucemia	47

mitochondrial dysfunction,[46] or parathyroid hormone activity de-regulation,[47] suggesting that autophagy increase is a common feature of diverse premature-aging syndromes (Table 1). This abnormal autophagy increase is commonly associated with a complex metabolic shift in glucose and lipid metabolism, involving the up-regulation of key genes for gluconeogenesis, glycogen accumulation, fatty acid synthesis and β-oxidation.[48-50] The metabolic shift is in turn linked to a substantial increase in the levels of hepatic glycogen and also to liver steatosis, as well as to significant changes in biochemical parameters, such as reduced levels of blood glucose, insulin, leptin and increased values of plasma adiponectin.[44] All these profound metabolic alterations likely lead to mTOR inhibition and up-regulation of LKB1-AMPK axis activity, both observed in progeroid murine models.[45] In fact, a decrease in blood glucose levels or an increase in circulating adiponectin leads to an in vivo induction of AMPK activity,[51] which in turn inhibits mTOR activity.[52] Given that both mTOR inhibition and AMPK activity increase have been shown to induce autophagy, the metabolic shift characteristic of progeroid mice likely underlies their unexpected autophagy increase. However, there could be additional mechanisms underlying the increase of autophagic activity in premature aging syndromes. In fact, it has been recently shown that nuclear blebs—a common feature of aged cells, also characteristic of cells from progeria patients—are degraded by autophagy, thus resulting in a modest but consistent increase in cellular autophagic activity.[53]

It is remarkable that the majority of the detected alterations in progeroid mice are associated with longer life-span rather than with the shortened longevity characteristic of these progeroid animals. As mentioned previously, autophagic activity is essential for dauer development and life-span extension in *C. elegans*[24] and a down-regulation of TOR-signaling extends life-span in both yeast and nematodes.[54,55] Similarly, AMPK over-expression promotes longevity in *C. elegans*[56] and the metabolic alterations such as hypoinsulinemia and hypoglycaemia found in progeroid mice, increase life-span in diverse model organisms.[32,57] In addition, many of the transcriptional alterations observed in key genes for glucose and lipid metabolism regulation resemble those observed in animals subjected to calorie restriction.[58,59]

Taken together, all these observations suggest that the different molecular defects leading to the development of premature aging trigger a complex and conserved metabolic response including many features tightly associated with life-span extension. At first sight, this fact seems intriguing. However, most of these features point towards a reduction in the metabolic activity of the organism, which inevitably leads to a reduction in cell division rate. This appears to be an adequate strategy to reduce the accumulation of cellular damage, as it has been reported that cell division drastically increases the rates of abnormal chromosome segregation and binucleation in cells from progeria patients.[60,61] Thus, it is reasonable to think that a normal growth rate would compromise somatic integrity in progeroid animals. In this case, a metabolic response aimed at re-allocating resources from growth to somatic preservation could be the best way to attenuate the consequences of the molecular alterations underlying progeroid syndromes. However, it is clear that this adaptive response fails to counteract the mentioned alterations, which irreversibly lead to the premature death observed in progeroid mice.

Conclusion

In summary, although it has been repeatedly shown that autophagic activity decreases with age, the paradoxical finding that autophagy is up-regulated in progeroid mice may also help to understand the mechanisms underlying the multiple tissue alterations observed in these syndromes. Although the observed metabolic shift could be beneficial, it could also be detrimental for the organism if overactivated. In this regard, we must take into account that although autophagy activation facilitates temporary adaptation to metabolic stress, this catabolic pathway may also lead to cell death when chronically activated.[5,62] This situation could contribute to the progressive muscular and cardiac wasting observed in both progeroid mice and progeria patients.[63,64] However, this hypothesis has to be tested using viable animal models with reduced autophagy or specific autophagy inhibitors, which are currently unavailable. These further studies will be helpful to clarify whether the observed increase of autophagic activity in progeroid mice helps to slow down the effects of the molecular alterations leading to premature aging, or by contrast, contributes to the development of the multiple pathologies observed in these mice. In this latter case, the autophagy pathway could be a future clinical target which may help to improve the prognosis of progeria patients.

Acknowledgements

We thank all members of our lab for their helpful comments on the manuscript and apologize for omission of relevant works due to space constraints. Our work is supported by grants from Ministerio de Ciencia e Innovación-Spain, European Union (FP7-Microenvimet) and Fundación M. Botín. The Instituto Universitario de Oncología is supported by Obra Social Cajastur and Acción Transversal del Cáncer RTICC.

References

1. Kirkwood TB. Understanding the odd science of aging. Cell 2005; 120:437-447.
2. Cuervo AM, Bergamini E, Brunk UT et al. Autophagy and aging: the importance of maintaining "clean" cells. Autophagy 2005; 1:131-140.
3. Muratani M, Tansey WP. How the ubiquitin-proteasome system controls transcription. Nat Rev Mol Cell Biol 2003; 4:192-201.
4. Ciechanover A. Proteolysis: from the lysosome to ubiquitin and the proteasome. Nat Rev Mol Cell Biol 2005; 6:79-87.
5. Shintani T, Klionsky DJ. Autophagy in health and disease: a double-edged sword. Science 2004; 306:990-995.
6. Marino G, Lopez-Otin C. Autophagy: molecular mechanisms, physiological functions and relevance in human pathology. Cell Mol Life Sci 2004; 61:1439-1454.
7. Terman A, Gustafsson B, Brunk UT. Autophagy, organelles and ageing. J Pathol 2007; 211:134-143.
8. Cuervo AM. Autophagy: in sickness and in health. Trends Cell Biol 2004; 14:70-77.

9. Hara T, Nakamura K, Matsui M et al. Suppression of basal autophagy in neural cells causes neurodegenerative disease in mice. Nature 2006; 441:885-889.
10. Komatsu M, Waguri S, Ueno T et al. Impairment of starvation-induced and constitutive autophagy in Atg7-deficient mice. J Cell Biol 2005; 169:425-434.
11. Del Roso A, Vittorini S, Cavallini G et al. Ageing-related changes in the in vivo function of rat liver macroautophagy and proteolysis. Exp Gerontol 2003; 38:519-527.
12. Donati A, Cavallini G, Paradiso C et al. Age-related changes in the regulation of autophagic proteolysis in rat isolated hepatocytes. J Gerontol A Biol Sci Med Sci 2001; 56:B288-B293.
13. Brunk UT, Jones CB, Sohal RS. A novel hypothesis of lipofuscinogenesis and cellular aging based on interactions between oxidative stress and autophagocytosis. Mutat Res 1992; 275:395-403.
14. Terman A. The effect of age on formation and elimination of autophagic vacuoles in mouse hepatocytes. Gerontology 1995; 41(Suppl 2):319-326.
15. Donati A, Recchia G, Cavallini G et al. Effect of aging and anti-aging caloric restriction on the endocrine regulation of rat liver autophagy. J Gerontol A Biol Sci Med Sci 2008; 63:550-555.
16. Cuervo AM. Autophagy and aging: keeping that old broom working. Trends Genet 2008; 24:604-612.
17. Corradetti MN, Guan KL. Upstream of the mammalian target of rapamycin: do all roads pass through mTOR? Oncogene 2006; 25:6347-6360.
18. Sarbassov DD, Ali SM, Sabatini DM. Growing roles for the mTOR pathway. Curr Opin Cell Biol 2005; 17:596-603.
19. Droge W, Kinscherf R. Aberrant insulin receptor signaling and amino acid homeostasis as a major cause of oxidative stress in aging. Antioxid Redox Signal 2008; 10:661-678.
20. Tsukada M, Ohsumi Y. Isolation and characterization of autophagy-defective mutants of Saccharomyces cerevisiae. FEBS Lett 1993; 333:169-174.
21. Thumm M, Egner R, Koch B et al. Isolation of autophagocytosis mutants of Saccharomyces cerevisiae. FEBS Lett 1994; 349:275-280.
22. Huang WP, Klionsky DJ. Autophagy in yeast: a review of the molecular machinery. Cell Struct Funct 2002; 27:409-420.
23. Hekimi S, Lakowski B, Barnes TM et al. Molecular genetics of life span in C. elegans: how much does it teach us? Trends Genet 1998; 14:14-20.
24. Melendez A, Talloczy Z, Seaman M et al. Autophagy genes are essential for dauer development and life-span extension in C. elegans. Science 2003; 301:1387-1391.
25. Hansen M, Chandra A, Mitic LL et al. A role for autophagy in the extension of lifespan by dietary restriction in C. elegans. PLoS Genet 2008; 4:e24.
26. Tavernarakis N, Pasparaki A, Tasdemir E et al. The effects of p53 on whole organism longevity are mediated by autophagy. Autophagy 2008; 4:870-873.
27. Simonsen A, Cumming RC, Brech A et al. Promoting basal levels of autophagy in the nervous system enhances longevity and oxidant resistance in adult Drosophila. Autophagy 2007; 4.
28. Kuma A, Hatano M, Matsui M et al. The role of autophagy during the early neonatal starvation period. Nature 2004; 432:1032-1036.
29. Komatsu M, Waguri S, Chiba T et al. Loss of autophagy in the central nervous system causes neurodegeneration in mice. Nature 2006; 441:880-884.
30. Zhang C, Cuervo AM. Restoration of chaperone-mediated autophagy in aging liver improves cellular maintenance and hepatic function. Nat Med 2008; 14:959-965.
31. Harrison DE, Strong R, Sharp ZD et al. Rapamycin fed late in life extends lifespan in genetically heterogeneous mice. Nature 2009; 460:392-395.
32. Russell SJ, Kahn CR. Endocrine regulation of ageing. Nat Rev Mol Cell Biol 2007; 8:681-691.
33. Cuervo AM. Calorie restriction and aging: the ultimate "cleansing diet". J Gerontol A Biol Sci Med Sci 2008; 63:547-549.
34. Boily G, Seifert EL, Bevilacqua L et al. SirT1 regulates energy metabolism and response to caloric restriction in mice. PLoS One 2008; 3:e1759.
35. Colman RJ, Anderson RM, Johnson SC et al. Caloric restriction delays disease onset and mortality in rhesus monkeys. Science 2009; 325:201-204.
36. Nakai A, Yamaguchi O, Takeda T et al. The role of autophagy in cardiomyocytes in the basal state and in response to hemodynamic stress. Nat Med 2007; 13:619-624.
37. Mathew R, Karp CM, Beaudoin B et al. Autophagy suppresses tumorigenesis through elimination of p62. Cell 2009; 137:1062-1075.
38. Mizushima N, Levine B, Cuervo AM et al. Autophagy fights disease through cellular self-digestion. Nature 2008; 451:1069-1075.
39. Scaffidi P, Misteli T. Lamin A-dependent nuclear defects in human aging. Science 2006; 312:1059-1063.

40. Haithcock E, Dayani Y, Neufeld E et al. Age-related changes of nuclear architecture in Caenorhabditis elegans. Proc Natl Acad Sci USA 2005; 102:16690-16695.
41. Navarro CL, Cau P, Levy N. Molecular bases of progeroid syndromes. Hum Mol Genet 2006; 15 Spec No 2:R151-R161.
42. Schumacher B, Garinis GA, Hoeijmakers JH. Age to survive: DNA damage and aging. Trends Genet 2008; 24:77-85.
43. Kudlow BA, Kennedy BK, Monnat RJ Jr. Werner and Hutchinson-Gilford progeria syndromes: mechanistic basis of human progeroid diseases. Nat Rev Mol Cell Biol 2007; 8:394-404.
44. Marino G, Ugalde AP, Salvador-Montoliu N et al. Premature aging in mice activates a systemic metabolic response involving autophagy induction. Hum Mol Genet. 2008;17(14):2196-211.
45. Marino G, Lopez-Otin C. Autophagy and aging: new lessons from progeroid mice. Autophagy 2008; 4:807-809.
46. Chen YF, Kao CH, Chen YT et al. Cisd2 deficiency drives premature aging and causes mitochondria-mediated defects in mice. Genes Dev 2009; 23:1183-1194.
47. Miao D, Su H, He B et al. Severe growth retardation and early lethality in mice lacking the nuclear localization sequence and C-terminus of PTH-related protein. Proc Natl Acad Sci USA 2008; 105:20309-20314.
48. Hoeijmakers JH. DNA damage, aging, and cancer. N Engl J Med 2009; 361:147-85.
49. van de Ven M, Andressoo JO, Holcomb VB et al. Adaptive stress response in segmental progeria resembles long-lived dwarfism and calorie restriction in mice. PLoS Genet 2006; 2:e192.
50. van der Pluijm I, Garinis GA, Brandt RM et al. Impaired genome maintenance suppresses the growth hormone—insulin-like growth factor 1 axis in mice with Cockayne syndrome. PLoS Biol 2006; 5:e2.
51. Long YC, Zierath JR. AMP-activated protein kinase signaling in metabolic regulation. J Clin Invest 2006; 116:1776-1783.
52. Hardie DG. The AMP-activated protein kinase pathway—new players upstream and downstream. J Cell Sci 2004; 117:5479-5487.
53. Park YE, Hayashi YK, Bonne G et al. Autophagic degradation of nuclear components in mammalian cells. Autophagy 2009; 5.
54. Vellai T, Takacs-Vellai K, Zhang Y et al. Genetics: influence of TOR kinase on lifespan in C. elegans. Nature 2003; 426:620.
55. Bonawitz ND, Chatenay-Lapointe M, Pan Y et al. Reduced TOR signaling extends chronological life span via increased respiration and upregulation of mitochondrial gene expression. Cell Metab 2007; 5:265-277.
56. Narbonne P, Roy R. Inhibition of germline proliferation during C. elegans dauer development requires PTEN, LKB1 and AMPK signalling. Development 2006; 133:611-619.
57. Kenyon C. The plasticity of aging: insights from long-lived mutants. Cell 2005; 120:449-460.
58. Anderson RM, Barger JL, Edwards MG et al. Dynamic regulation of PGC-1 alpha localization and turnover implicates mitochondrial adaptation in calorie restriction and the stress response. Aging Cell 2008; 7:101-111.
59. Gerhart-Hines Z, Rodgers JT, Bare O et al. Metabolic control of muscle mitochondrial function and fatty acid oxidation through SIRT1/PGC-1 alpha. EMBO J 2007; 26:1913-1923.
60. Dechat T, Shimi T, Adam SA et al. Alterations in mitosis and cell cycle progression caused by a mutant lamin A known to accelerate human aging. Proc Natl Acad Sci USA 2007; 104:4955-4960.
61. Cao K, Capell BC, Erdos MR et al. A lamin A protein isoform overexpressed in Hutchinson-Gilford progeria syndrome interferes with mitosis in progeria and normal cells. Proc Natl Acad Sci USA 2007; 104:4949-4954.
62. Maiuri MC, Zalckvar E, Kimchi A et al. Self-eating and self-killing: crosstalk between autophagy and apoptosis. Nat Rev Mol Cell Biol 2007; 8:741-752.
63. Merideth MA, Gordon LB, Clauss S et al. Phenotype and course of Hutchinson-Gilford progeria syndrome. N Engl J Med 2008; 358:592-604.
64. Pendas AM, Zhou Z, Cadinanos J et al. Defective prelamin A processing and muscular and adipocyte alterations in Zmpste24 metalloproteinase-deficient mice. Nat Genet 2002; 31:94-99.

Chapter 7

Regulation of Protein Turnover by Longevity Pathways

Tibor Vellai* and Krisztina Takács-Vellai

Abstract

Cellular homeostasis, which is needed for the cells to survive, requires a well-controlled balance in protein turnover. Both protein synthesis and degradation are influenced by distinct genetic pathways that control aging in divergent eukaryotic species. These conserved mechanisms involve the insulin/IGF-1 (Insulin-like Growth Factor-1), TGF-β (Transforming Growth Factor-β), JNK (c-Jun terminal kinase), RTK/Ras/MAPK (Receptor Tyrosine Kinase/Ras/Mitogen-Activated Protein Kinase) and TOR (kinase Target Of Rapamycin) signaling cascades and the mitochondrial respiratory system—each of them promotes protein synthesis; as well as the intracellular protein degradation machineries, including the ubiquitin-proteasome system and lysosome-mediated autophagy. In addition to providing building blocks for generation of new proteins and fuelling the cell with energy under starvation, the protein degradation processes eliminate damaged, nonfunctional proteins, the accumulation of which serves as the primary contributory factor to aging. Interestingly, a complex, intimate regulatory relationship exists between mechanisms promoting protein synthesis and those mediating protein degradation: under certain circumstances the former downregulate the latter. Thus, conditions that favor protein synthesis can enhance the rate at which damaged proteins accumulate. This may explain why genetic interventions and environmental factors (e.g., dietary restriction) that reduce protein synthesis, at least to tolerable levels, extend lifespan and increase resistance to cellular stress in various experimental model organisms of aging. In this chapter, the molecular mechanisms by which protein synthesis-promoting longevity pathways and protein degradation pathways interact with each other are discussed.

Protein Metabolism and Aging

Life is dependent on continuous generation, readjustment and renewal of cellular proteins, which become possible by processes of protein synthesis and degradation. The availability of newly produced proteins hence is a requirement for cellular homeostasis, a function known to be impaired during aging. Notably, a general decline in the rate of protein synthesis is a phenomenon that characterizes essentially all aging cells.[1] Recent genetic studies, however, indicate that lowering protein synthesis to tolerable levels that still enable cell growth and other vital cellular processes can significantly extend lifespan in the nematode *Caenorhabditis elegans*, a widely used invertebrate model of aging.[1-4] This finding was somehow unexpected as generation of new proteins should contribute to the rejuvenation and efficient working of the cells and eventually to the survival of the organism. To solve this apparent paradox, one should understand the molecular mechanisms by which protein synthesis and degradation are coupled in aging control. Furthermore, it also

*Corresponding Author: Tibor Vellai—Department of Genetics, Eötvös Loránd University, Pázmány Péter stny. 1/C. Budapest, Hungary, H-1117. Email: vellai@falco.elte.hu

Protein Metabolism and Homeostasis in Aging, edited by Nektarios Tavernarakis.

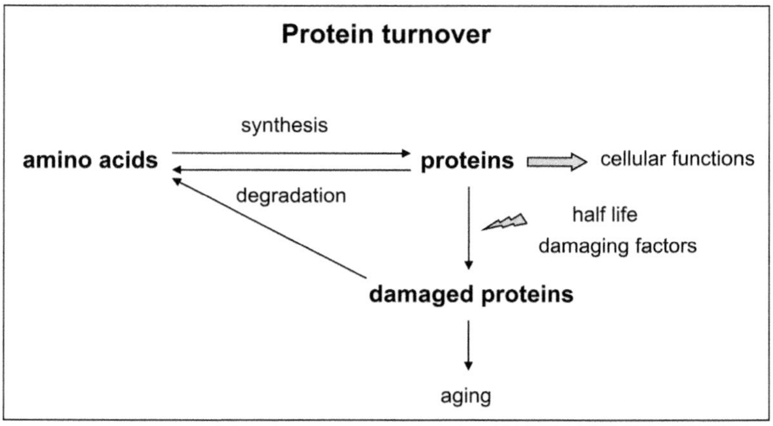

Figure 1. Turnover of intracellular proteins. Many cellular proteins are constantly degraded and resynthesized. Intracellular proteins that pass their life time or become damaged (e.g., misfolded) due to diverse cellular stress factors can form protein damages (cellular toxins), the accumulation of which contributes to the driving of the aging process. Elimination of such types of cellular damage occurs by the protein degradation processes. An age-dependent decline characterizes the rate of both protein synthesis and degradation.

remains to be elucidated whether lowered rate of protein synthesis at advanced ages is a cause, i.e., a contributory factor to, or consequence of the aging process. In general, elucidating how protein metabolism is linked to the aging process is a fundamental and fascinating issue in biology, with significant medical implications.

Protein turnover comprises the synthesis and degradation processes of proteins (Fig. 1). Translation of eukaryotic (messenger) mRNAs yields the short- and long-lived proteins, which can be damaged due to cellular stress or become intrinsically nonfunctional after passing their natural life time. For example, heat shock, a rapid elevation in temperature that occurs frequently under natural conditions, can cause misfolding, aggregation and cross-linking of various proteins. Reactive oxygen species (ROS)—which are small, short-lived and highly reactive molecules formed by incomplete one-electron reduction of oxygen and produced as a result of mitochondrial respiration, activity of specific enzymes or ionizing radiation—can also effectively damage (oxidize) proteins. Abnormal and thereby nonfunctional proteins generated by such effects can then be repaired by molecular chaperons or undergo degradation. The end-products of intracellular protein degradation processes are used to supply the cells with energy or as building blocks for protein synthesis, or further processed by the urea cycle removing the nitrogen in the form of urea. Damaged proteins are not simply unable to perform their biological activity, but can actively interfere with normal cellular functions. These factors, for example, can bind signaling proteins, thereby sending erroneous signals or blocking cell-cell communication. Thus, damaged proteins, together with other types of cellular damage, act as undesired cellular toxins.[5-7] Their effective removal from the cytoplasm is essential for the cells to operate persistently. In divergent eukaryotic species ranging from yeast to mammals, the lifelong, progressive accumulation of unrepaired cellular damage drives the aging process and leads to a decreased ability of the cell and, eventually of the organism, to survive.[5-7]

Locally, protein synthesis takes place on ribosomes (for reviewing the biochemical mechanism of eukaryotic mRNA translation, see ref. 1). Briefly, proteins are synthesized in the amino-to-carboxyl direction by the successive addition of amino acids to the carboxyl terminus of the growing peptide chains (Fig. 2). The amino acids are incorporated into the growing chain in activated form as aminoacyl-tRNAs. The linking of an amino acid to its specific tRNA is catalyzed by an *aminoacyl-tRNA synthetase* (AARS) and requires energy released from ATP hydrolysis. For each amino acid, there is usually one activating AARS enzyme and at least one kind of tRNA. Translation

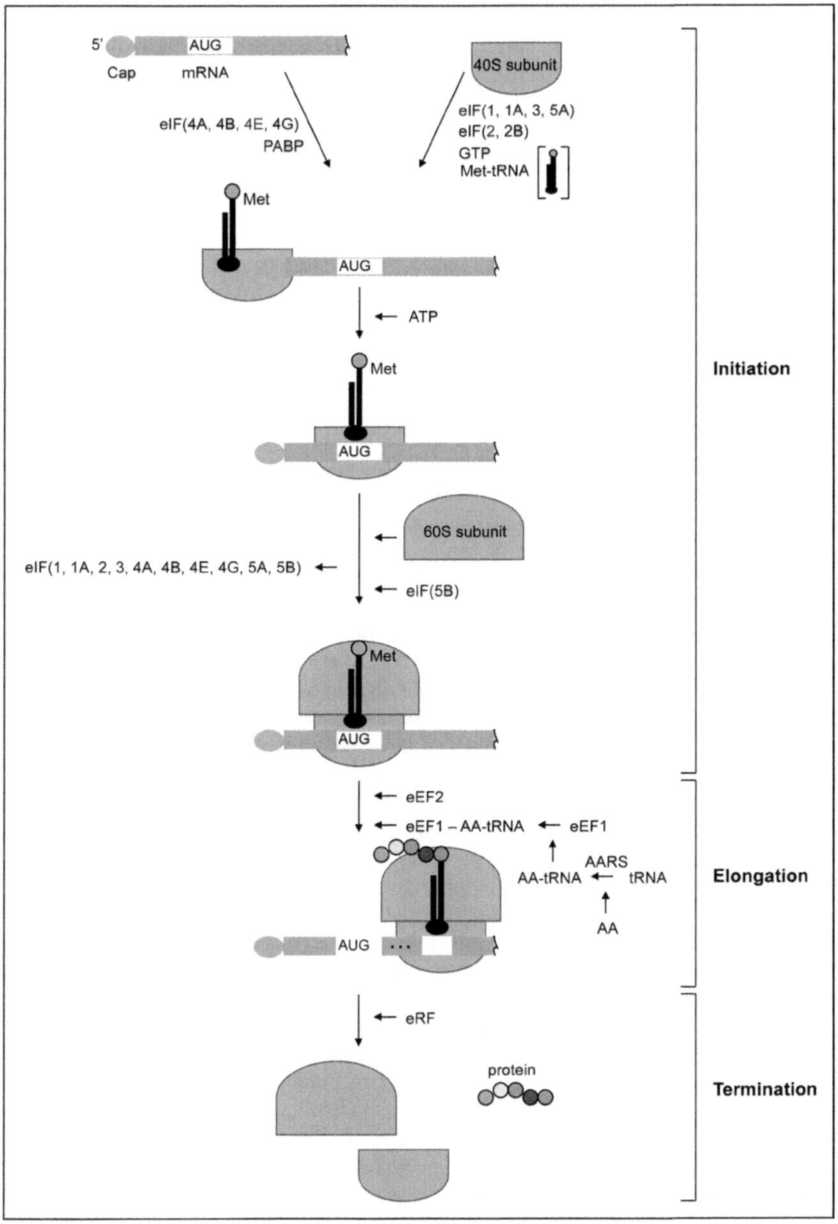

Figure 2. The main phases and components of mRNA translation in eukaryotes. mRNA translation consists of three main phases: initiation, elongation and termination. eIF denotes eukaryotic initiation factors, eEF stands for eukaryotic elongation factors, eRF indicates the eukaryotic release factor. Other abbreviations: GTP, guanosinetriphosphate; ATP, adenosinetriphosphate; PABP, poly-A binding protein; Cap, 7-monomethyl guanosine cap at the 5′ end of mRNA; Met-tRNA, methionyl transfer RNA; AA, aminoacyl; AARS, aminoacyl-tRNA synthetase; 40S and 60S, eukaryotic ribosomal subunits. The eIFs that play a crucial role in the global regulation of mRNA translation are indicated by red coloring. For details, see the introductory paragraph.

of mRNA molecules into proteins consists of three distinct, tightly controlled phases: initiation, elongation and termination. In eukaryotes, the initiation process involves a large number of initiation factors (eIFs), which dictate the formation of the so-called 43S pre-initiation complex (43S PIC) on the mRNA being translated. This complex integrates the activated 40S ribosomal subunit and initiator methyonil-tRNA (Met-tRNA) and joins with the 60S ribosomal subunit to form the 80S initiation complex. After formation of the 80S ribosome, elongation of the polypeptide chain starts and then proceeds with the appropriate amino acid-loaded tRNAs, which are synthesized by the corresponding AARSs. During the elongation process, the ribosome is translocated along the mRNA. Finally, mRNA translation is terminated by the eukaryotic release factor that reads a stop codon. Then, the ribosome dissociates from the mRNA, releasing the two ribosomal subunits.

The initiation process contains two crucial steps, where global mRNA translation is mostly regulated (for details, also see later chapters in the text). First, eIF2, which mediates binding of Met-tRNA to the 43S PIC, can be phosphorylated by eIF2B. This step influences the ability of eIF2B to convert GDP to GTP on eIF2, which is required for the incorporation of Met-tRNA into the 43S ribosomal subunit. Second, the recruitment of the 43S PIC on the 7-monomethyl guanosine cap at the 5′ end of all eukaryotic mRNAs is controlled by the cap-binding protein eIF4E, whose activity in turn can be modulated by the eIF4E-binding protein 4E-BP.

The assembly of new proteins requires a source of amino acids, which are generated by the digestion of proteins in the intestine and the degradation of proteins within the cell. The breakdown of intracellular proteins occurs by highly regulated degradative processes, including the ubiquitin-proteasome system and lysosome-mediated autophagy (Fig. 3A). Various proteins can become covalently modified with ubiquitin, which itself is a small protein and, after being tagged with chains of four or more ubiquitins, delivered to the ATP-dependent proteasome system for destruction.[8,9] Hence, ubiquitin is the tag that marks proteins for breakdown. Typically, the ubiquitin conjugation process involves three enzymes: ubiquitin-activating (E1), ubiquitin-conjugating (E2) and ubiquitin-ligase (E3) enzymes. The proteasome is a large, barrel-shaped, multiprotein protease complex that predominantly digests short-lived and misfolded proteins. Elimination of misfolded proteins can often be mediated by molecular chaperones, also called heat-shock or stress proteins, which either assist in the refolding of damaged proteins or sequester them for proteasome-mediated degradation.[10] Chaperones thus play a pivotal role in maintaining the conformational homeostasis of cytosolic proteins. Autophagy is the other main mechanism for intracellular protein degradation.[11,12] It is required for the bulk clearance of damaged and long-lived proteins. During autophagy, proteins, together with other cytosolic constituents, are delivered into lysosomes for breakdown that occurs by acidic proteases. Depending on the mechanism by which proteins are transported into lysosomes, three main types of autophagy can be distinguished: microautophagy, chaperon-mediated autophagy (CMA) and macroautophagy (for details, see the legend of Fig. 3B).[13] The products of autophagic breakdown are released back into the cytosol and utilized for the building of new proteins according to the actual needs of the cells. Therefore, autophagy is a vital mechanism for the renewal of intracellular proteins. It also plays a pivotal role in the cellular response to various stress conditions, such as amino acid deprivation, hypoxia, or elevated temperature, when fast reorganization of cellular functions is necessary.

In healthy cells, protein turnover (recycling) operates in a balanced way: unneeded, damaged proteins are effectively degraded by the cellular clearance mechanisms, providing building blocks and energy for the resynthesis process. Therefore, many cellular proteins are constantly degraded and resynthesized. However, the levels of both protein synthesis and degradation become decreased in aging cells. This could be the result of general deterioration in metabolic processes that accompanies aging. Alternatively, a developmental program mediated by signaling inputs or chromatin remodeling factors may adjust down the rate of protein metabolism over time. In addition, a complex regulatory interaction exists between longevity pathways promoting protein synthesis and molecular mechanisms underlying protein degradation. Nevertheless, intracellular protein damages inevitably accumulate as the (somatic) cells age and lowering protein synthesis can confer life span extension in certain eukaryotic organisms, probably through lowering the rate at which such types of damage accumulate.

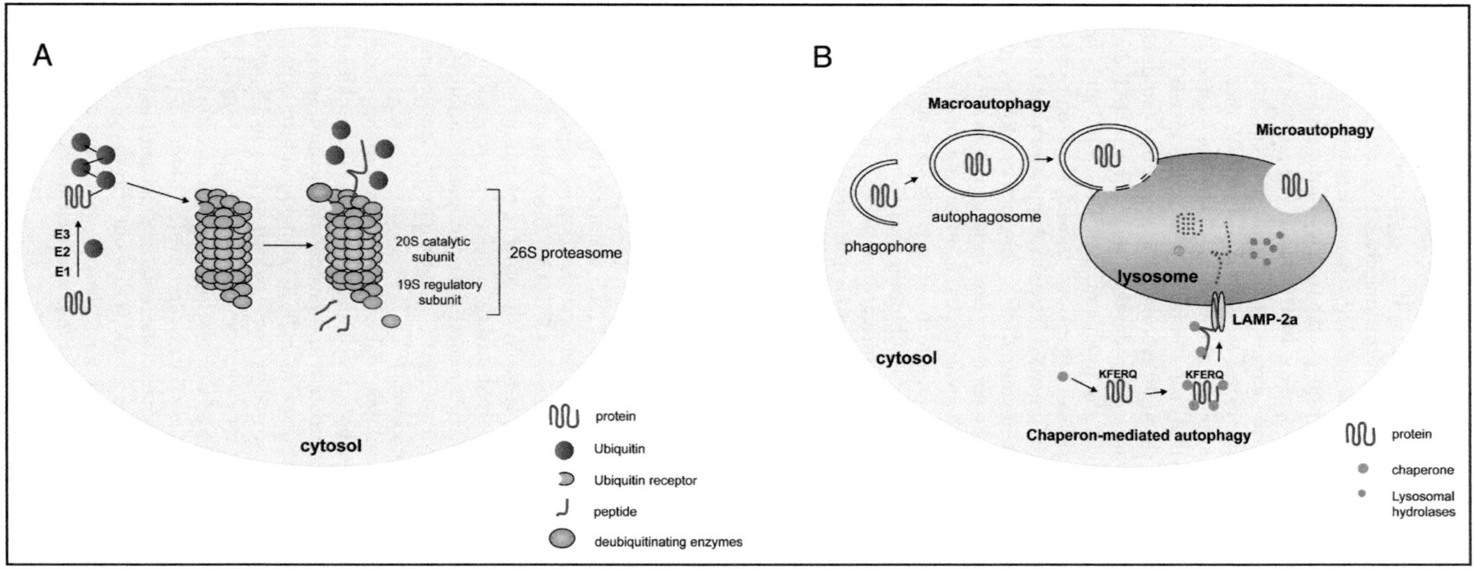

Figure 3. The main intracellular protein degradation processes. A) Many intracellular proteins can be degraded by the ubiquitin-proteasome system. The targeted protein is ubiquitinated by the subsequent action of ubiquitin-activating (E1), ubiquitin-conjugating (E2) and ubiquitin-ligase (E3) enzymes. After being tagged with chains of four or more ubiquitins, the protein is targeted to the proteasome for digestion into small pieces (peptides) by proteolysis. B) Autophagy involves the delivery of cellular constituents (including proteins) to lysosomes for degradation. The lumen of lysosomes contains the highest cellular concentration of proteases. Based on the mechanisms of cargo delivery to lysosomes, three main types of autophagy have been described: microautophagy, chaperon-mediated autophagy (CMA) and macroautophagy. In microautophagy, cytoplasmic material is sequestered through direct invagination of the lysosomal membrane. CMA selectively degrades proteins bearing a particular pentapeptide motif, KFERQ. Around 30% of soluble cytosolic proteins possess this motif. These proteins are recognized by a chaperone complex and translocated into the lysosome through a specific receptor called the lysosome-associated membrane protein (LAMP) 2A. Macroautophagy involves the formation of subcellular double-membrane-bound structures called autophagosomes to sequester cytoplasmic materials and deliver them into lysosomes. This process starts with the initiation of the formation of the isolation membrane, also termed phagophore. The growth of the phagophore terminates in the completion of the autophagosome. The subsequent fusion of the autophagosome with a lysosome forms an autolysosome, in which the enclosed material gets degraded. Only proteins are indicated, but other cellular constituents (macromolecules and organelles) can also be processed by autophagy.

Longevity Pathways That Promote Protein Synthesis

During the last three decades, a surprisingly large number of genes have been identified that regulate the aging process in divergent eukaryotic species. The recognition of their role in aging control was usually based upon the identification of the long-lived phenotype that their inactivation or overexpression cause (mutations that shorten life span generally have nothing to do with the aging process; rather they confer a nonspecific toxicity to the organism and thereby kill it prematurely). Combination of such genetic interventions (e.g., in the form of double mutant animals) often results in no additive increase in life span extension, rendering the corresponding genes into the same genetic (also called longevity) pathway. We mentioned *daf-2* and *age-1* as illustrative examples. These genes encode the nematode *Caenorhabditis elegans* insulin/IGF-1 receptor and class I phosphatidylinositol-3-kinase [PI3K (I)], respectively.[14,15] Lowering the activity of either of these genes through hypomorphic mutations can double the natural life span of worms.[15,16] Double mutant nematodes defective for both DAF-2 and AGE-1, however, do not live longer than either of the single mutants. Thus, *daf-2* and *age-1* act within a single signaling axis to control aging in this organism.[17] Indeed, the evolutionarily conserved insulin/IGF-1 hormonal system has been emerged as a central regulatory mechanism of aging in animal species ranging from worms to mammals.[16,18,19] In contrast, downregulating *daf-2* in *clk-1* mutant background confers an extreme increase in longevity as the double mutants live almost three-four times longer than the wild type (i.e., much longer than *daf-2* single mutants).[5,20] *clk-1*, loss-of-function mutations of which increase life-span moderately, encodes an enzyme that is necessary for the biosynthesis of ubiquinone, an integral part of the mitochondrial electron transport chain.[21] These genetic interactions imply that the insulin/IGF-1 signaling cascade and mitochondrial respiratory system function as two distinct, parallel acting longevity pathways.

Insulin/IGF-1 Signaling

Once the insulin/IGF-1 receptor is activated upon ligand binding, it further activates PI3K (I) (for reviews, see for example refs. 22, 23)(Fig. 4). This kinase converts phosphatidylinositol (3,4)-bisphosphate [PI(3,4)P_2] into phosphatidylinositol (3,4,5)-trisphosphate [PI(3,4,5)P_3], which is an important membrane-bound signaling molecule. The phosphatase and tensin homolog PTEN antagonizes PI3K (I) in this biochemical reaction. PI(3,4,5)P_3, either directly or through the 3-phosphoinositide-dependent protein kinase (PDK), mediates the activation of the serine-threonine kinase Akt/PKB (AKT8 virus proto-oncogen, also known as protein kinase B), which in turn represses the forkhead transcription factor FoxO by direct phosphorylation. Phosphorylated FoxO remains inactive in the cytoplasm. In the absence of inhibitory signal, FoxO becomes dephosphorylated and then transported into the nucleus, where it modulates the transcriptional activity of its target genes. Thus, a main mechanistic output of insulin/IGF-1 signaling is to inhibit FoxO. Consistently, the activity of FoxO is required for life span extension in insulin/IGF-1 signaling deficient nematodes and insects.[24-26]

Insulin/IGF-1 signaling controls multiple developmental, cellular and metabolic processes, including reproductive growth, aging, glucose metabolism, cell survival, growth and proliferation ,in divergent animal species.[27] Cell growth, differentiation and proliferation require an intense generation of new proteins. Thus, it is not surprising that insulin/IGF-1 signaling influences protein synthesis as well. A direct regulatory link is represented by FoxO, which, at least in *Drosophila*, appears to transcriptionally control *e4-BP* coding for an inhibitor of eIF4E (Figs. 2 and 4).[28] In addition, PDK and Akt/PKB trigger the kinase (S6K) that phosphorylates S6 protein involved in ribosome biogenesis.[22] Akt/PKB also downregulates the glycogen synthase kinase 3 (GSK3), whose activity in turns inhibits eIF2B.[29] To better understand the molecular mechanisms by which insulin/IGF-1 signaling affects mRNA translation, one should review our knowledge about the signaling network that comprises highly interconnected molecular pathways implicated in cell growth (Fig. 4). Remarkably, each of these pathways controls life span as well in a variety of organisms. We discuss them briefly below.

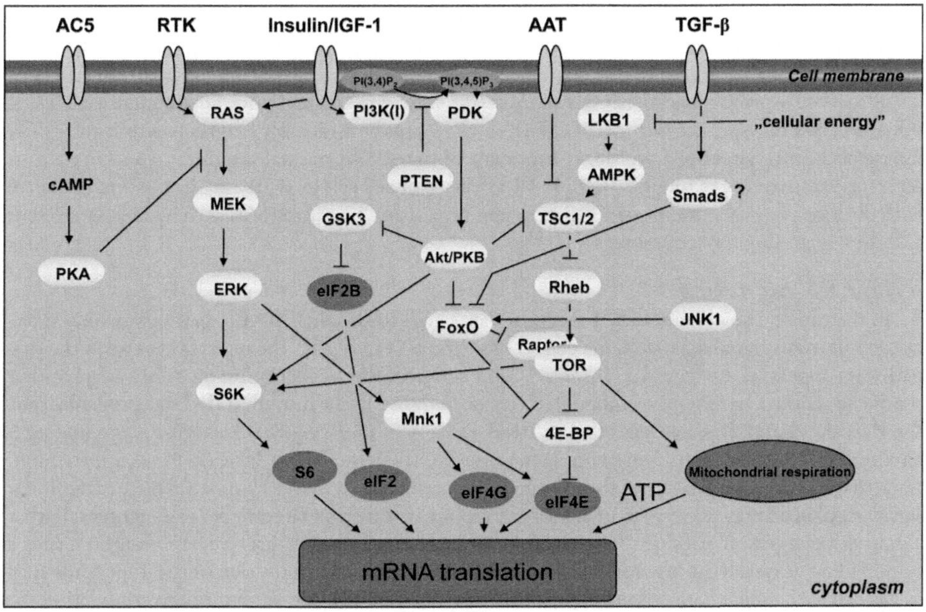

Figure 4. Model for the signaling network of longevity pathways linking protein synthesis to aging. The figure illustrates the mammalian counterparts of signaling and metabolic components involved in both aging and protein synthesis. Several interrelationships have only been demonstrated thus far in model organisms. Arrows (activations) and bars (inhibitions) indicate direct interactions that have been experimentally confirmed. Note that not all the indirect interactions have as yet necessarily been demonstrated. The mitochondrial respiratory chain representing a distinct longevity pathway is not shown here in details. "questionmark" indicates that the effects of TGF-β on FoxO may be SMAD-independent. S6K mediates ribosome biogenesis, whereas eIF2, eIF4E and eIF4G are required for Cap-dependent mRNA translation. RTK: receptor tyrosine kinase. For other abbreviations, see the text.

TOR Signaling

Akt/PKB activates the kinase target of rapamycin (TOR). This regulatory interaction occurs through inhibition of the tuberous sclerosis complexes 1 and 2 (TSC1/2) and Ras homolog enriched in brain (Rheb), which are upstream regulators of cellular energy-sensing TOR (Fig. 4).[30] TSC1/2, which is inhibited by amino acid transporters (AAT), also serve as sites where the adenosine monophosphate-activated protein kinase (AMPK) links the cellular AMP/ATP ratio and insulin/IGF-1 signaling to lifespan.[31] Knockdown of TOR activity in *C. elegans*, starting from the young adulthood onwards, leads to an extreme extension in life span.[32] Double mutant (epistasis) analysis in this organism indicates further that TOR and the insulin/IGF-1 cascade constitute a common signaling axis to control aging. In addition, the *C. elegans* FoxO-like protein DAF-16 directly represses the transcription of *Raptor*,[33] the mammalian counterpart of which encodes a regulatory associated protein of TOR.[34,35] Mutational or pharmacological inhibition of TOR similarly extends life span in yeast, *Drosophila* and the mouse.[36-38] Together, these data identify TOR signaling as an evolutionarily conserved longevity system. Accumulating evidence shows further that TOR regulates cell growth by altering gene transcription and mRNA translation in response to nutrient changes.[22,39] Two conserved effectors of TOR which are directly linked to the translational machinery involve eIF4E and S6K. As seen above and in Figure 2, eIF4E binds to the 7-methyl-guanosine Cap present in almost all eukaryotic mRNAs and interacts with the multidomain scaffold protein eIF4G.[22,40] eIF4G, in turn, cooperates with the poly(A)-binding protein PAPB. Combined interactions of eIF4E, eIF4G and PAPB enhance mRNA translation through bringing together the 5′ and 3′ ends of the mRNA. eIF4G also helps to recruit the 40S ribosomal subunit to the mRNA. TOR activity inhibits the eIF4E-binding protein 4E-BP by direct phosphorylation. As 4E-BP negatively regulates eIF4E, this phosphorylation function of TOR markedly promotes protein synthesis. Consistent with its regulatory dependence on TOR signaling, loss of a specific isoform of eIF4E in *C. elegans* considerably lengthens life span.[1-4] S6K kinase phosphorylates S6, a protein component of the 40S ribosomal subunit.[22] This kinase thus serves as another site through which TOR affects global protein synthesis. In good accordance with the long-lived phenotype of TOR deficient organisms, impairing S6K activity in yeast, worms and flies promotes organismal survival.[3,36,41]

RAS/ERK Signaling

In mammals, the insulin/IGF-1 receptor can signal to the Ras/MAPK pathway, which is another important signaling system in cell growth control (Fig. 4).[22,42] The interplay between the two pathways appears to be complex. Insulin/IGF-1 activates Ras by recruiting the Grb2 adaptor that binds to a unique tyrosine-phosphorylated consensus motif in the insulin/IGF-1 receptor substrate. On the other hand, Ras can activate Akt/PKB, thereby modulating FoxO activity. Ras signaling is also implicated in the regulation of life span in various models of aging. In yeast, for example, mutations that increase the activity of the mitogen-activated protein (MAP) kinase ERK2 (extracellular signal-regulated kinase 2) serve to increase both replicative life span (the number of daughter cells that a mother yeast cell can produce before senescence) and chronological life span (the length of time a nondividing yeast cell can survive).[43,44] Furthermore, mutational inactivation of the Type 5 adenylyl cyclase (AC5), which converts ATP to cAMP and inhibits Ras signaling through stimulating the cAMP-dependent protein kinase A (PKA), confers a life span extension by 30% in mice.[45] ERK1/2 kinases were found to stimulate S6K.[1] In addition, these kinases, together with p38 mitogen-activated protein kinase (p38MAPK), modulate the activity of the mitogen-activated protein kinase-interacting kinase (Mnk1), which, in turn, triggers eIF4E and eIF4G.[46] Based on these data, Ras signaling represents another regulatory mechanism that controls both protein synthesis and aging.

TGF-β Signaling

TGF-β signaling is also required for normal cell growth and thereby may have a role in protein synthesis.[47] The TGF-β receptors can act in Smad-dependent or -independent fashion. In *C. elegans*, similar to insulin/IGF-1 signaling, TGF-β activity regulates development (reproductive growth), cell size and life span.[48,49] Interestingly, each of these TGF-β functions is mediated, at least in part, by the FoxO-like transcription factor DAF-16.[48,50] For example, mutant worms with reduced activity of the TGF-β ligand DAF-7 develop as long-lived dauer larvae (dauer is a developmental diapausa persisting for more than 6 months), but die as disorganized dauers within 1-2 weeks if DAF-16 activity is absent.[50] As DAF-16 links insulin/IGF-1 signaling to protein synthesis and aging, TGF-β should also modulate both processes (Fig. 4). Indeed, the mammalian Type I TGF-β receptor can directly interact with and activate the protein phosphatase 2A (PP2A), which is an upstream stimulator of S6K.[51] Moreover, Type II TGF-β stimulates Ras/ERK signaling too. Together, it is likely that the TGF-β signal transduction system functions as a longevity pathway that indirectly promotes mRNA translation.

JNK Kinase-Mediated Signaling

In certain cellular settings, the c-Jun terminal kinase (JNK) is also under the control of the Type II TGF-β receptor.[47] This regulatory relationship is independent of Smad proteins. Furthermore, it was shown that genetic interventions that enhance JNK activity slow down the aging process of *C. elegans* and *Drosphila*.[52,53] In nematodes, JNK regulates life span by modulating phosphorylation and nuclear translocation of forkhead transcription factor DAF-16. Although it is not yet demonstrated, these data implicate that JNK signaling affects protein synthesis, probably through interacting with DAF-16.

Mitochondrial Respirtory Chain

Protein synthesis is a highly costly process. The proportion of cellular energy used for protein synthesis is estimated to be as high as 30-40% of total ATP and GTP.[54] Mitochondria, wherein the majority of cellular energy is generated in most eukaryotic cells, affect essentially all biochemical processes. Thus, mRNA translation should be tightly coupled to mitochondrial activity as well. The fact that mammalian TOR signaling plays a crucial role in the regulation of mitochondrial function strongly supports this view.[55] Furthermore, components of the mitochondrial respiratory chain constitute a well-established longevity pathway: lowering the rate of mitochondrial respiration was shown to increase longevity in worms.[56,57] This could be the result of decreased levels of ROS that mitochondria produce or a decline in protein synthesis, or both.

Together, numerous evolutionarily conserved molecular pathways, including the insulin/IGF-1, TOR, Ras/Erk, TGF-β and JNK signaling systems and the mitochondrial respiratory chain, that normally accelerate the aging process also—directly or indirectly—influence the cell's machinery for protein synthesis. These regulatory circuits function in a highly orchestrated, intertwined way (Fig. 4). However, some of the longevity pathways established, e.g., the Sirt1/2—HSF1 (eukaryotic heat-shock factor-1) signaling axis or signaling via the tumor suppressor p53 (reviewed in ref. 7), have not been mentioned thus far in this chapter. This is a consequence of the fact that no information is available yet on whether and if yes how, they control mRNA translation. Obviously, these additional longevity systems are inherently integrated into the global signaling network of aging,[7] and thereby estimated to influence protein synthesis. We conclude that in general, longevity pathways mediate signals that promote translation.

Interactions between Molecular Mechanisms Involved in Protein Synthesis and Degradation

Protein turnover, as well as the generation of protein damage, depends on the balance of protein synthesis and degradation processes. The more new proteins are synthesized, the more damage will be produced. Upon this consideration, one would expect that under favorable conditions (e.g., when nutrients are available) the activity of mechanisms that underlie protein breakdown also becomes elevated. Looking at autophagy as the main intracellular catabolic process, this is not the case. Its primary function is to provide macromolecular components (the products of lysosomal degradation, including amino acids and fatty acids) as an alternative energy source for fuelling the cells during starvation. Moreover, emerging evidence indicates that longevity pathways often downregulate the autophagic machinery (Fig. 5).[7,23] We discuss insulin/IGF-1 signaling as a prominent example. Nutrient supply stimulates the insulin/IGF-1 receptor, which in turn activates the nutrient sensing TOR kinase through modulating the downstream PI3K(I)-PDK-Akt/PKB kinase cascade, FoxO and TSC1/2 (Fig. 4). Activated TOR blocks the initial stages of autophagy. Alternatively, FoxO directly modulates the transcription of several key autophagy genes.[58] Genetic epistasis analysis in worms could reveal these connections more evidently: expression of the long-lived phenotype in insulin/IGF-1 or TOR signaling deficient mutant animals was shown to require elevated activity of autophagy genes.[59-61] Similar to autophagy, the proteasome system appears to function downstream of and is negatively regulated by insulin/IGF-1 signaling to control aging.[62] Certain components of the Ras/Erk pathway also act as direct negative regulators of autophagy.[44] These results are consistent with the general observation that increasing the activity of longevity pathways by environmental (e.g., nutrition and elevated temperature or oxygen levels) or intrinsic (e.g., growth hormones and mitogens) factors leads to the accumulation of unrepaired cellular damage. This causes acceleration of the aging process and shortening of life span.

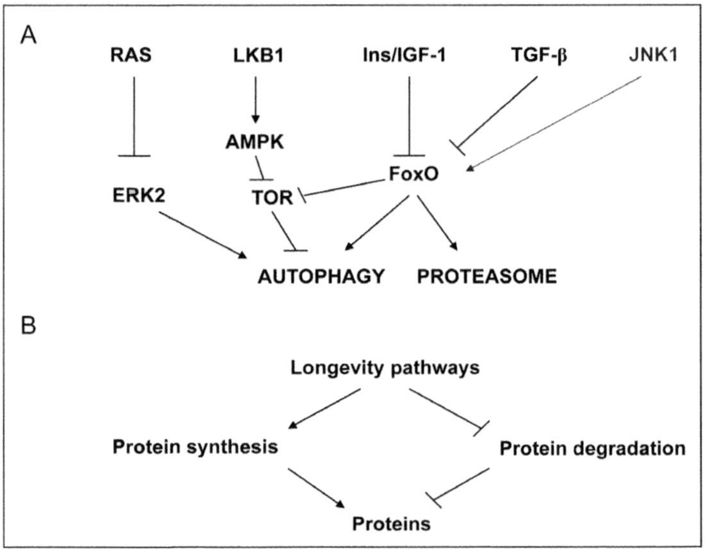

Figure 5. Interplay between molecular pathways mediating protein synthesis and degradation. A) Longevity pathways that promote protein synthesis downregulate the protein degradation machineries. This implies that conditions lowering protein synthesis to tolerable levels can lead to increased elimination of protein damages. Arrows indicate activations, bars indicate negative regulatory interactions. JNK1-mediated signaling is indicated by gray coloring as it acts as an anti-aging pathway. B) Protein turnover depends on the activity of protein synthesis and degradation processes. Molecular pathways that promote protein synthesis generally downregulate the degradation processes. It should be noted that the scheme is highly simplified as it lacks regulatory inputs that, for example, the proteasome system exerts on these molecular pathways.

Conclusion

As discussed above, longevity pathways generally promote protein synthesis (Fig. 5). Inactivation or downregulation of such pathways confers life span extension in diverse eukaryotic species. Consistently, decreased protein synthesis rates in *C. elegans* also lengthen life span.[1-4] One might argue that this longevity response is due to reduced energy consumption. However, reducing mRNA translation still extends life span in animals with lowered respiration. This suggests that protein synthesis affects aging independently of effects on energy consumption. Altering protein metabolism certainly has pleiotropic effects. However, we hypothesize that decreasing the rate of protein synthesis implies a lower incidence of the production of aberrant proteins. Damaged (misfolded, aggregated, crosslinked) proteins act as cellular toxins interfering with normal cellular functions. Such types of damage accumulate with a lower rate when growth rate is decreased. This assumption can be tested. For example, double mutant nematodes defective for both eIF4E and a key autophagy gene should be generated (inhibition of eIF4E lengthens whereas inactivation of autophagy shortens life span in this organism). If double mutants exhibit a short-lived phenotype, autophagy genes should mediate the effect that eIF4E has on aging. This would support the cellular damage theory of aging.

Acknowledgements

This work was supported by grants from the Hungarian Scientific Research Funds OTKA K68372 and NK78012 to T.V, OTKA PD75477 to K.T-V and the National Office for Research and Technology (TECH_08_A1/2-2008-0106) to T.V. T.V and K.T-V are grantees of the János Bolyai scholarship.

References

1. Tavernarakis N. Ageing and the regulation of protein synthesis: a balancing act? Trends Cell Biol 2008; 18:228-35.
2. Syntichaki P, Troulinaki K, Tavernarakis N. eIF4E function in somatic cells modulates ageing in Caenorhabditis elegans. Nature 2007; 445:922-6.
3. Hansen M, Taubert S, Crawford D et al. Lifespan extension by conditions that inhibit translation in Caenorhabditis elegans. Aging Cell 2007; 6:95-110.
4. Pan KZ, Palter JE, Rogers AN et al. Inhibition of mRNA translation extends lifespan in Caenorhabditis elegans. Aging Cell 2007; 6:111-9.
5. Hekimi S, Guarente L. Genetics and the specificity of the aging process. Science 2003; 299:1351-4.
6. Kirkwood TBL. A systematic look at an old problem. Nature 2008; 451:644-7.
7. Vellai T, Takács-Vellai K, Sass M et al. The regulation of aging—does autophagy underlie longevity? Trends Cell Biol 2009; in press.
8. Rubinsztein DC. The roles of intracellular protein-degradation pathways in neurodegeneration. Nature 2006; 443:780-6.
9. Hirsch C, Gauss R, Horn SC et al. The ubiquitylation machinery of the endoplasmic reticulum. Nature 2009; 485:453-60.
10. Sőti C, Sreedhar AS, Csermely P. Apoptosis, necrosis and cellular senescence: chaperone occupancy as a potential switch. Aging Cell 2003; 2:39-45.
11. Levine B, Kroemer G. Autophagy in the pathogenesis of disease. Cell 2008; 132:27-42.
12. Mizushima M, Levine B, Cuervo AM et al. Autophagy fights disease through cellular selfdigestion. Nature 2008; 451:1069-75.
13. Cuervo AM. Autophagy and aging: keeping that old broom working. Trends Genet 2008; 24:604-12.
14. Kimura KD, Tissenbaum HA, Liu Y et al. daf-2, an insulin receptor-like gene that regulates longevity and diapause in Caenorhabditis elegans. Science 1997; 277:942-6.
15. Morris JZ, Tissenbaum HA, Ruvkun G. A phosphatidylinositol-3-OH kinase family member regulating longevity and diapause in Caenorhabditis elegans. Nature. 1996; 382:536-9.
16. Kenyon C, Chang J, Gensch E et al. A C. elegans mutant that lives twice as long as wild type. Nature 1993; 366:461-4.
17. Dorman JB, Albinder B, Shroyer T et al. The age-1 and daf-2 genes function in a common pathway to control the lifespan of Caenorhabditis elegans. Genetics 1995; 141:1399-406.
18. Tatar M, Kopelman A, Epstein D et al. A mutant Drososphila insulin receptor homolog that extends lifespan and impairs neuroendocrine functions. Science 2001; 292:107-10.
19. Holzenberger M, Dupont J, Ducos B et al. IGF-1 receptor regulates lifespan and resistance to oxidative stress in mice. Nature 2003; 421:125-6.
20. Lakowski B, Hekimi S. The genetics of caloric restriction in Caenorhabditis elegans. Proc Natl Acad Sci USA 1998; 95:13091-6.
21. Miyadera H, Amino H, Hiraishi A et al. Altered quinone biosynthesis in the long-lived clk-1 mutants of Caenorhabditis elegans. J Biol Chem 2001; 276:7713-6.
22. Oldham S, Hafen E. Insulin/IGF and target of rapamycin signaling: a TOR de force in growth control. Trends Cell Biol 2003; 13:79-85.
23. Vellai T. Autophagy genes and ageing. Cell Death Differ 2009; 16:94-102.
24. Lin K, Dorman JB, Rodan A et al. daf-16: An HNF-3/forkhead family member that can function to double the life-span of Caenorhabditis elegans. Science 1997; 278:1319-22.
25. Ogg S, Paradis S, Gottlieb S et al. The Fork head transcription factor DAF-16 transduces insulin-like metabolic and longevity signals in C. elegans. Nature 1997; 389:994-9.
26. Hwangbo DS, Gersham B, Tu M-P et al. Drosophila dFOXO controls lifespan and regulates insulin signalling in brain and fat body. Nature 2004; 429:562-6.
27. Saltiel AR, Kahn CR. Insulin signalling and the regulation of glucose and lipid metabolism. Nature 2001; 414:799-806.
28. Jünger MA, Rintelen F, Stocker H et al. The Drosophila forkhead transcription factor FOXO mediates the reduction in cell number associated with reduced insulin signaling. J Biol 2003; 2:20.
29. Williams DD, Pavitt GD, Proud CG. Characterization of the initiation factor eIF2B and its regulation in Drosophila melanogaster. 2J Biol Chem 2003; 76:3733-3742.
30. Pan D, Dong J, Zhang Y et al. Tuberous sclerosis complex: from Drosophila to human disease. Trends Cell Biol 2004; 14:78-85.
31. Apfeld J, O'Connor G, McDonagh T et al. The AMP-activated protein kinase AAK-2 links energy levels and insulin-like signals to lifespan in C. elegans. Genes Dev 2004; 18:3004-9.
32. Vellai T, Takács-Vellai K, Zhang Y et al. Genetics: influence of TOR kinase on lifespan in C. elegans. Nature 2003; 426:620.

33. Jia K, Chen D, Riddle DL. The TOR pathway interacts with the insulin signaling pathway to regulate C. elegans larval development, metabolism and life span. Development 2004; 131:3897-906.
34. Kim DH, Sarbassov DD, Ali S et al. mTOR interacts with raptor to form a nutrient-sensitive complex that signals to the cell growth machinery. Cell 2002; 110:163-75.
35. Hara K, Maruki Y, Long X et al. Raptor, a binding partner of target of rapamycin (TOR), mediates TOR action. Cell 2002; 110:177-89.
36. Kapahi P, Zid BM, Harper T et al. Regulation of lifespan in Drosophila by modulation of genes in the TOR signaling pathway. Curr Biol 2004; 14:885-90.
37. Kaeberlein M, Powers III WR, Steffen KK et al. Regulation of yeast replicative life span by TOR and Sch9 in response to nutrients. Science 2005; 310:1193-96.
38. Harrison DE, Strong R, Sharp ZD et al. Rapamycin fed late in life extends lifespan in genetically heterogeneous mice. Nature 2009; 460:392-5.
39. Fafournoux P, Bruhat A, Jousse C. Amino acid regulation of gene expression. Biochem J 2000; 351:1-12.
40. Gingras AC, Raught B, Sonenberg N. eIF4 initiation factors: effectors of mRNA recruitment to ribosomes and regulators of translation. Annu Rev Biochem 1999; 68:913-63.
41. Wei M, Fabrizio P, Hu J et al. Life span extension by calorie restriction depends on Rim15 and transcription factors downstream of Ras/PKA, Tor and Sch9. PLoS Genet 2008; 4:e13.
42. Skolnik EY, Lee CH, Batzer A et al. The SH2/SH3 domain-containing protein GRB2 interacts with tyrosine-phosphorylated IRS1 and Shc: implications for insulin control of ras signalling. EMBO J 1993; 12:1929-36.
43. Fabrizio P, Pletcher SD, Minois N et al. Chronological aging-independent replicative life span regulation by Msn2/Msn4 and Sod2 in Saccharomyces cerevisiae. FEBS Lett 2004; 557:136-42.
44. Budovskaya YV, Stephan JS, Reggiori F et al. The Ras/cAMP-dependent protein kinase signaling pathway regulates an early step of the autophagy process in Saccharomyces cerevisiae. J Biol Chem 2004; 279:20663-71.
45. Yan L, Vatner DE, O'Connor JP et al. Type 5 adenylyl cyclase disruption increases longevity and protects against stress. Cell 2007; 130:247-58.
46. Pyronnet S, Dostie J, Sonenberg N. Suppression of cap-dependent translation in mitosis. Genes Dev 2001; 15:2083-93.
47. Derynck R, Zhang YE. Smad-dependent and Smad-independent pathways in TGF-b family signalling. Nature 2003; 425:577-84.
48. Shaw WM, Luo S, Landis J et al. The C. elegans TGF-β dauer pathway regulates longevity via insulin signalling. Curr Biol 2007; 17:1635-45.
49. Krishna S, Maduzia LL, Padgett RV. Specificity of TGFb signaling is conferred by distinct type I receptors and their associated SMAD proteins in Caenorhabditis elegans. Development 1999; 126:251-60.
50. Vellai T, Bicsák B, Tóth ML et al. Regulation of cell growth by autophagy. Autophagy 2008; 4:507-9.
51. Petritsch C, Beug H, Balmain A et al. TGF-b inhibits p70 S6 kinase via protein phosphatase 2A to induce G1 arrest. Genes Dev 2000; 14:3093-101.
52. Wang MC, Bohmann D, Jasper H. JNK extends life span and limits growth by antagonizing cellular and organism-wide responses to insulin signaling. Cell 2005; 121:115-25.
53. Oh SW, Mukhopadhyay A, Svrzikapa N et al. JNK regulates lifespan in Caenorhabditis elegans by modulating nuclear translocation of forkhead transcription factor DAF-16. Proc Natl Acad Sci USA 2005; 102:4494-9.
54. Buttgereit F, Brand MD. A hierarchy of ATP-consuming processes in mammalian cells. Biochem J 1995; 312:163-7.
55. Cunningham JT, Rodgers JT, Arlow DH et al. mTOR controls mitochondrial oxidative function through a YY1-PGC-1alpha transcriptional complex. Nature 2007; 450:736-40.
56. Dillin A, Hsu AL, Arantes-Oliveira N et al. Rates of behavior and aging specified by mitochondrial function during development. Science 2002; 298:2398-401.
57. Lee SS, Lee RY, Fraser AG et al. A systematic RNAi screen identifies a critical role for mitochondria in C. elegans longevity. Nat Genet 2003; 33:40-8.
58. Zhao J, Brault JJ, Schild A et al. FoxO3 co-ordinately activates protein degradation by the autophagic/lysosomal and proteasomal pathways in atrophying muscle cells. Cell Metab 2007; 6:472-83.
59. Meléndez A, Tallóczy Z, Seaman M et al. Autophagy genes are essential for dauer development and life span extension in C. elegans. Science 2003; 301:1387-91.
60. Tóth ML, Sigmond T, Borsos É et al. Longevity pathways converge on autophagy genes to regulate life span in Caenorhabditis elegans. Autophagy 2008; 4:330-8.
61. Hansen M, Chandra A, Mitic LL et al. A role for autophagy in the extension of lifespan by dietary restriction in C. elegans. PLoS Genet 2008; 4:e24.
62. Ghazi A, Henis-Korenblit S, Kenyon C. Regulation of Caenorhabditis elegans life span by a proteasomal E3 ligase complex. Proc Natl Acad Sci USA 2007; 104:5947-52.

Chapter 8

Protein Metabolism and Lifespan in *Caenorhabditis elegans*

Geert Depuydt, Jacques R. Vanfleteren and Bart P. Braeckman*

Abstract

Lifespan of the versatile model system *Caenorhabditis elegans* can be extended by a decrease of insulin/IGF-1 signaling, TOR signaling, mitochondrial function, protein synthesis and dietary intake. The exact molecular mechanisms by which these modulations confer increased life expectancy are yet to be determined but increased stress resistance and improved protein homeostasis seem to be of major importance. In this chapter, we explore the interactions among several genetic pathways and cellular functions involved in lifespan extension and their relation to protein homeostasis in *C. elegans*. Several of these processes have been associated, however some relevant data are conflicting and further studies are needed to clarify these interactions. In mammals, protein homeostasis is also implicated in several neurodegenerative diseases, many of which can be modeled in *C. elegans*.

Introduction

C. elegans *as a Model Organism*

Caenorhabditis elegans is a free-living soil nematode that was introduced in the mid 1960's by Sydney Brenner as a potent model organism to study animal development and neurology. Since its introduction, this anatomically simple nematode, whose body plan includes muscles, nervous system and digestive tract, has become a valuable and widely used research tool with a fully sequenced and well annotated genome.[1-4] *C. elegans* is principally a hermaphrodite, but males arise spontaneously at a frequency of about 0.1%. The ability to carefully orchestrate the crossing of hermaphrodites with males is especially useful for genetic studies. No inbreeding depression is present in *C. elegans,* which eliminates major complications when analyzing the genetics of life history traits such as lifespan.[5,6] The rapid lifecycle (three days), large progeny (± 200) and short lifespan (two to three weeks) have been very advantageous for its use in aging (Fig. 1).[7] *C. elegans* is easily cultured in large numbers and worm strains can be conserved indefinitely in liquid nitrogen. A single wild-type individual has an invariable body plan consisting of 959 somatic nuclei and the body is transparent during all life stages. These traits allow for detailed in vivo cellular studies during development and morphogenesis. Over the years, many powerful molecular genetic tools have become available for *C. elegans* research such as random mutagenesis, RNA mediated interference (RNAi) and transgenesis of fluorescent marker proteins for expression studies. For a large number of genes, well-characterized mutants are available at the Caenorhabditis Genetics Center (http://biosci.umn.edu/CGC/). Other valuable research resources include Wormbase (http://www.wormbase.org), the *C. elegans* model organism database, Wormbook, a comprehensive reference

*Corresponding Author: Bart P. Braeckman—Department of Biology, Ghent University, K.L. Ledeganckstraat 35, B-9000 Ghent, Belgium. Email: bart.braeckman@ugent.be

Protein Metabolism and Homeostasis in Aging, edited by Nektarios Tavernarakis.

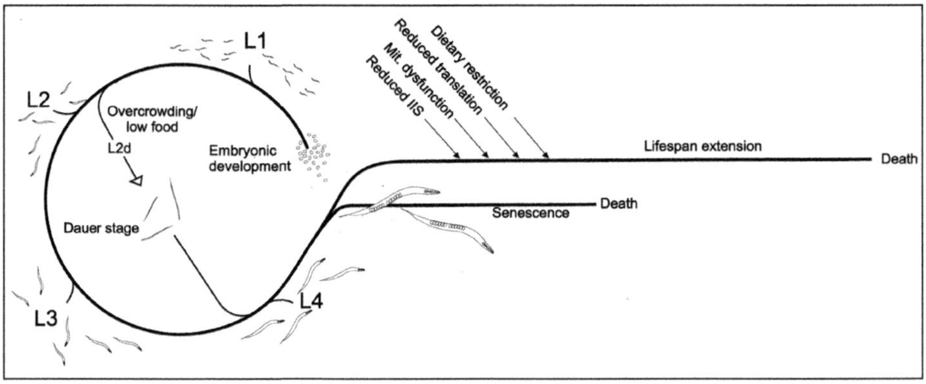

Figure 1. *C. elegans* lifecycle. The lifecycle can be broadly divided into an embryonic and a postembryonic development to adulthood. The latter consists of four larval stages (L1-L4) each separated by a molt. The total lifespan of *C. elegans* is approximately two weeks and highly dependent on ambient temperature and feeding conditions. In harsh conditions larvae will develop into an alternative L3: the dauer ("enduring") stage. Dauer animals can survive for over three months without food and are stress resistant. When re-exposed to food, these dauers resume development, reproduce and age normally.

book freely available online (http://www.wormbook.org) and RNAi feeding libraries that jointly cover approximately 86% of the genes in the *C. elegans* genome.[8,9]

Several genetic factors and environmental interventions have been shown to extend normal lifespan in *C. elegans*, thus providing an insight into the molecular mechanisms that govern aging. For example, reduced activity of the *C. elegans* insulin/insulin-like growth factor (IGF-1) tyrosine kinase receptor DAF-2 (DAuer Formation abnormal-2) or the downstream phosphatidylinositol-3-OH kinase (PI3K), AGE-1, during adulthood, causes nuclear translocation of the forkhead (FOXO) transcription factor ortholog DAF-16/FOXO, thereby inducing several longevity promoting genes involved in stress resistance.[10-14] Likewise, steroid hormone signaling from the reproductive system affects lifespan through DAF-16/FOXO.[15] Dietary restriction (DR), the substantial decrease in food uptake without essential nutrient deprivation, extends *C. elegans* lifespan as well. The identification of DR-specific transcription factors in *C. elegans* such as PHA-4/FOXA and SKN-1/NRF2 may, if conserved, help to identify the mechanisms by which DR enhances longevity in vertebrates.[16,17] Inhibition of the nutrient sensor TOR[18] and reduced mitochondrial respiration[19] have also been found to extend lifespan in *C. elegans*, independently of DAF-16/FOXO. Recently, molecular processes that affect lifespan downstream of the signaling pathways, such as autophagy and mRNA translation, are starting to be characterized in *C. elegans*.

The Protein Homeostasis and Longevity Hypotheses

Error Catastrophe and Oxidative Damage Accumulation

In 1963, Orgel[20] advanced the error catastrophe hypothesis of aging in which he suggested that, as an animal ages, there is an increased frequency of errors at the transcriptional and/or translational level of protein synthesis resulting in the synthesis of aberrant protein molecules and accumulation of partially or totally inactive enzymes. Orgel further hypothesized that once a critical level of faulty proteins is reached, death occurs. Indeed, in the 1970's the accumulation of catalytically incompetent forms of enzymes was reported for a variety of organisms (for a review see refs. 21-23), including nematodes.[24-30] However, several groups later showed that the effect of aging on the fidelity of protein synthesis is surprisingly small,[31-35] thus contradicting Orgel's hypothesis. Still, altered proteins do accumulate (within cells/tissues) with age and in 1956, Harman[36,37] had already suggested that senescence is the result of the accumulation of damaged macromolecules due to constant attack by free radicals. The free radical theory of aging remains one of the most widely accepted hypotheses as to why organisms age.

Protein Turnover Hypothesis

Whereas powerful repair mechanisms exist for damaged DNA,[38] only limited protein damage can be repaired.[39] Cells deal with irreversibly damaged proteins by replacing them through protein turnover. The potentially widespread deleterious effects of the accumulation of aberrant proteins on normal cellular function could contribute significantly to the aging process.[48,48a,48b] The amount of protein damage that accumulates with age is extensive. In aged individuals approximately 40%-50% of all cellular proteins suffer some form of oxidative damage.[40] In the 1970's and 1980's many investigators measured protein synthesis and protein degradation as a function of age in various tissues and organisms[41] including nematodes.[26,42-47] Overall, these studies showed that the average protein half-life increases significantly with age. These observations support an attractive alternative to the error catastrophe theory in that the increased "dwell time" of proteins (i.e., decreased protein turnover) with age results in an increased probability that proteins become post-translationally altered (oxidation, deamidation, glycation, racemization, isomerization),[48,49] denaturated[22] or misfolded[50] and accumulate instead of being replaced. The age-related decline in protein turnover is also thought to render cells less responsive to external stimuli and could explain why stress resistance in animals diminishes with age.[51]

Dietary Restriction, TOR Signaling and Protein Homeostasis

Dietary Restriction

One of the most exciting observations in the area of aging research is that dietary restriction (DR) consistently increases the lifespan and overall health in a wide variety of species from yeast, worms and flies to rodents by slowing down the aging process. Not only do these animals grow older, they also show a delayed onset of a wide range of age-related pathologies such as cancer, diabetes and cardiovascular disease in the case of mammals.[52,53] Possibly, the decrease of nutrient intake improves mitochondrial respiration and therefore, decreases free radical production, effectively slowing down the age-related accumulation of oxidized proteins.[54] In accordance, several studies have shown that moderate DR leads to lower levels of oxidative damage to proteins[55-58] and macromolecules in general.[54,59] In addition, DR also confers a general protective effect against proteotoxicity associated with misfolding or aggregation of proteins in *C. elegans*[60] and mice.[61] Of note, the ability of DR to significantly slow down the age-related decline in protein turnover generally accompanies enhanced survival in metazoans.[62-66] Although the intensive study of DR over seven decades has resulted in a thorough description of its effects on physiology, senescence and age-related disease, the genetic pathways governing this lifespan increase and the molecular processes involved are only beginning to be understood.

TOR

Classic genetic analysis in yeast, worms and flies has assigned the highly conserved serine/threonine kinase TOR as a downstream effector of the longevity response to DR.[67-69] TOR integrates nutritional and hormonal cues to modulate cell growth and proliferation.[70] Reduced TOR signaling was shown to extend lifespan in yeast,[68,71] *Drosophila*[67] and worms.[18,69] Deletion of the *C. elegans* TOR homolog, *let-363*/TOR and its binding partner *daf-15*/Raptor results in arrested L3 larvae that are distinct from dauers.[18,72,73] *let-363*/TOR(-) arrested larvae have a more than doubled lifespan compared to wild-type controls and a comparable lifespan extension is found when *let-363*/TOR RNAi treatment is initiated at the first day of adulthood.[18,69] No further increase in lifespan is observed in long-lived *daf-2* mutant worms, although deletion of *daf-16*/FOXO cannot suppress the longevity phenotype associated with *let-363*/TOR(RNAi), suggesting that these two pathways act in parallel or may converge downstream of *daf-16*/FOXO to extend lifespan.[18] Heterozygote *daf-15*(−/+) nematodes were also reported to be long lived, but lifespan extension through *daf-15* is dependent on functional DAF-16/FOXO.[74] In contrast, no obvious lifespan extension was found in balanced *let-363(h111)* heterozygous mutants or in *let-363; daf-15* double heterozygotes.[74] Mair and Dillin noted that *let-363*/TOR is part of an operon that also contains a transcription initiation factor TFIIIB (B0261.1) and a mitochondrial ribosomal subunit (B0261.4).[53] Collateral inhibition of gene sequences downstream the RNAi gene target within the same operon has been described.[75] Downregulation of TFIIIB is associated with larval arrest and lethality and RNAi of the mitochondrial ribosomal subunit B0261.4 was reported to extend lifespan independently of *daf-16*/FOXO.[53,76-78] Therefore, dissection of the genetic pathway(s) downstream of *let-363*/TOR remains a challenge.

Reduced Protein Synthesis Extends Lifespan

TOR regulates several processes that affect global protein turnover, including translation and autophagy, but also amino acid synthesis, ribosome biogenesis and transcription of ribosomal genes.[70] Therefore, these processes present likely candidates to be involved in lifespan extension by dietary restriction. An important downstream target of mammalian TOR is ribosomal S6 kinase 1 (p70S6K).[79] Phosphorylation of the 40S ribosomal S6 protein by p70S6K leads to a general increase in translation capacity, although the exact mechanism is still subject to debate.[80] The impact of translation regulation on lifespan has been particularly well characterized in *C. elegans* (for an overview, see Table 1).[81] Depletion of *rsks-1*/S6K, the *C. elegans* homolog of mammalian p70S6K and numerous translation initiation factors, including polyA-binding proteins, has been found to extend nematode lifespan.[69,82-87] Reducing the level of any of a large number of ribosomal proteins or genes involved in ribosome biogenesis, such as *nol-5*, *exos-3* and RNA polymerase I, also extend adult lifespan.[69,88] Down-regulation of several transfer RNA (tRNA) synthetases has a beneficial effect on lifespan as well.[88] Interestingly, inhibiting mitochondrial translation also extends lifespan, independently of *daf-16*/FOXO.[77]

The mitochondrial genome encodes only 12 polypeptides, all of which are components of the electron transport chain and ATP synthase.[89] Thus, defects in the mitochondrial translation machinery are likely to compromise mitochondrial electron transport and ATP synthesis. Inactivation of the mitochondrial leucyl-tRNA synthetase gene (*lrs-2*) or the mitochondrial ribosomal subunit B0261.4 results in disorganized, swollen and sometimes fused mitochondria along with substantially lowered ATP content and respiration rates. These long-lived worms are able to withstand heat-stress and hydrogen peroxide as well as or even better than control worms but they are hypersensitive to the NADPH-dependent superoxide generating herbicide paraquat, possibly due to elevated levels of NADPH in *lrs-2*(RNAi) worms.[77]

The mechanism by which reduced protein translation extends lifespan in *C. elegans* remains unanswered to date. Given that TOR controls protein homeostasis, it is tempting to assume that inhibition of translation would extend longevity by the same pathway as reduction of TOR-signaling. However, the genetic studies mentioned above have led to seemingly contradictory conclusions regarding the relationship between translation factors and TOR and insulin/IGF-1 signaling (IIS).

Lowering the Rate of Translation Induces Increased Stress Resistance

The *daf-16*/FOXO transcription factor is a major regulator of the genetic response to starvation, heat and oxidative stress.[82] Indeed, long-lived IIS mutants display concomitant increase of stress resistance. In general, reduced protein synthesis significantly increases resistance to thermal- and oxidative stress compared to wild type. For example, RNAi induced reduction in the rate of protein synthesis in the short-lived, paraquat hypersensitive *mev-1(kn1)* mutant, which lacks a component of complex II of the mitochondrial electron transport chain, confers a marked boost in paraquat resistance[83,85] and also extends lifespan of *mev-1(kn1)* in one,[83] but not in another report.[85] General suppression of translation by treatment with cycloheximide or by knockdown of genes encoding aminoacyl-tRNA synthetases was shown to rescue animals from hypoxia-induced cell death.[90] Moreover, resistance to hypoxia and the rate of translation were strongly inversely related. Interestingly, maximal resistance to hypoxia by suppression of translation requires the unfolded protein response (UPR) for its full phenotypic expression. Hypoxia produces misfolded proteins that accumulate in the lumen of the endoplasmic reticulum (ER) and thereby activate the UPR.[91] In mammalian cells, activation of the UPR results in the phosphorylation of the eukaryotic translation initiation factor eIF2α by the protein kinase-like endoplasmic reticulum kinase (PERK), thereby inhibiting translation of the majority of genes and cell-cycle progression.[92,93] In contrast, eIF2α phosphorylation also induces translation of approximately one-third of the UPR-dependent genes including, the activating transcription factor ATF4 (yeast GCN4). The PERK, eIF2α, ATF4 regulatory axis induces expression of genes involved in amino acid biosynthesis and transport, anti-oxidative stress response and proapoptotic functions.[94] It is proposed that under protein stress a reduction in translation, elicited by the UPR, reduces the unfolded protein load to a level manageable by the UPR.[90,95]

Protein Synthesis and Reproduction

Reduced protein synthesis and TOR signaling significantly affect growth and fecundity (Table 1), suggesting that longevity and reproduction may be coupled and that the relocation of resources away from reproduction in favor of maintenance and repair could extend lifespan.[96] Tohyama interpreted the failure of eIF2Bδ RNAi treatment to extend the lifespan of germline defective *glp-4*(*bn2*)I nematodes as an example of such trade-offs.[85] However, a lack of germline (*glp-4*(*bn2*)I) did not suppress the effect of *ife-2*/eIF4G deficiency on animal lifespan[83] and Hansen et al (2007)[69] reported significant lifespan extension of sterile (*fer-15*(*b26*)II; *fem-1*(*hc17*)III) animals treated with ribosomal protein RNAi. Furthermore, none of the RNAi treatments that inhibit translation further increases the lifespan of *daf-2* mutants (Table 1), whereas gonad ablation doubles the long lifespan of *daf-2* mutants.[97] Finally, lifespan extension following germ line removal is dependent on *daf-16*/FOXO,[98] whereas several translation-inhibiting RNAi treatments (e.g., ribosomal subunits) extend lifespan independently of *daf-16*/FOXO.[69] Therefore, reduced brood size per se cannot explain lifespan extension of translation deficient worms.

Protein Synthesis and Mitochondrial Dysfunction

Reduced mitochondrial respiration in *C. elegans* can extend lifespan.[19,77,99] To test whether reduced protein synthesis extends lifespan by reducing respiration rates, respiratory mutants *isp-1* (*iron-sulfur protein-1*, a subunit of the mitochondrial complex III) and *clk-1* (*clock abnormality-1*, a gene required for ubiquinone biosynthesis),[100] were treated with RNAi of ribosomal proteins, *rsks-1*/S6K, *ife-2*/eIF4E, *ifg-1*/eIF4G.[69,83,84] All RNAi clones tested further extended the lifespan of these mutants. Therefore, decreased respiration and translation probably extend lifespan independently.

Translation Initiation

Complex Interaction between Translation Initiation Factors, TOR Signaling and DR

The FOXA/HNF-3-related transcription factor PHA-4 is required for DR induced longevity in *C. elegans*. Moreover, *let-363*/TOR and *rsks-1*/S6K antagonize *pha-4*/FOXA to control postembryonic development.[16,101] Inversely, PHA-4/FOXA activity is required for lifespan extension and even survival of the animal in response to decreased or inactivated TOR signaling.[101] Surprisingly, in contrast to TOR and *rsks-1*/S6K, *ife-2*/eIF4E does not rely on PHA-4/FOXA activity to prolong lifespan. In addition, *rsks-1*/S6K and *let-363*/TOR, but not *ife-2*/eIF4E, can suppress *pha-4*/FOXA associated larval lethality. These data suggest that *ife-2*/eIF4E and possibly additional transcription initiation factors might be separated from the *let-363*/TOR, *rsks-1*/S6K, *pha-4*/FOXA regulatory axis. Indeed, depletion of *ife-2*/eIF4E further extends lifespan in a *let-363*/TOR deficient background.[69,83] In addition to *ife-2*/eIF4E, depletion of *ifg-1*/eIF4G was found to further extend the long lifespan of *rsks-1*/S6K(RNAi) and *let-363*/TOR(RNAi) nematodes and *iftb-1*/eIF2β RNAi further extends the lifespan of *rsks-1*(*sv31*) mutants.[69,84]

Table 1. Overview of translation factors that have been cited to influence *C. elegans* lifespan

C. elegans Gene	Name	Short Description	Method of Inactivation	Abnormal Dauer/ Larval Arrest	Lifespan Extension: N2	*let-363/* TOR	*daf-16/* FOXO	*daf-2; daf-16*	*daf-2*	*eat-2*	*sir-2.1*	Mitochondria (*clk-1* or *isp-1*)	Thermo-tolerance	Oxidative Stress Resistance	Reduced Fecundity	References
B0261.2	*let-363/ mTOR*	DNA-dependent protein kinase	*let-363* (*h111*); *let-363* (*h114*); *let-363* (*h131*), RNAi (hat./ad.)	+	+		+		–	+ [82] – [69]	+		+		+	18, 69, 72, 74, 82, 101
C10C5.6	*daf-15/* Raptor	Guanine nucleotide binding protein MIP1	*daf-15*(*m81*), RNAi (hat.)	+	+		–		+						+	73, 74
Y47D3A.16	*rsks-1/* p70S6K	Ribosomal protein S6 kinase	*rsks-1*(*sv31*), *rsks-1*(*ok1255*), RNAi (hat./ad.)	–	+ [69, 84] – [83]	+ [69] – [84]	+		+ [84] – [69]	+	+	+	+ [69] – [84]	–	+ [69] – [72]	69, 72, 83, 84, 101
T01H8.1	*rskn-1/* p90S6K	Ribosomal protein S6 kinase	RNAi (ad.)		–											83
T27F7.3	eIF1	Translation initiation factor 1	RNAi (ad.)	+	+		+									87
Y37E3.10	eIF2α	Translation initiation factor 2, alpha subunit	RNAi (hat.)	+	+										+	72
K04G2.1	*iftb-1/* eIF2β	Translation initiation factor 2, beta subunit	RNAi (hat./ad.)	+ [72] – [82]	+		–		–	+	+		+[d]		+	69, 72, 82

continued on next page

Table 1. Continued

					Lifespan Extension											
C. elegans Gene	Name	Short Description	Method of Inactivation	Abnormal Dauer/ Larval Arrest	N2	*let-363*/ TOR	*daf-16*/ FOXO	*daf-2; daf-16*	*daf-2*	*eat-2*	*sir-2.1*	Mitochondria (*clk-1* or *isp-1*)	Thermo-tolerance	Oxidative Stress Resistance	Reduced Fecundity	References
Y39G10AR.8	eIF2γ	Translation initiation factor 2, gamma subunit	RNAi (ad.)	+	+			+	+	+			+	+	+	88
F11A3.2	eIF2Bδ	Translation initiation factor 2B, delta subunit	RNAi (ad.)	+	+		–	–		+			+[e]	+	+	85
D2085.3	eIF2Bε	Translation initiation factor 2B, epsilon subunit	RNAi (ad.)	+	+			+	+	+			–	–	+	88
C27D11.1	*egl-45*/ eIF3a	Translation initiation factor 3, subunit a	RNAi (ad.)	+	+		+									87
D2013.7	*eif-3.f*/ eIF3f	Translation initiation factor 3, subunit f	RNAi (ad.)	+	+		+	+	+	+			+	+	+	87, 88
Y54E2A.11	*eif-3.B*/ eIF3b	Translation initiation factor 3, subunit b	RNAi (ad.)	+	+		+									87

continued on next page

Table 1. Continued

C. elegans Gene	Name	Short Description	Method of Inactivation	Abnormal Dauer/ Larval Arrest	Lifespan Extension: N2	*let-363*/ TOR	*daf-16*/ FOXO	*daf-2; daf-16*	*daf-2*	*eat-2*	*sir-2.1*	Mitochondria (*clk-1* or *isp-1*)	Thermo-tolerance	Oxidative Stress Resistance	Reduced Fecundity	References
Y65B4A.6	eIF-4AI	ATP-depende RNA helicase FAL1, involved in rRNA maturation	RNAi (hat.)	–											+	72
F57B9.6	*inf-1*/ eIF4A2, eIF4A1	Translation init. factor 4F, helicase subunit 4A	RNAi (ad.)		+		+									87
F53A2.6	*ife-1*/ eIF4E (a)	Translation initiation factor 4F cap-binding subunit	RNAi (ad.)		+ [84] – [83]											83, 84
R04A9.4	*ife-2*/ eIF4E	Translation initiation factor 4F cap-binding subunit	*ife-2(ok306)*, RNAi	–[b]	+	+	+ [83] – [69]		+	+		+	+ [69][d] – [83]	+	–	69, 83, 101
B0348.6	*ife-3*/ eIF4E	Translation initiation factor 4F cap-binding subunit	RNAi (hat.)		–											83

continued on next page

Table 1. Continued

					Lifespan Extension											
C. elegans Gene	Name	Short Description	Method of Inactivation	Abnormal Dauer/ Larval Arrest	N2	*let-363*/ TOR	*daf-16*/ FOXO	*daf-2; daf-16*	*daf-2*	*eat-2*	*sir-2.1*	Mitochondria (*clk-1* or *isp-1*)	Thermo-tolerance	Oxidative Stress Resistance	Reduced Fecundity	References
C05D9.5	*ife-4*/ eIF4E	Translation initiation factor 4F cap-binding subunit	*ife-4(ok320)*, RNAi (hat.),		–											83
Y57A10A.30	*ife-5*/ eIF4E	Translation initiation factor 4F cap-binding subunit	RNAi (hat.)		–											83
M110.4	*ifg-1*/ eIF4G	Translation initiation factor 4F, ribosome/ mRNA-bridging subunit 4G	RNAi (hat./ad.)	+ [72] – [82]	+	+	+ [82][c], [84,87] – [69, 82]		+[84] – [69]	+	+	+	+ [69] – [84]	–	+	69, 72, 82, 84, 87
T05G5.10	*iff-1*/ eIF5A	Translation initiation factor 5A	RNAi (ad.)		+		–				+					86
C23G10.3	*rps-3*	40s ribosomal protein S3	RNAi (ad.)	+	+		+									87
Y71A12B.1	*rps-6*	40s ribosomal protein S6	RNAi (ad.)		+								+			69
F42C5.8	*rps-8*	40s ribosomal protein S8	RNAi (ad.)	+	+		+									87

continued on next page

Table 1. Continued

C. elegans Gene	Name	Short Description	Method of Inactivation	Abnormal Dauer/ Larval Arrest	Lifespan Extension								Thermo-tolerance	Oxidative Stress Resistance	Reduced Fecundity	References
					N2	*let-363*/ TOR	*daf-16*/ FOXO	*daf-2; daf-16*	*daf-2*	*eat-2*	*sir-2.1*	Mitochondria (*clk-1* or *isp-1*)				
D1007.6	*rps-10*	40s ribosomal protein S10	RNAi (ad.)		+											69
F40F11.1	*rps-11*	40s ribosomal protein S11	RNAi (ad.)		+		+									69, 87
F36A2.6	*rps-15*	40s ribosomal protein S15	RNAi (hat./ad.)	+	+		+		–	+	+	+	+		+	69
F53A3.3	*rps-22*	40s ribosomal protein S22	RNAi (hat./ad.)		+		+		–	+	+	+				69
F39B2.6	*rps-26*	40s ribosomal protein S26	RNAi (ad.)		+											69
B0041.4	*rpl-4*	60S ribosomal protein L4	RNAi (ad.)		+		+									69
R151.3	*rpl-6*	60S ribosomal protein L6	RNAi (ad.)		+											69
R13A5.8	*rpl-9*	60S ribosomal protein L9	RNAi (ad.)		+											69
C09D4.5	*rpl-19*	60S ribosomal protein L19	RNAi (ad.)	+	+		+	+	+	+			+	+	+	69
C27A2.2	*rpl-22*	60S ribosomal protein L22	RNAi not specified		+											178

continued on next page

Table 1. Continued

					Lifespan Extension											
C. elegans Gene	Name	Short Description	Method of Inactivation	Abnormal Dauer/ Larval Arrest	N2	*let-363*/ TOR	*daf-16*/ FOXO	*daf-2; daf-16*	*daf-2*	*eat-2*	*sir-2.1*	Mitochondria (*clk-1* or *isp-1*)	Thermo-tolerance	Oxidative Stress Resistance	Reduced Fecundity	References
Y106G6H.3	*rpl-30*	60S ribosomal protein L30	RNAi (ad.)		+		+									69
Y106G6H.2	*pab-1*/ PABPC1	PolyA binding protein (RRM superfamily)	RNAi (ad.)		+											84
F18H3.3	*pab-2*/ PABP-1	PolyA binding protein (RRM superfamily)	RNAi (ad.)		+											84
C17E4.5	*pabp-2*/ PABPN1	Nuclear polyA Binding protein	RNAi (ad.)		–											84
R166.5	*mnk-1*	MAP kinase iegrating kinase (MNK) homolog	RNAi (hat.)		–											83
W01B11.3	*nol-5*/ Nop58p/ Nop5p	Ribosome bio-genesis protein -p58p/-p5p	RNAi (ad.)	+	+			+	+	+			+	–	+	88
F59C6.4	*exos-3*	3′–5′ exoribonuclease subunit Rrp40 involved in rRNA processing	RNAi (ad.)	+	+			+	+	+			+	–	+	88

continued on next page

Table 1. Continued

C. elegans Gene	Name	Short Description	Method of Inactivation	Abnormal Dauer/ Larval Arrest	Lifespan Extension								Thermo-tolerance	Oxidative Stress Resistance	Reduced Fecundity	References
					N2	*let-363*/ TOR	*daf-16*/ FOXO	*daf-2*; *daf-16*	*daf-2*	*eat-2*	*sir-2.1*	Mitochondria (*clk-1* or *isp-1*)				
Y48G1A.4	/	Nucleolar protein involved in 40S ribosome biogenesis	RNAi (ad.)	+	+			+	+	+			+	+	+	88
C47D12.6	*trs-1*	Threonyl-tRNA syhetase	RNAi (ad.)	+	+			+	+	+			+	–	+	88
F22D6.3	*nrs-1*	Asparaginyl-tRNA syhetase	RNAi (ad.)	+	+			+	+	+			+	–	+	88
F26F4.10	*rrt-1*	Arginyl-tRNA synthetase	RNAi (hat./ad.)	+	+									+	+	90
ZK524.3	*lrs-2*	Mitochondrial leucyl-tRNA synthetase	RNAi (hat.)	–	+		+						+	+ –[f]	+	77
C25A1.7a	*irs-2*	Mitochondrial isoleucyl-tRNA synthetase	RNAi (hat.)	–	–		–									77
C48E7.2	/	RNA polymerase III subunit	RNAi (ad.)	+	+			+	+	+			+	+	–	88
F14B4.3	/	RNA polymerase I subunit	RNAi (ad.)	+	+			+	+	+			–	–	+	88

continued on next page

Table 1. Continued

					Lifespan Extension											
C. elegans Gene	Name	Short Description	Method of Inactivation	Abnormal Dauer/ Larval Arrest	N2	*let-363*/ TOR	*daf-16*/ FOXO	*daf-2; daf-16*	*daf-2*	*eat-2*	*sir-2.1*	Mitochondria (*clk-1* or *isp-1*)	Thermo-tolerance	Oxidative Stress Resistance	Reduced Fecundity	References
K01C8.6	/	Mitochondrial ribosomal protein L10	RNAi (ad.)	+	+			+	+	+			–	–	+	88
F59A3.3	/	Mitochondrial ribosomal protein L24	RNAi (ad.)	+	+			+	+	+			+	+	+	88
B0511.8	*tag-264*	Mitochondrial 28S ribosomal protein S30	RNAi (ad.)	+	+			+	+	+			+	+	+	88
B0261.4	/	Mitochondrial ribosomal protein	RNAi (hat.)		+		+						+	+ –[f]		77

'+' and '–' represent 'lifespan extension' and 'no lifespan extension', respectively. (ad., adult; hat., hatching). [a]The *C. elegans* genome encodes five eIF4E isoforms (*ife-1* to *-5*) that differ in cap-binding specificity and anatomical expression. *ife-2*, *-3* and *-5* are expressed in the germline, whereas *ife-2* and *-4* are specifically expressed in the soma.[178] [b]*ife-2(ok306)* mutants grown at 25°C suffer from extensive embryonic lethality (92%) compared to mutants grown at 20°C (6%).[83] [c]Initiation of RNAi treatment either from hatching or at young adult results in a *daf-16*-dependent and *daf-16*-independent lifespan extension, respectively.[82] [d]*daf-16*-dependent.[69] [e]Only after at least four days of RNAi treatment.[85] [f]Increased resistance to hydrogen peroxide, increased sensitivity for paraquat.[77]

TOR RNAi in young *eat-2* mutant adults –a genetic model for dietary restriction in *C. elegans*– does not further extend lifespan.[69,102] If DR-induced longevity would only depend on reduced protein synthesis by TOR-inhibition one would expect that reducing translation by any other means in animals subjected to DR would probably not further extend lifespan. However, several groups have consistently shown that reduced protein synthesis further extends lifespan of *eat-2* deficient worms.[69,83,85,88] Likewise, inhibition of either *ifg-1*/eIF4G or *rsks-1*/S6K significantly increases survival under starvation conditions.[84]

In *C. elegans*, the NAD^+-dependent histone deacetylases SIR-2 (sirtuin-2) is also not required for lifespan extension by reduced protein synthesis.[69,84,86] Indeed, RNAi of several translation initiation factors, ribosomal proteins, *let-363*/TOR and *rsks-1*/S6K were all able to extend the lifespan of short-lived *sir-2* mutants. Although, SIR-2 activity has been associated with the response of DR in yeast, worms and flies,[103-106] the role of sirtuins in DR is strongly contested within the field of aging research.[53,69] Another major output of TOR signaling is the inhibition of autophagy induction. Autophagy and maybe not protein synthesis, might embody a major process that drives DR induced longevity.

Interaction between Translation Initiation Factors with IIS

Like TOR, *rsks-1*/S6K and ribosomal proteins extend lifespan of *daf-16*/FOXO mutants, but not those of *daf-2* mutants.[69] Surprisingly, unlike data on ribosomal proteins, several groups have found a genetic interaction between translation initiation factors and *daf-16*/FOXO.[69,82,85,86] However, an equal number of reports did not find such interactions.[83,84,87,88] Unfortunately, these contradictions are not easily resolved by obvious differences in the experimental procedures. For example, the Johnson group reported that RNAi-induced depletion of *iftb-1*/eIF2β or *ifg-1*/eIF4G throughout development extends lifespan independently of *daf-16*/FOXO, whereas RNAi inhibition initiated at young adulthood requires *daf-16*/FOXO to extend lifespan.[82] Therefore, inhibition of translation initiation factors could have differential effects depending on life-stage. However, two other groups reported a *daf-16*/FOXO independent lifespan extension when young adult worms were treated with *ifg-1*/eIF2β RNAi.[87,88] Also, the use of different *daf-16*/FOXO alleles seems inadequate to explain conflicting results among research groups (Table 1). For example, intact DAF-16/FOXO has been shown to be required for lifespan extension of *ifg-1*/eIF4G RNAi worms in one study[69] but was found dispensable in another.[84] Inhibition of the translational machinery is capable to affect DAF-16/FOXO nuclear localization and by consequence induces the expression of several known *daf-16*/FOXO reporter genes, such as *sod-3, hsp-16.2, hsp70* and *hsp90*.[82,85,87] Although these findings do not prove a direct regulatory interaction between components of the translation machinery and insulin signaling, they do suggest that a *daf-16*/FOXO-dependent stress response could account for some the observed lifespan extensions by reduced protein synthesis. However, the group of Ruvkun showed that although RNAi targeting protein synthesis does not require functional DAF-16/FOXO to extend lifespan, inactivation of some of these genes can induce nuclear localization of DAF-16/FOXO and consequently drive expression of *sod-3*.[87]

A Model for Translation Inhibition Induced Longevity

Hansen et al (2007)[69] reported that depletion of ribosomal proteins extends lifespan of *daf-16*/FOXO mutants, but not *daf-2* mutants. In contrast, in their hands, lifespan extension by RNAi of all translation initiation factors tested (*iftb-1*/eIF2β, *ife-2*/eIF4E and *ifg-1*/eIF4G) was completely dependent on DAF-16/FOXO and even slightly shortened the lifespan of the long-lived *daf-2* mutant. These findings suggest that inhibition of translation initiation not only engages DAF-16/FOXO but also blocks all of the DAF-16/FOXO-independent mechanisms of lifespan extensions that result from reduced protein synthesis.[69] In an attempt to resolve these issues, Hansen et al (2007)[69] proposed a switch-like model, in which the activation of a DAF-16/FOXO-dependent pathway suppresses the operation of the DAF-16/FOXO-independent pathway (i.e., lifespan extension by reduced protein synthesis per se), although the molecular mechanism underlying such interaction has not been specified. Others have suggested that a negative feedback loop

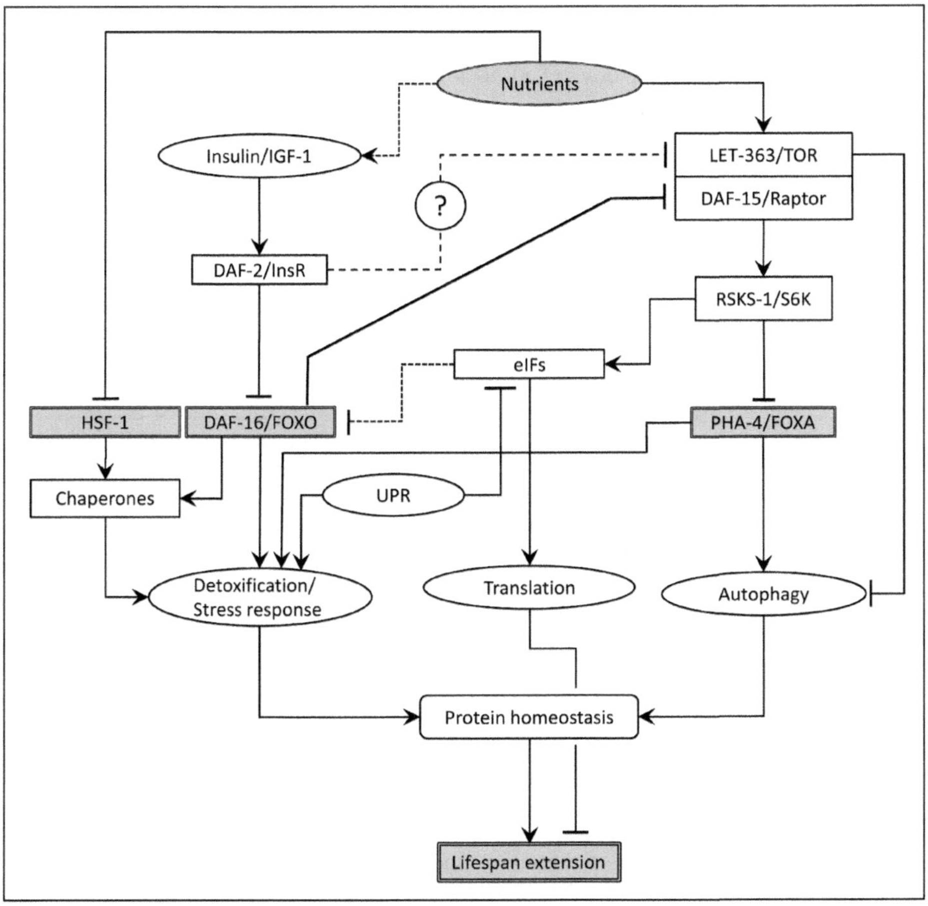

Figure 2. Interactions among insulin/IGF-1-signaling and TOR-signaling pathway and their relation to protein homeostasis and longevity. Arrows depict positive regulation and capped lines denote negative regulation. Dashed lines represent predicted interactions.

between *daf-16*/FOXO and TOR-signaling could explain the claims of genetic interaction between translation initiation factors and *daf-16/FOXO* (Fig. 2). Indeed, *daf-16*/FOXO negatively regulates *daf-15* and thus TOR-signaling, as described by Jia et al (2004).[74] Therefore, inactivation of *daf-16*/FOXO may increase TOR activity in *C. elegans*, which is predicted to boost translation and thereby suppress the long life phenotype of *ife-2*/eIF4E mutants. Secondly, this negative feedback loop would also explain the lack of additive effect in the long-lived double mutant *let-363; daf-2*, although *let-363*/TOR appears largely independent of IIS. Indeed, constitutive activity of DAF-16/FOXO in *daf-2* mutants could decrease TOR activity in such a way that lifespan of *daf-2* single mutants resembles that of *daf-2; let-363* double mutants.[101] In agreement with such a model, *daf-2* mutants do show significantly more reduced protein synthesis rates with age compared to wild-type (own unpublished results) and also display a DR-phenotype.[107] Alternatively, *ife-2*/eIF4E or other translation initiation factors impinge on *daf-16*/FOXO more directly, or induce a general stress response that activates DAF-16/FOXO. Translation initiation factors have been associated in mediating specific stress responses. eIF2α is a well known phosphorylation target in response to various stresses including, amino acid starvation, ER-stress and viral infection in yeast and mammals, causing general suppression of protein synthesis and upregulation of autophagy.[95,108-110] In *S. Pombe* one of two eIF4E homologs is also associated with the stress response by modulating the translational efficiency of a subset of mRNAs.[111]

HSF-1 Mediated Defense against Proteotoxicity

The *C. elegans* transcription factor HSF-1 is a highly conserved, ubiquitously expressed transcriptional regulator of stress inducible heat shock genes, many of which encode molecular chaperones and components of the protein degradation machinery. Inhibition of HSF-1 function has been shown to lead to a decreased lifespan and an accelerated aging phenotype in *C. elegans,* whereas overexpression of HSF-1 results in a lifespan extension dependent on the presence of DAF-16/FOXO. Conversely, lifespan extension via reduced insulin/IGF-1 signaling requires HSF-1 activity next to DAF-16/FOXO.[112,113] Overexpression of individual molecular chaperones under HSF-1 transcriptional control, including members of the Hsp70 and small heat shock protein (*shsp*) gene families such as HSP-16, in *Drosophila* and *C. elegans* have been shown to extend lifespan, whereas knockdown of individual *shsp* genes shorten the lifespan of long-lived *daf-2* mutants.[112,114-116] The existence of sequences identical to both DAF-16/FOXO and HSF-1 consensus binding sites upstream of many *shsp* genes[112] and the increased *shsp* expression in IIS pathway mutants[11] indicate that HSF-1 and DAF-16/FOXO function in concert to extend lifespan, at least in part, by increasing *shsp* gene expression. sHSPs are known to form large oligomers that bind to unfolded proteins and prevent them from aggregating.[117,118] In aging animals this may prevent oxidized or otherwise damaged proteins from aggregating before they can be refolded or degraded. An important role for HSF-1 in the protection against proteotoxicity has been observed in transgenic nematodes engineered to express human proteins associated with neurodegenerative diseases such as Alzheimer's, Huntington's and Parkinson's disease.[119,120] A common trait with most of these age-associated neuropathologies is the aggregation of specific proteins that is believed to be toxic and to underlie the disease pathology (i.e., proteotoxicity). A progressive paralysis with age is seen in transgenic nematodes expressing beta amyloid peptide $A\beta_{42}$ (Alzheimer's disease), polyglutamine tract expansion Q_{35}YFP (Huntington's disease) or an aggregation prone GFP (GFP-degron) in the body wall muscles.[119,121] Both reduced IIS and DR were shown to protect extensively against proteotoxicity in all models tested ($A\beta_{42}$, Q_{35}YFP and GFP-degron) and these animals remained largely paralysis-free or showed a significantly delayed onset of paralysis.[60,122,123] A similar improvement in the progression of neurodegeneration in mouse models has been observed.[61] Both reduced IIS and DR were found to require functional HSF-1 to confer increased protection against proteotoxicity.[60,122] An interesting observation is that, although animals subjected to DR or decreased IIS show increased protection against proteotoxicity, no apparent difference was found in either size or number of Q_{35}YFP or $A\beta_{42}$ proteotoxic aggregates.[60,122] Cohen and colleagues showed that the transcriptional output of HSF-1 and DAF-16/FOXO due to reduced *daf-2* signaling differently affects the fate of an aggregation-prone protein. HSF-1 was found to prevent the formation of aggregation-prone peptides into toxic oligomers, likely by elevating the levels of protecting molecular chaperones, whereas DAF-16/FOXO seems to enhance the formation of large, inert aggregates from toxic oligomers. Although, both activities seem paradoxal, Cohen et al suggest a hierarchical mechanism in which low molecular weight oligomers are transformed into high molecular weight aggregates of lower toxicity by the action of DAF-16/FOXO, only after the preferred route of efficient disaggregation of toxic oligomers by action of HSF-1 is overtaxed under these pathological conditions. These findings seem to be in accordance with several lines of evidence that inclusion bodies are not the main cause of toxicity and probably represent a cellular protective response.[124,125]

In mammalian and human Alzheimer brains elevated levels of several small heat shock proteins (shsp) have been observed and colocalize with β-amyloid (Aβ) protein depositions.[126] In a transgenic *C. elegans* Alzheimer's disease model, overexpression of HSP-16.2, a *C. elegans* chaperone protein homologous to αB crystalline under control of HSF-1, suppresses proteotoxicity associated with human $A\beta_{42}$.[127] *C. elegans* does not naturally produce the Aβ-peptide, so the interaction of HSP-16.2 and Aβ likely results from a general function of HSP-16.2 to bind (potentially toxic) intermediate multimers and prevent the formation of toxic oligomer species.

Although reduced IIS and DR both are capable of eliciting enhanced protection against age-related proteotoxicity in a *hsf-1*-dependent manner, the mechanism by which they do so seems to be different. Whereas IIS requires the activity of both *daf-16*/FOXO and *hsf-1*, DR only requires *hsf-1*, but *daf-16*/FOXO is expendable.[60] Hsu and colleagues showed that dietary restriction (*eat-2* mutants) extends lifespan of *hsf-1* treated animals, which suggests DR acts in parallel to *hsf-1* to modulate longevity.[112] In a hypothesis put forward by Steinkraus et al, DR and *hsf-1* could act in parallel to promote longevity, both contributing to improved protein homeostasis.[60] While *hsf-1* regulates the expression of molecular chaperones, DR could contribute to protein homeostasis by enhancing protein degradation (e.g., autophagy) and/or modulating mRNA translation, in effect enhancing protein turnover. Another possibility put forward is that HSF-1 functions together with one or more other transcription factors in conferring the phenotypes associated with DR. Two such transcription factors required for most regimes of DR, *skn-1*/NRF2 and *pha-4*/FOXA are attractive candidates for such co-operation.[60] Interestingly, in yeast, activation of *hsf-1* results in reduced TOR activity.[128] Perhaps in *C. elegans* a similar mechanism is conserved in which HSF-1 can inhibit TOR-signaling resulting in several shared phenotypes between reduced IIS and DR. Further studies are necessary to clarify the precise relationship between *hsf-1* and DR and TOR-signaling.

Autophagy

Autophagy Regulates Aging in C. elegans

Autophagy

In *C. elegans* there is clear evidence that longevity is closely linked to the capacity of the cell to regulate autophagic activity, another downstream target of TOR-signaling (Table 2). Macroautophagy (hereafter referred to as autophagy) is a major cellular catabolic process conserved in all eukaryotic cells that involves the degradation and recycling of cellular material and organelles by the lysosomal system.[129] Together with the ubiquitin-proteasome system (UPS), autophagy is responsible for the basal turnover of cellular material within the cell, but during cellular stress such as starvation, autophagic cellular self digestion is strongly upregulated in an attempt to maintain cellular energy and reclaim homeostasis. For example, in the yeast *Saccharomyces cerevisiae*, autophagy is essential for survival during starvation and sporulation.[130]

Autophagy Is Required for Longevity in IIS Mutants

Meléndez was the first to show that in *C. elegans* autophagy genes are essential for completion of normal dauer morphogenesis and that *bec-1* (the *C. elegans* ortholog of the yeast and mammalian autophagy genes ATG6/VPS30 and *beclin 1*, respectively) is required for the longevity phenotype of *daf-2* mutant animals.[131] *daf-2(-)* animals have markedly increased levels of autophagic activity, implying that insulin signaling negatively regulates the induction of autophagy.[131] Besides *bec-1*, inactivation of other autophagy genes including, *atg-7*, *atg-18*, *lgg-1*, *unc-51* and *lgg-3* were also shown to negatively affect the longevity phenotype of *daf-2* mutants, emphasizing the importance of autophagy in the regulation of worm lifespan.[132-134] In *C. elegans,* reduced IIS is thought to regulate autophagic induction, in part by inhibiting the DAF-16/FOXO-dependent suppression of *daf-15*/Raptor and thus silencing TOR kinase signaling.[74] Arrested L3 larvae deficient of TOR and long lived *daf-15*/Raptor heterozygotes indeed have elevated levels of autophagy which is also required for their long lifespan.[134,135] Therefore, autophagy may be a major target of the TOR kinase signaling pathway in modulating the lifespan of *C. elegans* along with normal larval development and dauer formation.

Autophagy and DR

In yeast, worms and flies dietary restriction extends lifespan, at least in part, by downregulating TOR-signaling.[67,68,135] The survival of wild-type worms during starvation was found to be dependent on autophagy and autophagic activity is increased in several feeding-defective *C. elegans* (*eat-*) mutants as well as in worms subjected to DR.[134-137] This suggests a role for autophagy in DR-induced longevity. Indeed, autophagy genes were found to be epistatic to the long lifespan of both *eat-2(ad1113)* and *eat-2(ad1116)* mutant worms.[133-135] The long lifespan of *eat-2(ad1116)* mutants requires the activity of the FOXA transcription factor *pha-4*/FOXA which was shown to be a downstream target of *let-363*/TOR, *rsks-1*/S6K-signaling and essential for lifespan extension in response to DR.[16,101] Hansen and colleagues showed that induction of autophagy in response to DR requires PHA-4/FOXA activity, indicating that autophagy is a transcriptionally regulated response to food limitation. Unfortunately, the genes downstream of PHA-4/FOXA that regulate autophagy have not yet been identified.[135] Taken together these results imply that IIS, TOR and DR extend lifespan, at least in part, by converging on autophagy to promote longevity in *C. elegans*.

Autophagy and p53

Autophagy was also shown to mediate the lifespan extending effect of loss-of-function in *cep-1*, the *C. elegans* ortholog of the tumor suppressor protein p53.[138,139] In *Drosophila* expression of dominant negative forms of the p53 ortholog Dmp53 in adult neurons extends lifespan by inhibiting insulin signaling, suggesting that *cep-1* might affect aging in *C. elegans* to some extent by endocrine signaling.[140] Indeed, there are some indications that also in worms, lifespan extension by *cep-1* requires functional DAF-16/FOXO.[139]

Interaction between DR, IIS and Autophagy

Despite the seemingly rejuvenating effect of autophagy on overall organismal physiology, autophagy is neither sufficient nor required to extend lifespan in *C. elegans*. Hansen and colleagues showed that *daf-16*/FOXO is dominant over autophagy to extend lifespan of *daf-2* mutants, in spite of increased autophagy in the absence of *daf-16*/FOXO. However, direct measurements of functional autophagic activity in *daf-2*(-); *daf-16* (-) double mutants have not been performed.[135] Moreover, Pan et al and Hansen et al reported that long-lived mitochondrial respiratory chain mutants *clk-1* and *isp-1* do not require autophagy for their long lifespan and inhibiting protein synthesis in well-fed animals extends lifespan in the absence of autophagy.[84,135] In contrast, Tóth et al found that lifespan extension of respiration impaired *clk-1* and *atp-3* required autophagy.[134] Taken together, these observations suggest that additional processes beyond autophagy must be involved in lifespan extension. Finally, excessive activation of autophagy can also negatively affect survival of *C. elegans* during starvation as it damages tissues critical for survival, such as pharyngeal muscle cells.[137]

The finding by Hansen that reduced IIS induces autophagy independently of DAF-16/FOXO, but requires both DAF-16/FOXO and increased autophagy to extend lifespan is intriguing in several ways. First, it contrasts with dietary restriction where autophagy[135] but not DAF-16/FOXO[102,141] is required to extend lifespan, which suggests that autophagy modulates lifespan in IIS mutants, at least in part, by a different mechanism compared to DR. Secondly, these results imply that reduced protein synthesis in response to DR is not sufficient to extend lifespan in the absence of autophagy, whereas inhibiting translation in well-fed worms is.[135] Hansen and colleagues therefore suggest that disrupting protein synthesis in well fed animals might trigger a novel, lifespan extending pathway that is not triggered by dietary restriction/TOR.[135] Indeed, the lifespan of *eat-2* mutants is further extended by direct protein synthesis inhibition but not by TOR-inhibition.[69,88] However, further work is needed to explore these suppositions in more detail. Finally, these results suggest that cross-talk may exist between IIS and TOR-signaling upstream of DAF-16/FOXO in *C. elegans*.

Autophagy and Neuropathology

Besides starvation, autophagy is also upregulated by other forms of stress, such as the accumulation of intracellular protein aggregates, associated with aging and numerous neurodegenerative disorders.[129] In fact, the pathological accumulation of autophagosomes and autophagosome-like

Table 2. Overview of autophagy genes implicated in* C. elegans *aging

					Required for Lifespan Extension						
Yeast Ortholog	***C. elegans* Gene**	**Name**	**Method of Inactivation**	**Abnormal Dauer/ Larval Arrest**	**Reduced Lifespan**	**IIS**	**DR**	***let-363*/TOR**	**Reduced Protein Synthesis**	**ETC**	**References**
ATG6/ VPS30	T19E7.3	*bec-1*	*bec-1 (ok691)*, RNAi (hat./ad.)	+	–	+	+	+	–	+ [134] – [135]	131, 132, 134, 135, 137, 179
ATG1	Y60A3A.1	*unc-51*	*unc-51(e369)*, *unc-51 (e1189)*, RNAi (hat.)	+	+		+	+		+	131, 134
ATG4	ZK792.1	/	RNAi (hat.)	–							131
ATG7	M7.5	*atg-7*	RNAi (hat./ad.)	+	+	+	+				72, 131, 132
ATG8	C32D5.9	*lgg-1*	RNAi (hat.)	+ (a)	+						131, 134
ATG8	ZK593.6	*lgg-2*	RNAi (hat.)	– (a)							131
AUT10	F41E6.13	*atg-18*	RNAi (hat.)	+							131
ATG12	B0336.8	*lgg-3*	RNAi (hat./ad.)	+	+	+					72, 132
ATG17	R07G3.3	*npp-21*	RNAi (hat.)	–							131
VPS34	B0025.1	*vps-34*	RNAi (ad.)		–		+		–	+ [134] – [135]	134, 135
ATG18	F41E6.13	*atg-18*	*atg18 (gk378)*		+			+		+	134
ATG9	T22H9.2	*atg-9*	RNAi (hat.)		+						134

'+' and '–' denote 'required' and 'not required' for lifespan extension, respectively. (ad., adult; hat., hatching). [a]lethal.

structures in brains of patients with central nervous system (CNS) disorders such as Alzheimer's, Parkinson's and Huntington's disease raises the question as to the (etiologic) role for autophagic (dys-)function in these disease conditions.[142-144] Therefore, the effect of autophagy on the degradation of protein aggregates was investigated in cell- and animal models using pharmacological inducers and inhibitors of autophagy.[145] For example, autophagy was shown to be involved in the clearance of polyglutamine (polyQ) aggregates in vitro and treatment of mice and *Drosophila* with rapamicyn (a potent inhibitor of TOR activity) protected against Huntington's disease.[146,147]

Huntington and Alzheimer

In *C. elegans* models for neurodegenerative disorders, autophagy genes were found to protect from, or significantly delay, the onset of proteotoxicity.[148-150] Genetic impairment of autophagy results in increased size and number of polyglutamine (polyQ) aggregates in neurons and muscles, indicating an essential in vivo role for autophagy in the suppression of aberrant polyQ protein induced proteotoxicity.[148,149] Abnormal accumulation of immature autophagosomal vacuoles in Alzheimer brains suggests that autophagic activity may be severely compromised in this disease.[151] Transgenic expression of human β-amyloid peptide (Aβ) in *C. elegans* muscle also causes accumulation of autophagosomal vesicles, thereby mimicking Alzheimer brains.[150] Reduced IIS in this strain was shown to promote autophagic degradation of Aβ protein accompanied by a significant reduction in the accumulation of autophagosomes and a markedly delayed onset of worm paralysis. The favorable effect of reduced IIS on Aβ toxicity therefore seems to originate from promoting proper completion of autophagy and consequent degradation of Aβ.

Parkinson's Disease

Parkinson's disease is characterized by the progressive loss of dopaminergic nigrostratial neurons in the brain, along with the accumulation of proteins in so called Lewy bodies. Genomic multiplication of α-synuclein (α-syn) results in Parkinson's disease and is central to the formation of Lewy bodies.[152] Besides α-syn, torsinA also accumulates in Lewy bodies acting as a molecular chaperone-like protein together with heat shock proteins to suppress aggregation.[153,154] Also, a specific mutation in torsinA is linked to the autosomal-dominant heritable movement disorder Oppenheim's dystonia.[155] Using *C. elegans* as a model organism to examine the role of torsin activity, overexpression of both human and nematode torsin proteins results in a dramatic reduction of polyQ-dependent protein aggregation.[156] In addition, overexpression of *tor-2*, one of three *C. elegans* homologues of human torsinA, ameliorates the formation of α-syn aggregates in nematode body-wall muscle cells.[157] A genome-wide RNAi screen in worms overexpressing human α-syn identified 80 *C. elegans* genes of which expression influences the formation of α-syn inclusions.[158] Approximately half of these have an established human ortholog. Quality control and vesicle-trafficking genes expressed in the ER/golgi complex and vesicular compartments were overrepresented and several molecular modulators of lifespan were identified. A similar systematic RNAi screen identified worm othologs of known familial Parkinson's disease genes such as *pdr-1*/parkin, *djr1.1*/DJ-1, *pink-1*/PINK1, *nhr-6*/NURR1 and also *tor-2*. Several autophagy-related genes were identified, such as *unc-51*, *atg-7*, *vps-34*, *lgg-3* and *lgg-1*. A highly conserved protein that demonstrated high neuroprotective capacity was *C. elegans* VPS41. In *S. cerevisiae*, VPS41 is involved in trafficking from the trans-Golgi to the vacuole/lysosome.[159] Together these results suggest a significant role for lysosomal function in maintaining homeostatic balance in response to excessive α-syn. Microarray of α-syn overexpressing nematodes also revealed upregulation of seven genes that function in the UPS along with 35 genes involved in mitochondrial function.[160] Inhibition of the proteasome and dysfunctional mitochondria have indeed been linked to the formation of α-syn inclusions and degeneration of dopaminergic neurons in mammals.[161,162]

Autophagy and Neuronal Cell Death

Many studies have reported an increase in markers of autophagy following hypoxic-ischemia in cardiac myocytes and cerebral neurons.[163,164] Both cell protective and cell death promoting properties have been attributed to autophagy in different models and experimental contexts, obscuring the functional significance of autophagy under these conditions. In *C. elegans* it was found that blocking autophagy is deleterious for the survival of *C. elegans* following a severe anoxic insult and worsens the cellular pathology seen after anoxia.[165] Interestingly, mutation in both apoptotic- or ion-channel-mediated necrotic cell death pathways abrogates the anoxia-hypersensitive phenotype in autophagy-deficient nematodes. This suggests a cytoprotective role for autophagy following anoxic insult by preventing apoptotic and necrotic cell death in *C. elegans*.[165,166] In contrast, a considerable body of literature reports that macroautophagy is also a cell death (promoting) mechanism.[167] For example, in *C. elegans*, gain of function mutations in genes that encode specific ion channel subunits such as the degenerins DEG-1, MEC-4 and the acetylcholine receptor subunit DEG-3 lead to toxic ion-channel induced necrotic cell death of a subset of neurons that can be substantially prevented by inactivation of autophagy.[168,169] The role for autophagy in the survival of neuronal cells under pathological conditions thus seems to be dependent on the type of insult and specific cellular environment.[170]

Proteasome Function in Proteotoxicity and Longevity

The ubiquitin proteasome system (UPS) degrades oxidatively damaged and misfolded proteins and prevents their accumulation in insoluble aggregates.[171] In addition, the attachment of ubiquitin molecules by ubiquitin ligases marks specific proteins for degradation by the proteasome. Ubiquitin ligases regulate several cellular processes by controlling the concentration of key regulatory proteins in diverse molecular pathways.[172] For example, the ubiquitin ligase E3 can modulate the stability of DAF-16/FOXO thereby determining worm lifespan.[173,174]

Impairment of general proteasome function, by depleting several structural and enzymatic subunits, shortens *C. elegans* lifespan[173,175] and causes the nuclear localization of the SKN-1/NRF2, a functional ortholog of the mammalian transcription factor NRF2, which is involved in the oxidative stress response.[176]

AIRAP (arsenic-inducible proteasomal 19S regulatory particle-associated protein) modulates the 26S proteasome to counteract proteotoxicity induced by an environmental toxin.[177] Worms lacking *aip-1*, a homolog of mammalian AIRAP, develop normally but exhibit a shortened lifespan.[175] In addition, *aip-1*(-) promotes the accumulation Q_{35}-YFP in SDS-resistant aggregates and accelerates $A\beta_{1\text{-}42}$ induced paralysis. Therefore, *aip-1* links cellular resistance to proteotoxicity and the maintenance of normal lifespan.

Conclusion

A strong correlation between macromolecular damage and age has been observed in many species, including *C. elegans*. This damage likely occurs through the interaction of naturally produced free radicals with their immediate intracellular environment. The constant turnover of proteins through autophagy and the proteasome constitutes the main mechanism by which a cell can clean itself from protein damage, thereby maintaining proteome integrity. As the protein translation and degradation capacity of cells diminishes with age, the turnover of aberrant proteins becomes compromised. In turn, this contributes to the general deterioration of cellular homeostasis seen during senescence. In *C. elegans*, dietary restriction and reduced insulin/IGF-1 signaling rely upon the induction of autophagic activity to extend lifespan. Also, in several age-related neurodegenerative disease models in *C. elegans*, elevated levels of autophagy protect from or delay the formation of toxic protein aggregates. Similarly, the expression of molecular chaperones, many of which are under control of insulin/IGF-1 signaling, promotes longevity and offers protection against aggregation-induced proteotoxicity. These findings advocate a major role for protein homeostasis in senescence and age-related diseases.

Increased protein turnover is by definition associated with increased protein translation. However, in *C. elegans*, a reduction in protein synthesis has a beneficial effect on lifespan. Interaction between translation initiation and the insulin/IGF-1 signaling or other stress-response pathways may be, at least in part, responsible for the observed lifespan extension and increased stress resistance. Unraveling the exact genetic interplay and mechanism by which reduced protein synthesis confers longevity constitutes the focus of current research efforts in several *C. elegans* aging laboratories. Similarly, autophagy has only recently claimed a central role in the aging process. Therefore, several questions remain open for future research endeavors. For example, mitophagy, the selective engulfment and degradation of mitochondria by lysosomes, could, if conserved, have a major impact on the aging process. Also, the effect of recycling of amino-acids and/or other degradation products on lifespan extension is currently unknown. This recycling process is a hallmark for animal survival during prolonged periods of starvation. Moreover, the molecular mechanisms underlying recycling itself are still poorly understood in general and deserve further attention.

References

1. Riddle DL. C. elegans II. Cold Spring Harbor: Cold Spring Harbor laboratory; 1997.
2. Wood WB. The Nematode Caenorhabditis elegans. Cold Spring Harbor: Cold Spring Harbor laboratory; 1988.
3. Consortium TCeS. Genome sequence of the nematode C. elegans: a platform for investigating biology. Science 1998; 282(5396):2012-2018.
4. Schwarz EM. Genomic classification of protein-coding gene families (September 23, 2005), WormBook, ed. The C. elegans Research Community, WormBook, doi/10.1895/wormbook.1.29.1, http://www.wormbook.org.
5. Johnson TE, Wood WB. Genetic analysis of life-span in Caenorhabditis elegans. Proc Natl Acad Sci USA 1982; 79(21):6603-6607.
6. Johnson TE, Hutchinson EW. Absence of strong heterosis for life span and other life history traits in Caenorhabditis elegans. Genetics 1993; 134(2):465-474.
7. Kenyon C. The nematode Caenorhabditis elegans. Science 1988; 240(4858):1448-1453.
8. Kamath RS, Fraser AG, Dong Y et al. Systematic functional analysis of the Caenorhabditis elegans genome using RNAi. Nature 2003; 421(6920):231-237.
9. Rual JF, Ceron J, Koreth J et al. Toward improving Caenorhabditis elegans phenome mapping with an ORFeome-based RNAi library. Genome Res 2004; 14(10B):2162-2168.
10. Murphy CT, McCarroll SA, Bargmann CI et al. Genes that act downstream of DAF-16 to influence the lifespan of Caenorhabditis elegans. Nature 2003; 424(6946):277-283.
11. McElwee J, Bubb K, Thomas JH. Transcriptional outputs of the Caenorhabditis elegans forkhead protein DAF-16. Aging Cell 2003; 2(2):111-121.
12. Kenyon C, Chang J, Gensch E et al. A C-Elegans Mutant That Lives Twice as Long as Wild-Type. Nature 1993; 366(6454):461-464.
13. Friedman DB, Johnson TE. A Mutation in the Age-1 Gene in Caenorhabditis-Elegans Lengthens Life and Reduces Hermaphrodite Fertility. Genetics 1988; 118(1):75-86.
14. Kimura KD, Tissenbaum HA, Liu YX et al. daf-2, an insulin receptor-like gene that regulates longevity and diapause in Caenorhabditis elegans. Science 1997; 277(5328):942-946.
15. Arantes-Oliveira N, Apfeld J, Dillin A et al. Regulation of life-span by germ-line stem cells in Caenorhabditis elegans. Science 2002; 295(5554):502-505.
16. Panowski SH, Wolff S, Aguilaniu H et al. PHA-4/Foxa mediates diet-restriction-induced longevity of C. elegans. Nature 2007; 447(7144):550-555.
17. Bishop NA, Guarente L. Two neurons mediate diet-restriction-induced longevity in C. elegans. Nature 2007; 447(7144):545-549.
18. Vellai T, Takacs-Vellai K, Zhang Y et al. Genetics: influence of TOR kinase on lifespan in C. elegans. Nature 2003; 426(6967):620.
19. Dillin A, Hsu AL, Arantes-Oliveira N et al. Rates of behavior and aging specified by mitochondrial function during development. Science 2002; 298(5602):2398-2401.
20. Orgel LE. The maintenance of the accuracy of protein synthesis and its relevance to ageing. Proc Natl Acad Sci USA 1963; 49:517-521.
21. Rothstein M. Aging and the alteration of enzymes: a review. Mech Ageing Dev 1975; 4(5-6):325-338.
22. Rothstein M. The formation of altered enzymes in aging animals. Mech Ageing Dev 1979; 9(3-4):197-202.
23. Stadtman ER. Protein modification in aging. J Gerontol 1988; 43(5):B112-120.
24. Gershon H, Gershon D. Detection of inactive enzyme molecules in ageing organisms. Nature 1970; 227(5264):1214-1217.
25. Reiss U, Rothstein M. Age-related changes in isocitrate lyase from the free living nematode, Turbatrix aceti. J Biol Chem 1975; 250(3):826-830.
26. Zeelon P, Gershon H, Gershon D. Inactive enzyme molecules in aging organisms. Nematode fructose-1,6-diphosphate aldolase. Biochemistry 1973; 12(9):1743-1750.
27. Reznick AZ, Gershon D. Age related alterations in purified fructose-1,6-diphosphate aldolase from the nematode Turbatrix aceti. Mech Ageing Dev 1977; 6(5):345-353.

28. Sharma HK, Gupta SK, Rothstein M. Age-related alteration of enolase in the free-living nematode, Turbatrix aceti. Arch Biochem Biophys 1976; 174(1):324-332.
29. Gupta SK, Rothstein M. Phosphoglycerate kinase from young and old Turbatrix aceti. Biochim Biophys Acta 1976; 445(3):632-644.
30. Sharma HK, Rothstein M. Age-related changes in the properties of enolase from Turbatrix aceti. Biochemistry 1978; 17(14):2869-2876.
31. Parker J, Flanagan J, Murphy J et al. On the accuracy of protein synthesis in Drosophila melanogaster. Mech Ageing Dev 1981; 16(2):127-139.
32. Johnson TE, McCaffrey G. Programmed aging or error catastrophe? An examination by two-dimensional polyacrylamide gel electrophoresis. Mech Ageing Dev 1985; 30(3):285-297.
33. Fleming JE, Quattrocki E, Latter G et al. Age-dependent changes in proteins of Drosophila melanogaster. Science 1986; 231(4742):1157-1159.
34. Takahashi R, Goto S. Fidelity of aminoacylation by rat-liver tyrosyl-tRNA synthetase. Effect of age. Eur J Biochem 1988; 178(2):381-386.
35. Vanfleteren JR, De Vreese A. Analysis of the proteins of aging Caenorhabditis elegans by high resolution two-dimensional gel electrophoresis. Electrophoresis 1994; 15(2):289-296.
36. Harman D. Aging: a theory based on free radical and radiation chemistry. J Gerontol 1956; 11(3):298-300.
37. Harman D. Free radicals in aging. Mol Cell Biochem 1988; 84(2):155-161.
38. Balajee AS, Bohr VA. Genomic heterogeneity of nucleotide excision repair. Gene 2000; 250(1-2):15-30.
39. Grune T, Reinheckel T, Davies KJ. Degradation of oxidized proteins in mammalian cells. FASEB J 1997; 11(7):526-534.
40. Stadtman ER, Starke-Reed PE, Oliver CN et al. Protein modification in aging. EXS 1992; 62:64-72.
41. Van Remmen H, Ward WF, Sabia RV et al. Gene expression and protein degradation. In: Masoro, EJ, ed. Handbook of Physiology, Section 11: Aging. New York: Oxford University Press, 1995:171-234.
42. Sharma HK, Prasanna HR, Lane RS et al. The effect of age on enolase turnover in the free-living nematode, Turbatrix aceti. Arch Biochem Biophys 1979; 194(1):275-282.
43. Reznick AZ, Gershon D. The effect of age on the protein degradation system in the nematode Turbatrix aceti. Mech Ageing Dev 1979; 11(5-6):403-415.
44. Egilmez NK, Rothstein M. The effect of aging on cell-free protein synthesis in the free-living nematode Turbatrix aceti. Biochim Biophys Acta 1985; 840(3):355-363.
45. Prasanna HR, Lane RS. Impaired Protein-Synthesis in Aged Turbatrix-Aceti. Federation Proceedings 1977; 36(3):292-292.
46. Prasanna HR, Lane RS. Protein degradation in aged nematodes (Turbatrix aceti). Biochem Biophys Res Commun 1979; 86(3):552-559.
47. Sarkis GJ, Ashcom JD, Hawdon JM et al. Decline in protease activities with age in the nematode Caenorhabditis elegans. Mech Ageing Dev 1988; 45(3):191-201.
48. Ryazanov AG, Nefsky BS. Protein turnover plays a key role in aging. Mech Ageing Dev 2002; 123(2-3):207-213.
48a. Brunk UT, Terman A. The mitochondrial-lysosomal axis theory of aging: accumulation of damaged mitochondria as a result of imperfect autophagocytosis. Eur J Biochem 2002; 269(8):1996-2002.
48b. Stroikin Y, Dalen H, Brunk UT, et al. Testing the "garbage" accumulation theory of ageing: mitotic activity protects cells from death induced by inhibition of autophagy. Biogerontology. 2005;6(1):39-47.
49. Tavernarakis N, Driscoll M. Caloric restriction and lifespan: a role for protein turnover? Mech Ageing Dev 2002; 123(2-3):215-229.
50. Gafni A. Altered protein metabolism in aging. Annu Rev Gerontol Geriatr 1990; 10:117-131.
51. Richardson A, Cheung HT. Current concepts: I. The relationship between age-related changes in gene expression, protein turnover and the responsiveness of an organism to stimuli. Life Sci 1982; 31(7):605-613.
52. Masoro EJ. Overview of caloric restriction and ageing. Mech Ageing Dev 2005; 126(9):913-922.
53. Mair W, Dillin A. Aging and survival: the genetics of life span extension by dietary restriction. Annu Rev Biochem 2008; 77:727-754.
54. Gredilla R, Barja G. Minireview: the role of oxidative stress in relation to caloric restriction and longevity. Endocrinology 2005; 146(9):3713-3717.
55. Forster MJ, Sohal BH, Sohal RS. Reversible effects of long-term caloric restriction on protein oxidative damage. J Gerontol a-Biol 2000; 55(11):B522-B529.
56. Lass A, Sohal BH, Weindruch R et al. Caloric restriction prevents age-associated accrual of oxidative damage to mouse skeletal muscle mitochondria. Free Radical Bio Med 1998; 25(9):1089-1097.
57. Hepple RT, Qin M, Nakamoto H et al. Caloric restriction optimizes the proteasome pathway with aging in rat plantaris muscle: implications for sarcopenia. American Journal of Physiology-Regulatory Integrative and Comparative Physiology 2008; 295(4):R1231-R1237.

58. Youngman LD, Park JY, Ames BN. Protein oxidation associated with aging is reduced by dietary restriction of protein or calories. Proc Natl Acad Sci USA 1992; 89(19):9112-9116.
59. Sohal RS, Weindruch R. Oxidative stress, caloric restriction and aging. Science 1996; 273(5271):59-63.
60. Steinkraus KA, Smith ED, Davis C et al. Dietary restriction suppresses proteotoxicity and enhances longevity by an hsf-1-dependent mechanism in Caenorhabditis elegans. Aging Cell 2008; 7(3):394-404.
61. Martin B, Mattson MP, Maudsley S. Caloric restriction and intermittent fasting: two potential diets for successful brain aging. Ageing Res Rev 2006; 5(3):332-353.
62. Goto S, Takahashi R, Araki S et al. Dietary restriction initiated in late adulthood can reverse age-related alterations of protein and protein metabolism. Ann N Y Acad Sci 2002; 959:50-56.
63. Ishigami A, Goto S. Effect of dietary restriction on the degradation of proteins in senescent mouse liver parenchymal cells in culture. Arch Biochem Biophys 1990; 283(2):362-366.
64. Dhahbi JM, Mote PL, Wingo J et al. Caloric restriction alters the feeding response of key metabolic enzyme genes. Mech Ageing Dev 2001; 122(10):1033-1048.
65. Lewis SE, Goldspink DF, Phillips JG et al. The effects of aging and chronic dietary restriction on whole body growth and protein turnover in the rat. Exp Gerontol 1985; 20(5):253-263.
66. Aksenova MV, Aksenov MY, Carney JM et al. Protein oxidation and enzyme activity decline in old brown Norway rats are reduced by dietary restriction. Mech Ageing Dev 1998; 100(2):157-168.
67. Kapahi P, Zid BM, Harper T et al. Regulation of lifespan in Drosophila by modulation of genes in the TOR signaling pathway. Curr Biol 2004; 14(10):885-890.
68. Kaeberlein M, Powers RW, 3rd, Steffen KK et al. Regulation of yeast replicative life span by TOR and Sch9 in response to nutrients. Science 2005; 310(5751):1193-1196.
69. Hansen M, Taubert S, Crawford D et al. Lifespan extension by conditions that inhibit translation in Caenorhabditis elegans. Aging Cell 2007; 6(1):95-110.
70. Wullschleger S, Loewith R, Hall MN. TOR signaling in growth and metabolism. Cell 2006; 124(3):471-484.
71. Powers RW 3rd, Kaeberlein M, Caldwell SD et al. Extension of chronological life span in yeast by decreased TOR pathway signaling. Genes Dev 2006; 20(2):174-184.
72. Long X, Spycher C, Han ZS et al. TOR deficiency in C. elegans causes developmental arrest and intestinal atrophy by inhibition of mRNA translation. Curr Biol 2002; 12(17):1448-1461.
73. Hara K, Maruki Y, Long X et al. Raptor, a binding partner of target of rapamycin (TOR), mediates TOR action. Cell 2002; 110(2):177-189.
74. Jia K, Chen D, Riddle DL. The TOR pathway interacts with the insulin signaling pathway to regulate C. elegans larval development, metabolism and life span. Development 2004; 131(16):3897-3906.
75. Bosher JM, Dufourcq P, Sookhareea S et al. RNA interference can target pre-mRNA: consequences for gene expression in a Caenorhabditis elegans operon. Genetics 1999; 153(3):1245-1256.
76. Simmer F, Moorman C, van der Linden AM et al. Genome-wide RNAi of C. elegans using the hypersensitive rrf-3 strain reveals novel gene functions. PLoS Biol 2003; 1(1):E12.
77. Lee SS, Lee RY, Fraser AG et al. A systematic RNAi screen identifies a critical role for mitochondria in C. elegans longevity. Nat Genet 2003; 33(1):40-48.
78. Fraser AG, Kamath RS, Zipperlen P et al. Functional genomic analysis of C. elegans chromosome I by systematic RNA interference. Nature 2000; 408(6810):325-330.
79. Hay N, Sonenberg N. Upstream and downstream of mTOR. Genes Dev 2004; 18(16):1926-1945.
80. Ruvinsky I, Sharon N, Lerer T et al. Ribosomal protein S6 phosphorylation is a determinant of cell size and glucose homeostasis. Genes Dev 2005; 19(18):2199-2211.
81. Rhoads RE et al. Mechanism and regulation of translation in C. elegans (January 28, 2006), WormBook, ed. The C. elegans Research Community, WormBook, doi/10.1895/wormbook.1.63.1, http://www.wormbook.org.
82. Henderson ST, Bonafe M, Johnson TE. daf-16 protects the nematode Caenorhabditis elegans during food deprivation. J Gerontol A Biol Sci Med Sci 2006; 61(5):444-460.
83. Syntichaki P, Troulinaki K, Tavernarakis N. eIF4E function in somatic cells modulates ageing in Caenorhabditis elegans. Nature 2007; 445(7130):922-926.
84. Pan KZ, Palter JE, Rogers AN et al. Inhibition of mRNA translation extends lifespan in Caenorhabditis elegans. Aging Cell 2007; 6(1):111-119.
85. Tohyama D, Yamaguchi A, Yamashita T. Inhibition of a eukaryotic initiation factor (eIF2Bdelta/F11A3.2) during adulthood extends lifespan in Caenorhabditis elegans. FASEB J 2008; 22(12):4327-4337.
86. Hamilton B, Dong Y, Shindo M et al. A systematic RNAi screen for longevity genes in C. elegans. Genes Dev 2005; 19(13):1544-1555.
87. Curran SP, Ruvkun G. Lifespan regulation by evolutionarily conserved genes essential for viability. PLoS Genet 2007; 3(4):e56.
88. Chen D, Pan KZ, Palter JE et al. Longevity determined by developmental arrest genes in Caenorhabditis elegans. Aging Cell 2007; 6(4):525-533.

89. Okimoto R, Macfarlane JL, Clary DO et al. The mitochondrial genomes of two nematodes, Caenorhabditis elegans and Ascaris suum. Genetics 1992; 130(3):471-498.
90. Anderson LL, Mao X, Scott BA et al. Survival from hypoxia in C. elegans by inactivation of aminoacyl-tRNA synthetases. Science 2009; 323(5914):630-633.
91. Koumenis C, Naczki C, Koritzinsky M et al. Regulation of protein synthesis by hypoxia via activation of the endoplasmic reticulum kinase PERK and phosphorylation of the translation initiation factor eIF2alpha. Mol Cell Biol 2002; 22(21):7405-7416.
92. Harding HP, Zhang Y, Ron D. Protein translation and folding are coupled by an endoplasmic-reticulum-resident kinase. Nature 1999; 397(6716):271-274.
93. Lee AH, Iwakoshi NN, Anderson KC et al. Proteasome inhibitors disrupt the unfolded protein response in myeloma cells. Proc Natl Acad Sci USA 2003; 100(17):9946-9951.
94. Malhotra JD, Kaufman RJ. The endoplasmic reticulum and the unfolded protein response. Semin Cell Dev Biol 2007; 18(6):716-731.
95. Kaufman RJ. Regulation of mRNA translation by protein folding in the endoplasmic reticulum. Trends Biochem Sci 2004; 29(3):152-158.
96. Kirkwood TB, Austad SN. Why do we age? Nature 2000; 408(6809):233-238.
97. Arantes-Oliveira N, Berman JR, Kenyon C. Healthy animals with extreme longevity. Science 2003; 302(5645):611.
98. Hsin H, Kenyon C. Signals from the reproductive system regulate the lifespan of C. elegans. Nature 1999; 399(6734):362-366.
99. Feng J, Bussiere F, Hekimi S. Mitochondrial electron transport is a key determinant of life span in Caenorhabditis elegans. Dev Cell 2001; 1(5):633-644.
100. Ewbank JJ, Barnes TM, Lakowski B et al. Structural and functional conservation of the Caenorhabditis elegans timing gene clk-1. Science 1997; 275(5302):980-983.
101. Sheaffer KL, Updike DL, Mango SE. The Target of Rapamycin pathway antagonizes pha-4/FoxA to control development and aging. Curr Biol 2008; 18(18):1355-1364.
102. Lakowski B, Hekimi S. The genetics of caloric restriction in Caenorhabditis elegans. Proc Natl Acad Sci USA 1998; 95(22):13091-13096.
103. Wang Y, Tissenbaum HA. Overlapping and distinct functions for a Caenorhabditis elegans SIR2 and DAF-16/FOXO. Mech Ageing Dev 2006; 127(1):48-56.
104. Lin SJ, Defossez PA, Guarente L. Requirement of NAD and SIR2 for life-span extension by calorie restriction in Saccharomyces cerevisiae. Science 2000; 289(5487):2126-2128.
105. Rogina B, Helfand SL. Sir2 mediates longevity in the fly through a pathway related to calorie restriction. Proc Natl Acad Sci USA 2004; 101(45):15998-16003.
106. Wood JG, Rogina B, Lavu S et al. Sirtuin activators mimic caloric restriction and delay ageing in metazoans. Nature 2004; 430(7000):686-689.
107. Gems D, Sutton AJ, Sundermeyer ML et al. Two pleiotropic classes of daf-2 mutation affect larval arrest, adult behavior, reproduction and longevity in Caenorhabditis elegans. Genetics 1998; 150(1):129-155.
108. Dever TE. Translation initiation: adept at adapting. Trends Biochem Sci 1999; 24(10):398-403.
109. Talloczy Z, Jiang W, Virgin HWt et al. Regulation of starvation- and virus-induced autophagy by the eIF2alpha kinase signaling pathway. Proc Natl Acad Sci USA 2002; 99(1):190-195.
110. Py BF, Boyce M, Yuan J. A critical role of eEF-2K in mediating autophagy in response to multiple cellular stresses. Autophagy 2009; 5(3):393-396.
111. Ptushkina M, Malys N, McCarthy JE. eIF4E isoform 2 in Schizosaccharomyces pombe is a novel stress-response factor. EMBO Rep 2004; 5(3):311-316.
112. Hsu AL, Murphy CT, Kenyon C. Regulation of aging and age-related disease by DAF-16 and heat-shock factor. Science 2003; 300(5622):1142-1145.
113. Garigan D, Hsu AL, Fraser AG et al. Genetic analysis of tissue aging in Caenorhabditis elegans: a role for heat-shock factor and bacterial proliferation. Genetics 2002; 161(3):1101-1112.
114. Yokoyama K, Fukumoto K, Murakami T et al. Extended longevity of Caenorhabditis elegans by knocking in extra copies of hsp70F, a homolog of mot-2 (mortalin)/mthsp70/Grp75. Febs Letters 2002; 516(1-3):53-57.
115. Walker GA, Lithgow GJ. Lifespan extension in C-elegans by a molecular chaperone dependent upon insulin-like signals. Aging Cell 2003; 2(2):131-139.
116. Tatar M, Khazaeli AA, Curtsinger JW. Chaperoning extended life. Nature 1997; 390(6655):30.
117. Horwitz J. Alpha-crystallin can function as a molecular chaperone. Proc Natl Acad Sci USA 1992; 89(21):10449-10453.
118. Clark JI, Muchowski PJ. Small heat-shock proteins and their potential role in human disease. Curr Opin Struct Biol 2000; 10(1):52-59.
119. Link CD. Transgenic invertebrate models of age-associated neurodegenerative diseases. Mech Ageing Dev 2001; 122(14):1639-1649.

120. Brignull HR, Morley JF, Morimoto RI. The stress of misfolded proteins: C. elegans models for neurodegenerative disease and aging. Adv Exp Med Biol 2007; 594:167-189.
121. Link CD, Fonte V, Hiester B et al. Conversion of green fluorescent protein into a toxic, aggregation-prone protein by C-terminal addition of a short peptide. J Biol Chem 2006; 281(3):1808-1816.
122. Cohen E, Bieschke J, Perciavalle RM et al. Opposing activities protect against age-onset proteotoxicity. Science 2006; 313(5793):1604-1610.
123. Morley JF, Brignull HR, Weyers JJ et al. The threshold for polyglutamine-expansion protein aggregation and cellular toxicity is dynamic and influenced by aging in Caenorhabditis elegans. Proc Natl Acad Sci USA 2002; 99(16):10417-10422.
124. Ross CA, Poirier MA. Opinion: What is the role of protein aggregation in neurodegeneration? Nat Rev Mol Cell Biol 2005; 6(11):891-898.
125. Arrasate M, Mitra S, Schweitzer ES et al. Inclusion body formation reduces levels of mutant huntingtin and the risk of neuronal death. Nature 2004; 431(7010):805-810.
126. Fonte V, Kapulkin V, Taft A et al. Interaction of intracellular beta amyloid peptide with chaperone proteins. Proc Natl Acad Sci USA 2002; 99(14):9439-9444.
127. Fonte V, Kipp DR, Yerg J, 3rd et al. Suppression of in vivo beta-amyloid peptide toxicity by overexpression of the HSP-16.2 small chaperone protein. J Biol Chem 2008; 283(2):784-791.
128. Bandhakavi S, Xie H, O'Callaghan B et al. Hsf1 activation inhibits rapamycin resistance and TOR signaling in yeast revealed by combined proteomic and genetic analysis. PLoS ONE 2008; 3(2):e1598.
129. Levine B, Klionsky DJ. Development by self-digestion: Molecular mechanisms and biological functions of autophagy. Dev Cell 2004; 6(4):463-477.
130. Tsukada M, Ohsumi Y. Isolation and characterization of autophagy-defective mutants of Saccharomyces cerevisiae. FEBS Lett 1993; 333(1-2):169-174.
131. Melendez A, Talloczy Z, Seaman M et al. Autophagy genes are essential for dauer development and life-span extension in C. elegans. Science 2003; 301(5638):1387-1391.
132. Hars ES, Qi H, Ryazanov AG et al. Autophagy regulates ageing in C. elegans. Autophagy 2007; 3(2):93-95.
133. Jia KL, Levine B. Autophagy is required for dietary restriction-mediated life span extension in C-elegans. Autophagy 2007; 3(6):597-599.
134. Toth ML, Sigmond T, Borsos E et al. Longevity pathways converge on autophagy genes to regulate life span in Caenorhabditis elegans. Autophagy 2008; 4(3):330-338.
135. Hansen M, Chandra A, Mitic LL et al. A role for autophagy in the extension of lifespan by dietary restriction in C. elegans. PLoS Genet 2008; 4(2):e24.
136. Morck C, Pilon MC. elegans feeding defective mutants have shorter body lengths and increased autophagy. BMC Dev Biol 2006; 6:39.
137. Kang C, You YJ, Avery L. Dual roles of autophagy in the survival of Caenorhabditis elegans during starvation. Genes Dev 2007; 21(17):2161-2171.
138. Tavernarakis N, Pasparaki A, Tasdemir E et al. The effects of p53 on whole organism longevity are mediated by autophagy. Autophagy 2008; 4(7):870-873.
139. Arum O, Johnson TE. Reduced expression of the Caenorhabditis elegans p53 ortholog cep-1 results in increased longevity. J Gerontol A Biol Sci Med Sci 2007; 62(9):951-959.
140. Bauer JH, Chang C, Morris SN et al. Expression of dominant-negative Dmp53 in the adult fly brain inhibits insulin signaling. Proc Natl Acad Sci USA 2007; 104(33):13355-13360.
141. Houthoofd K, Braeckman BP, Johnson TE et al. Life extension via dietary restriction is independent of the Ins/IGF-1 signalling pathway in Caenorhabditis elegans. Exp Gerontol 2003; 38(9):947-954.
142. Nixon RA, Yang DS, Lee JH. Neurodegenerative lysosomal disorders: a continuum from development to late age. Autophagy 2008; 4(5):590-599.
143. Martinez-Vicente M, Cuervo AM. Autophagy and neurodegeneration: when the cleaning crew goes on strike. Lancet Neurol 2007; 6(4):352-361.
144. Cherra SJ, Chu CT. Autophagy in neuroprotection and neurodegeneration: A question of balance. Future Neurol 2008; 3(3):309-323.
145. Jaeger PA, Wyss-Coray T. All-you-can-eat: autophagy in neurodegeneration and neuroprotection. Mol Neurodegener 2009; 4(1):16.
146. Ravikumar B, Vacher C, Berger Z et al. Inhibition of mTOR induces autophagy and reduces toxicity of polyglutamine expansions in fly and mouse models of Huntington disease. Nat Genet 2004; 36(6):585-595.
147. Iwata A, Christianson JC, Bucci M et al. Increased susceptibility of cytoplasmic over nuclear polyglutamine aggregates to autophagic degradation. Proc Natl Acad Sci USA 2005; 102(37):13135-13140.
148. Jia K, Hart AC, Levine B. Autophagy genes protect against disease caused by polyglutamine expansion proteins in Caenorhabditis elegans. Autophagy 2007; 3(1):21-25.
149. Khan LA, Yamanaka T, Nukina N. Genetic impairment of autophagy intensifies expanded polyglutamine toxicity in Caenorhabditis elegans. Biochem Biophys Res Commun 2008; 368(3):729-735.

150. Florez-McClure ML, Hohsfield LA, Fonte G et al. Decreased insulin-receptor signaling promotes the autophagic degradation of beta-amyloid peptide in C. elegans. Autophagy 2007; 3(6):569-580.
151. Nixon RA, Wegiel J, Kumar A et al. Extensive involvement of autophagy in Alzheimer disease: an immuno-electron microscopy study. J Neuropathol Exp Neurol 2005; 64(2):113-122.
152. Singleton AB, Farrer M, Johnson J et al. alpha-Synuclein locus triplication causes Parkinson's disease. Science 2003; 302(5646):841.
153. Shashidharan P, Good PF, Hsu A et al. TorsinA accumulation in Lewy bodies in sporadic Parkinson's disease. Brain Res 2000; 877(2):379-381.
154. McLean PJ, Kawamata H, Shariff S et al. TorsinA and heat shock proteins act as molecular chaperones: suppression of alpha-synuclein aggregation. J Neurochem 2002; 83(4):846-854.
155. Ozelius LJ, Hewett JW, Page CE et al. The early-onset torsion dystonia gene (DYT1) encodes an ATP-binding protein. Nat Genet 1997; 17(1):40-48.
156. Caldwell GA, Cao S, Sexton EG et al. Suppression of polyglutamine-induced protein aggregation in Caenorhabditis elegans by torsin proteins. Hum Mol Genet 2003; 12(3):307-319.
157. Hamamichi S, Rivas RN, Knight AL et al. Hypothesis-based RNAi screening identifies neuroprotective genes in a Parkinson's disease model. Proc Natl Acad Sci USA 2008; 105(2):728-733.
158. van Ham TJ, Thijssen KL, Breitling R et al. C. elegans model identifies genetic modifiers of alpha-synuclein inclusion formation during aging. PLoS Genet 2008; 4(3):e1000027.
159. Rehling P, Darsow T, Katzmann DJ et al. Formation of AP-3 transport intermediates requires Vps41 function. Nat Cell Biol 1999; 1(6):346-353.
160. Vartiainen S, Pehkonen P, Lakso M et al. Identification of gene expression changes in transgenic C. elegans overexpressing human alpha-synuclein. Neurobiol Dis 2006; 22(3):477-486.
161. McNaught KS, Jenner P. Proteasomal function is impaired in substantia nigra in Parkinson's disease. Neurosci Lett 2001; 297(3):191-194.
162. McNaught KS, Mytilineou C, Jnobaptiste R et al. Impairment of the ubiquitin-proteasome system causes dopaminergic cell death and inclusion body formation in ventral mesencephalic cultures. J Neurochem 2002; 81(2):301-306.
163. Gustafsson AB, Gottlieb RA. Eat your heart out: Role of autophagy in myocardial ischemia/reperfusion. Autophagy 2008; 4(4):416-421.
164. Chu CT. Eaten alive: autophagy and neuronal cell death after hypoxia-ischemia. Am J Pathol 2008; 172(2):284-287.
165. Samokhvalov V, Scott BA, Crowder CM. Autophagy protects against hypoxic injury in C. elegans. Autophagy 2008; 4(8):1034-1041.
166. Takacs-Vellai K, Vellai T, Puoti A et al. Inactivation of the autophagy gene bec-1 triggers apoptotic cell death in C. elegans. Curr Biol 2005; 15(16):1513-1517.
167. Codogno P, Meijer AJ. Autophagy and signaling: their role in cell survival and cell death. Cell Death Differ 2005; 12 Suppl 2:1509-1518.
168. Toth ML, Simon P, Kovacs AL et al. Influence of autophagy genes on ion-channel-dependent neuronal degeneration in Caenorhabditis elegans. J Cell Sci 2007; 120(Pt 6):1134-1141.
169. Samara C, Syntichaki P, Tavernarakis N. Autophagy is required for necrotic cell death in Caenorhabditis elegans. Cell Death Differ 2008; 15(1):105-112.
170. Vellai T, Toth ML, Kovacs AL. Janus-faced autophagy: a dual role of cellular self-eating in neurodegeneration? Autophagy 2007; 3(5):461-463.
171. Jung T, Grune T. The proteasome and its role in the degradation of oxidized proteins. IUBMB Life 2008; 60(11):743-752.
172. Sarikas A, Xu X, Field LJ et al. The cullin7 E3 ubiquitin ligase: a novel player in growth control. Cell Cycle 2008; 7(20):3154-3161.
173. Ghazi A, Henis-Korenblit S, Kenyon C. Regulation of Caenorhabditis elegans lifespan by a proteasomal E3 ligase complex. Proc Natl Acad Sci USA 2007; 104(14):5947-5952.
174. Li W, Gao B, Lee SM et al. RLE-1, an E3 ubiquitin ligase, regulates C. elegans aging by catalyzing DAF-16 polyubiquitination. Dev Cell 2007; 12(2):235-246.
175. Yun C, Stanhill A, Yang Y et al. Proteasomal adaptation to environmental stress links resistance to proteotoxicity with longevity in Caenorhabditis elegans. Proc Natl Acad Sci USA 2008; 105(19):7094-7099.
176. Kahn NW, Rea SL, Moyle S et al. Proteasomal dysfunction activates the transcription factor SKN-1 and produces a selective oxidative-stress response in Caenorhabditis elegans. Biochem J 2008; 409(1):205-213.
177. Stanhill A, Haynes CM, Zhang Y et al. An arsenite-inducible 19S regulatory particle-associated protein adapts proteasomes to proteotoxicity. Mol Cell 2006; 23(6):875-885.
178. Keiper BD, Lamphear BJ, Deshpande AM et al. Functional characterization of five eIF4E isoforms in Caenorhabditis elegans. J Biol Chem 2000; 275(14):10590-10596.

Chapter 9

Mitochondrial Protein Quality Control Systems in Aging and Disease

Karin Luce, Andrea C. Weil and Heinz D. Osiewacz*

Abstract

Preserving the integrity of proteins, biomolecules prone to molecular damage, is a fundamental function of all biological systems. Impairments in protein quality control (PQC) may lead to degenerative processes, such as aging and various disorders and diseases. Fortunately, cells contain a hierarchical system of pathways coping protein damage. Specific molecular pathways detect misfolded proteins and act either to unfold or degrade them. Degradation of proteins generates peptides and amino acids that can be used for remodelling of impaired pathways and cellular functions. At increased levels of cellular damage whole organelles can be removed via autophagy, a process that depends on the activity of lysosomes. In addition, cells may undergo apoptosis, a form of programmed cell death, which in single-cellular and lower multicellular organisms can lead to death of the individual.

Molecular damage of cellular compartments is mainly caused by reactive oxygen species (ROS). ROS is generated via different cellular pathways and frequently arises in the mitochondrial electron transport chain as a by-product of oxygenic energy transduction. Consequently, mitochondrial proteins are under high risk to become damaged. Perhaps for this reason mitochondria contain a very efficient PQC system that keeps mitochondrial proteins functional as long as damage does not reach a certain threshold and the components of this system themselves are not excessively damaged. The mitochondrial PQC system consists of chaperones that counteract protein aggregation through binding and refolding misfolded polypeptides and of membrane-bound and soluble ATP-dependent proteases that are involved in degradation of damaged proteins. During aging and in neurodegenerative diseases components of this PQC system, including Lon protease present in the mitochondrial matrix, become functionally impaired.

In this chapter we summarise the current knowledge of cellular quality control systems with special emphasis on the role of the mitochondrial PQC system and its impact on biological aging and disease.

Introduction

Cellular proteins may become damaged at any point during a cell's life cycle, resulting in misfolding and aggregation, which eventually leads to cellular dysfunction, degenerative disorders, disease and aging. Living systems contain various molecular pathways to repair, stabilize and refold affected proteins. These pathways control the levels of damaged and misfolded proteins, thus keeping the cells and tissues functional. However, if nonfunctional proteins accumulate wihin the cell (e.g., due to an increased production of damaged proteins), they are detected and degraded by a proteolytic system.[1]

*Corresponding Author: Heinz D. Osiewacz—Johann Wolfgang Goethe University, Faculty for Biosciences and Cluster of Excellence Macromolecular Complexes, Institute of Molecular Biosciences, Max-von-Laue-Str. 9, 60438 Frankfurt, Germany.
Email: osiewacz@bio.uni-frankfurt.de

Protein Metabolism and Homeostasis in Aging, edited by Nektarios Tavernarakis.

Table 1. Overview of known mitochondrial chaperones and proteases involved in protein quality control and the maintenance of mitochondrial functions

Component	Name	Localization	Function
Protease	Lon	Matrix	Degradation/processing
Protease	Clp (Protease Subunit)	Matrix	Degradation
Protease	i-AAA	Inner membrane	Degradation/processing
Protease	m-AAA	Inner membrane	Degradation/processing
Protease	Rhomboid	Inner membrane	Processing
Protease	HtrA2/Omi	Intermembrane space	Processing
Chaperone	Hsp60	Matrix	Folding/refolding
Chaperone	Hsp70	Matrix	Folding/refolding
Chaperone	Hsp78	Matrix	Folding/refolding
Chaperone	ClpX (Hsp100)	Matrix	Folding/refolding
Chaperone	Prohibitins	Inner membrane	Binding/processing

In the cytoplasm of all eukaryotes the proteasome represents the major proteolytic protein quality control (PQC) system. Degradation by the proteasome requires ATP to drive the unfolding and translocation of a polypeptide substrate into a chamber of sequestered proteolytic active sites.[2] Substrates are either recognized by the proteasome through post-translational ubiquitination of one or more lysine residues, a process requiring several enzymatic steps, or reach the proteasome directly by chaperone-mediated delivery. In addition to the cytoplasm, the proteasome is also active in the nucleus, but is not seen in any other cellular organelles. Damaged proteins in these compartments can undergo proteasome independent degradation, or may be still be degraded by the proteasome, if transported to a site where proteasome is present. This mechanism is utilised for proteins of the endoplasmatic reticulum (ER) that require the activity of an ER-associated degradation (ERAD) system.[3,4]

Damaged proteins can also be degraded by autophagy, a process defined as the main mechanism of selective removal of whole organelles via the activity of lysosomes.[5] In recent years it has become clear that removal of mitochondria by autophagy, termed 'mitophagy', is the main mechanism of mitochondrial protein turnover.[5] The activation of mitochondrial permeability transition and the subsequent loss of the membrane potential is a signal for the induction of mitophagy. However, before mitophagy becomes induced, a threshold level of intra-mitochondrial protein damage and protein aggregation needs to be reached. Below this level a number of chaperones and ATP-dependent proteases control protein homeostasis in the mitochondria and thus prevent or, at least, delay the induction of mitophagy (Table 1).

Two major pathways are known to remove damaged proteins from mitochondria and avoid irreversible protein aggregation. Firstly, chaperones recognize misfolded proteins, bind and try to refold them into their native form thus reconstituting protein function. However, if damage is excessive and refolding fails to occur, proteases of the AAA family (**A**TPases **a**ssociated with a variety of cellular **a**ctivities) eliminate misfolded proteins by degrading them into small fragments and amino acids (Fig. 1), which are transported out of mitochondria and are then available for de novo protein synthesis in the cytoplasm.

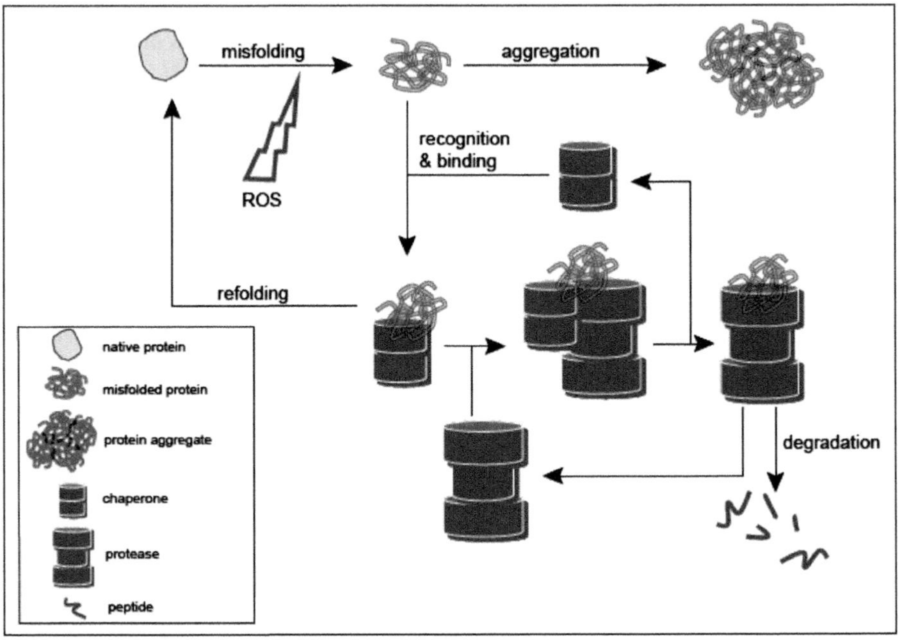

Figure 1. Functional model of chaperones and proteases and their interactions responsible for mitochondrial PQC. Misfolded proteins are recognized by and bound to mitochondrial chaperones. Chaperones work by rescuing misfolded polypeptides and rendering their natural task. If refolding into the native protein conformation fails, substrate polypeptides are translocated to a mitochondrial protease in a chaperone-dependent mechanism. Subsequently misfolded proteins are degraded by proteases into small peptide fragments.

If the proteolytic system is overwhelmed by excessive protein damage the defective mitochondria can either be functionally restored by fusion with intact mitochondria or parts of them with highly damaged proteins separated from the less damaged part of the mitochondrial network.[6,7] Parts with excessive damage are subsequently removed by autophagy, a process preventing the release of pro-apoptotic proteins from damaged mitochondria and avoiding cellular death.[8]

It thus appears that an efficient, hierarchical protein quality control system is active in mitochondria. Efficiency of this system appears to be crucial in particular because mitochondria, as the powerhouse of the eukaryotic cell, are a major generator of reactive oxygen species (ROS) which are able to damage all types of biomolecules including proteins, lipids and nucleic acids. It remains an important issue to elucidate how the system is regulated and what determines the limits of damage this system can successfully handle in order to protect biological systems from degeneration. Part of the answer seems to be linked to multiple functions of the individual components of the PQC system.

Mitochondrial Chaperones Are Necessary for Regulated Mitochondrial PQC

Mitochondrial chaperones, proteins that assist in protein folding, are able to remove damaged proteins through an energy-driven ATP-dependent mechanism which requires recognition, binding and refolding of damaged proteins and the subsequent release of the reconstituted and functional proteins. Up to now, a number of major chaperones involved in mitochondrial

PQC have been identified. They belong to the Hsp60, Hsp70 and Hsp100 family of heat shock proteins.

Mitochondrial Hsp70 (mtHsp70) is a major component of the mitochondrial import machinery. It binds to newly imported mitochondrial proteins from the TOM/TIM complex and is involved in the subsequent folding of the imported proteins into its active conformation.[9] Organisms lacking mtHsp70 are not viable, demonstrating the importance of this protein.[10] Additionally, mtHsp70 is involved in efficient degradation of damaged proteins by proteases. Work on *Saccharomyces cerevisiae* revealed this function of mtHsp70 and of a cochaperone, Mdj1, for degradation of misfolded proteins by the matrix protease Pim1, the yeast homologue of the human mitochondrial Lon protease.[11] Using a hybrid protein of cytochrome b2 fused to a mutated form of cytosolic mouse DHFR (dihydrofolate reductase), mtHsp70 and Mdj1 were demonstrated to be required for the proteolytic breakdown identifying them as essential components of the proteolytic machinery within the mitochondrial matrix. Both chaperones are sufficient for preventing irreversible protein aggregation and allowing efficient proteolysis via Pim1. Additional data revealed that- apart from Lon- proteolysis of misfolded proteins via the membrane bound m-AAA protease is also dependent on mtHsp70.[12]

The physiological function of Mcx1, the mitochondrial homologue of the bacterial ClpX/Hsp100 chaperone, is still not clear.[13] Mcx1 localizes to the soluble matrix fraction of mitochondria and is peripherally associated with the inner mitochondrial membrane. Under normal conditions, deletion of Mcx1 has no obvious phenotype. However, experiments discovering genes required for mitochondrial biogenesis in *S. cerevisiae* identified an increased frequency of spontaneous loss of the mitochondrial DNA (mtDNA) indicating an impact of Mcx1 on mtDNA stability.[14] In prokaryotes, ClpX forms one of the known ATP-dependent regulatory subunits of the Clp protease, is required for N-rule substrates, proteins bearing an amino-terminal destabilizing amino acid and promotes substrate unfolding for proteolysis.[15,16] Under heat shock conditions, Mcx1 accumulates in mitochondria suggesting a role in heat shock response by refolding damaged proteins.

Hsp78, a mitochondrial homologue of the bacterial Hsp100/ClpB chaperone, is involved in mitochondrial PQC as well. Importantly, Hsp78 can substitute for chaperone functions of mutated mtHsp70.[17] Under heat shock conditions, amounts of Hsp78 are strongly increased in mitochondria.[13] Hsp78 is crucial for maintenance of respiratory competence and for mitochondrial genome integrity under severe temperature stress (thermo tolerance). Given that mitochondrial protein synthesis is a thermo sensitive process, reactivation of mitochondrial protein synthesis after heat stress strongly depends on the presence of Hsp78.[18] Recently, Hsp78 was demonstrated to be responsible for resolubilization of protein aggregates generated by heat stress in vivo.[19] Especially aggregated mtHsp70 was shown to be resolubilized by Hsp78 demonstrating a unique role of the chaperone in mitochondria. Thus, resolubilization of protein aggregates in mitochondria is dependent on Hsp78.

Another mitochondrial chaperone involved in mitochondrial PQC is Hsp60, which is responsible for proper folding of proteins imported into mitochondria.[20] In yeast, deletion of hsp60 is lethal.[21] In humans, Hsp60 has been linked to neurodegenerative diseases. Mutations in the gene, coding for Hsp60, have been identified in patients with an autosomal dominant form of hereditary spastic paraplegia (SPG13), a neurodegenerative disorder characterized by a progressive paraparesis of the lower limps.[22] The mutated Hsp60 protein displays a reduced chaperonin activity, a decreased ATPase activity and exhibits a strongly reduced capacity to promote folding of e.g., denaturated malate dehydrogenase.[23] Furthermore, mutation of Hsp60 is accompanied by a decreased activity of mitochondrial matrix proteases, highlighting the close relationship and interaction between mitochondrial chaperones and proteases.[24]

Under (cellular) conditions of accumulating protein damage (e.g., stress, diseases and aging), chaperone activity is very often "overcharged" and mitochondria accumulate irreversible protein aggregates. To prevent this dangerous process, these organelles contain proteases that degrade misfolded and damaged proteins and recycle amino acids into the pool for protein synthesis.

Role of the Mitochondrial Proteases in Maintaining Mitochondrial Functions

Chloroplasts and mitochondria, both organelles of an endosymbiontic origin, contain several PQC proteases that are highly conserved from prokaryotes to higher eukaryotes. In mitochondria these proteases are present in all sub-compartments (Fig. 2). The mitochondrial matrix contains two soluble protease systems, Lon and Clp protease.[25,26] In the inner mitochondrial membrane three membrane-integrated proteases, the i-AAA, m-AAA and rhomboid protease PARL/Pcp1 are present.[27-29] Additionally, the intermembrane space harbours the HtrA2/Omi protease and a number of different oligopeptidases.[30-32]

The mechanism(s) by which mitochondrial proteases recognize and degrade misfolded polypeptide chains is rather unclear and speculative. There are three possible theories to describe this process:

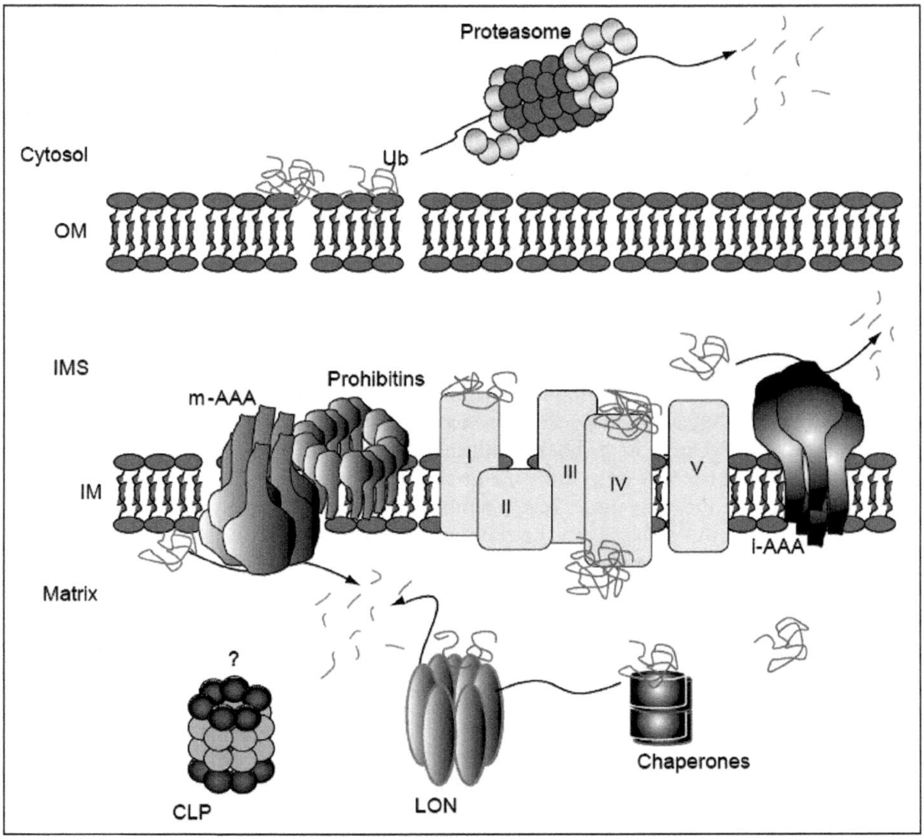

Figure 2. Scheme displaying key proteases involved in mitochondrial PQC. In the mitochondrial matrix Lon and Clp protease are present. While the molecular function of Clp is still unclear, Lon preferably degrades soluble misfolded polypeptides in an ATP-dependent mechanism. In the inner mitochondrial membrane, the i-AAA and the m-AAA proteases are involved in degradation and processing of respiratory chain subunits. The activity of the m-AAA protease appears to be strongly regulated by prohibitins. Damaged proteins of the outer mitochondrial membrane are probably ubiquitinated and applied to the ubiquitin-proteasome degradation system. OM: outer mitochondrial membrane; IMS: intermembrane space; IM: inner mitochondrial membrane.

i. Similar to the ubiquitin-proteasome system in the cytoplasm, substrate polypeptides may be marked with ubiquitin making them accessible for proteases by an ubiquitin recognition site.
ii. Chaperones may bind to damaged proteins and deliver these substrates to the proteases via binding of the chaperone to the protease.
iii. Proteases may contain specific substrate-binding sites for distinct recognition.

Until now no ubiquitin-conjugation system has been identified in mitochondria. However, recent data demonstrate the ubiquitination of a restricted set of mitochondrial proteins.[33] Furthermore it was found, that a subunit of the F_1F_0-ATPase is ubiquitinated, retranslocated from the inner mitochondrial membrane to the outer membrane and degraded by the proteasome in the cytoplasm under the control of the cytosolic chaperone Hsp90. Similar observations were reported for the pro-apoptotic intermembrane space protein endoG.[34] However, at the moment it is unclear how mitochondrial proteins are recognized, marked with ubiquitin and translocated through the mitochondrial membranes. At this time, it is rather theoretical and highly speculative how important this way of protein degradation for removal of damaged mitochondrial proteins is.

Mitochondrial protein degradation by chaperone-mediated binding and proteolysis or by direct substrate-binding of the proteases appears to be of greater relevance. This type of proteolysis is mediated by both membrane bound proteases like i-AAA and m-AAA and the soluble Lon and Clp proteases in the matrix which are indispensable for efficient removal of damaged mitochondrial proteins (Fig. 2). Point mutations or deletion of one of the genes coding for these proteases leads to severe phenotypes, neurodegenerative diseases (ND) and death. Some proteases contain specific substrate-binding sites, others not. For example, the i-AAA protease recognizes misfolded polypeptides by conserved helices in the proteolytic and AAA domains.[35] In contrast, proteolysis via Lon and m-AAA protease is dependent on the chaperone mtHSP70, indicating a chaperone-mediated substrate delivery. In the following paragraphs the most important mitochondrial specific proteases will be characterized and their molecular functions discussed in more detail.

Membrane-Bound AAA Proteases

Probably, the most fundamental function of the inner mitochondrial membrane is the energy conversion via oxidative phosphorylation.[36] To sustain proper activity of the respiratory chain, assembly of nuclear and mitochondrial encoded subunits of the complexes as well as the degradation of excess or damaged and non-assembled subunits need to be well coordinated. Two ATP-dependent membrane-bound proteases are known to be involved in the energetically unfavourable dislocation of non-assembled inner-membrane proteins.[37-39] Both are members of the highly conserved AAA family.[40] In yeast and humans, the i-AAA protease forms a homo-oligomeric complex of Yme1 subunits and exposes its catalytic domain to the intermembrane space.[37,41] The m-AAA protease is oriented towards the mitochondrial matrix.[42] In yeast, this complex is hetero-oligomeric, composed of Yta10 and Yta12. In humans, the orthologue consists of homo- or hetero-oligomeric complexes containing paraplegin and Afg3L2 or Afg3L2 only.[29,43,44] The activity of the m-AAA protease is regulated by prohibitins, a eukaryotic protein family with potential roles in senescence and tumor suppression.[45]

Both proteases exert versatile functions including the proteolytic degradation of polypeptides but also processing and translocation of substrate proteins. Most information about the function of the proteases has been obtained from *S. cerevisiae*. Using various reporter proteins, overlapping substrate specificities have been demonstrated.[37,38,42] However, so far no endogenous substrate has been identified which can be degraded by both proteases. Apart from the proteolytic activity, both AAA proteases display chaperone-like activity since they can bind unfolded polypeptides and suppress their aggregation.[29,38,39]

The loss of function of these two AAA proteases leads to severe phenotypes including neurodegenerative diseases in mammals. However, in most cases the molecular basis of these phenotypes is still unclear. Yeast cells lacking Yme1 are respiratory deficient (*petite*) and show an increased rate of DNA escape from mitochondria to the nucleus.[27] Moreover, cells require Yme1

Table 2. Identified mitochondrial substrate proteins of mitochondrial proteases

Protease	Substrate	Function	Reference
Lon	aconitase	Tricarboxylic acid cycle	95-97
Lon	Ilv1	Isoleucine biosynthesis	97
Lon	Ilv2	Isoleucine and valine biosynthesis	97
Lon	Lsc1	Tricarboxylic acid cycle	97
Lon	Lys4	Lysine biosynthesis	97
Lon	Cox IV	Respiratory chain subunit (complex IV)	85
Lon	Cox V	Respiratory chain subunit (complex IV)	85
Lon	Cox II*	Respiratory chain subunit (complex IV)	85
Lon	MPPa	Processing of imported proteins	48
Lon	StAR	Cholesterol transfer protein	128
m-AAA	MrpL32*	Subunit of mitochondrial ribosomes	65
m-AAA	Opa1*	Mitochondrial fusion	108
m-AAA	Ccp1	ROS scavenging	68
m-AAA	Cox I	Respiratory chain subunit (complex IV)	61
m-AAA	Cob	Respiratory chain subunit (complex III)	61
m-AAA	Subunit 6	ATP synthase (complex V)	29,62,64
m-AAA	Subunit 8	ATP synthase (complex V)	29,64
m-AAA	Subunit 9	ATP synthase (complex V)	29,62,64
i-AAA	Cox II	Respiratory chain subunit (complex IV)	41,49,50,69
i-AAA	Opa1*	Mitochondrial fusion	111,110
i-AAA	Nde1	External NADH dehydrogenase	51
i-AAA	Prohibitins	Cell proliferation and cristae morphogenesis	31

*Substrates are only proteolytically processed, not degraded.

for respiratory growth at high temperatures and display a growth defect on glucose-containing media at low temperatures.[27] Deletion of yme1 also leads to an altered mitochondrial morphology with an accumulation of punctuate and grossly swollen mitochondria.[46] Since it was found that yeast mitochondria lacking Yme1 contain considerably increased phosphatidylserine decarboxylase 1 (Psd1) activity and accordingly an increased level of phosphatidylethanolamine, the altered protein/lipid composition might be an explanation for the abnormal mitochondrial morphology.[47,48] One substrate of the i-AAA protease has been identified as the mitochondrially encoded Cox2 protein, a subunit of cytochrome c oxidase (Cox) (Table 2).[41,49,50] Non-assembled Cox2 has been shown to be rapidly degraded by Yme1 but remains stable in Δyme1 yeast cells. In cells lacking Yme1 the increased stability of nonfunctional Cox2 is supported by the drastically reduced number of peptides derived from proteolysis of Cox2.[51] Deletion of Yme1 also results in the complete loss of peptides derived from degradation of Nde1, an external NADH dehydrogenase.[51]

Peptides generated by the proteolytic breakdown of mitochondrial proteins are believed to be involved in signalling pathways between mitochondria and the nucleus. Therefore, deletion of yme1 in yeast leads to the up-regulation of nuclear genes with function in mitochondrial gene expression and the biogenesis of the respiratory chain.[52] This signalling pathway would allow the re-adjustment of the reduced activity of the F_1F_0-ATPase observed in the absence of the i-AAA protease.[52-55] So far, only little is known about the human i-AAA protease Yme1L1. It is speculated that Yme1L1 is also involved in the pathogenesis of the neurodegenerative disorder hereditary spastic paraplegia (HSP).[56] The protein is ubiquitously expressed and localized to mitochondria.[56] Expression of the human homologue in Δyme1 yeast cells restores growth on nonfermentable carbon sources at elevated temperatures pointing to a functional conservation.[57]

As observed for Yme1-deficient yeast cells, cells lacking either Yta10 or Yta12 or expressing a proteolytically inactive variant of the m-AAA protease are respiratory deficient.[55,58-61] In the absence of Yta10 or Yta12 the activities of the cytochrome c oxidase, the bc_1-complex and the ATP synthase are reduced.[58,59] This is consistent with the finding that the expression of the mitochondrial genes cox1 (subunit 1 of the cytochrome c oxidase) and cob (cytochrome b of the cytochrome bc_1-complex) are under control of proteolytic processing by the m-AAA protease.[61,62] Interestingly, similar observations have been reported for the soluble matrix protease Lon, pointing to overlapping functions not solely between the integral membrane-bound proteases but also between those proteases and soluble proteases.[63] Furthermore, the m-AAA protease is involved in the assembly of the subunits of the ATP synthase.[29,62,64] An additional reason for the respiratory deficiency might be the impaired processing of MrpL32, a subunit of mitochondrial ribosomes.[65] Processing of the N-terminal targeting sequence is essential for ribosomal assembly. Thus, translation of the mitochondrial encoded proteins is affected demonstrating a crucial role of the m-AAA protease in the control of mitochondrial protein synthesis. Interestingly, this role introduces the possibility of a negative-feedback regulation. The accumulation of nonnative proteins, for example in stressed or aged cells, might compete with MrpL32 for binding to the m-AAA protease. Hence, processing of MrpL32 is impaired resulting in reduced mitochondrial translation. In mammals the loss of function of Paraplegin or Afg3L2 results in the neurodegenerative disease HSP demonstrating the importance of this protease for mitochondrial PQC as well.[22,66]

Recently it has been shown that membrane-bound AAA proteases are involved in the ATP-dependent translocation of proteins.[67,68] In yeast, maturation of the nuclear encoded cytochrome c peroxidase 1 (Ccp1) depends on the successive action of the m-AAA and rhomboid protease Pcp1. The m-AAA protease is required for dislocation and positioning of the precursor form of Ccp1 which can then be cleaved by the rhomboid protease. A similar function has been proposed for the i-AAA protease.[67] Mitochondrial import of the human polynucleotide phosphorylase (PNPase) depends on Yme1 which binds to the precursor form and mediates its translocation across the outer membrane. It is tempting to speculate that more substrates exist which depend on the ATP-dependent dislocation instead of the proteolytic activity of the AAA proteases. This is supported by lethality of the simultaneous deletion of both the i-AAA and the m-AAA protease in yeast cells which can be restored by the expression of a proteolytically inactive form of Yta10 in Δyme1 yeast strains.[38,69]

Apart from the AAA proteases a number of additional proteases exist in mitochondrial membranes performing mainly processing functions. The role of mammalian mitochondrial HtrA2/Omi in protein quality control is still unclear. The precursor form of the serine protease HtrA2/Omi is temporarily anchored to the inner membrane where it undergoes proteolytic maturation and is released into the intermembrane space. Loss of its protease activity results in susceptibility toward Parkinson's disease, in which mitochondrial dysfunction and impairment of protein folding and degradation are key features.[30]

The rhomboid protease is an evolutionary conserved intramembrane serine protease.[70] The yeast rhomboid Pcp1 has two known substrates, Ccp1 and Mgm1.[71-73] Both proteins span the inner membrane once and are completely or partially released after cleavage by Pcp1. As mentioned before, maturation of Ccp1 also depends on the activity of the m-AAA protease suggesting the existence of additional substrates which require the coordinated activity of several proteases for maturation.

The fact that the loss of AAA proteases causes severe phenotypes in yeast and leads to axonal degeneration in mice as well as in humans underlines the importance of these proteases in PQC to maintain mitochondrial function.[66,74]

Soluble Matrix Proteases

In the mitochondrial matrix a wide range of metabolic functions need to be coordinated. In this subcompartment, the mtDNA as well as the citric acid cycle and other important pathways like lipid synthesis are present. The regulated transcription and translation as well as the removal of damaged enzymes and regulatory proteins need to be tightly controlled. Two soluble protease complexes, Lon and Clp, are part of the corresponding machinery.

Clp protease is an ATP-dependent serine protease catalyzing unfolding and degradation of damaged and specifically tagged proteins.[75] The protease is composed of the proteolytic subunit ClpP and the chaperones ClpX or ClpA. The functional form of ClpP is a tetradecamer forming a stable double ring structure. Each ring contains seven ClpP subunits. The chaperones ClpA and ClpX are members of the AAA protein family, a group of unfoldases that catalyze restructure and translocation of proteins and other biomolecules. They form ATP-dependent hexameric ring structures. The active protease is formed by stacking the chaperone hexamer to the ends of ClpP, resembling a proteasome-like structure.[16,76] While the ClpAP complex is only present in prokaryotes, ClpXP is highly conserved within mitochondria of eukaryotes, except the lower eukaryote *S. cerevisiae* which lacks the proteolytic subunit ClpP.[77] In prokaryotes, ClpXP degrades short lived regulatory proteins and is an important component of the prokaryotic PQC and heat shock response.[78] While most eukaryotes contain the protease and chaperone subunit within mitochondria, almost nothing is known regarding their specific biological function and their substrate binding and specifity. Interestingly, inactive ClpP leads to the complete loss of the mitochondrial unfolded protein response (mtUPR) during heat stress in *Caenorhabditis elegans*.[79] UPR is a mechanism involved in restoring normal cellular functions of the cell after heat stress via the activation of signalling pathways leading to increased transcription of nuclear genes coding for molecular chaperones involved in protein folding. Inhibition of the proteolytic activity of Clp rapidly blocks induction of genes of the hsp60 and hsp70 family, suggesting that Clp protease has a direct role in mtUPR signalling to the nucleus. Noticeable, like ClpP, the membrane-bound i-AAA protease seems to be involved in human mtUPR as well.[80] However, the identity of Clp protease substrates for mtUPR is still missing. Moreover, other functions and substrates of Clp need to be investigated more extensively.

The best characterized and probably most important matrix protease involved in mitochondrial PQC is Lon, a highly conserved ATP-dependent serine/lysine protease of the AAA^+ family.[81] The ATPase and the protease domain are located on the same polypeptide chain, whereas ATP hydrolysis and proteolytic activity are tightly coupled. Lon is only active in a homo-oligomeric state. In yeast mitochondria, a heptameric-ring structure was demonstrated by electron microscopy.[82] Apart from its proteolytic function, Lon is also involved in protein complex assembly and in the maintenance of mitochondrial DNA.[83-85] In yeast, proteolysis mediated by Lon is required for the expression of mitochondrial intron-containing transcripts.[63,86] Lon is part of the mitochondrial nucleoid, a complex maintaining mtDNA, as well.[63,86] Additionally, human Lon binds to single-stranded DNA oligonucleotides corresponding to a TG-rich region that overlaps the heavy strand- and light-strand promoters of human mtDNA.[63,87] Interestingly, binding of Lon to mtDNA stimulates the ATPase activity as well as the protease activity.[63] Thus, Lon may be responsible to regulate both, protein turnover and mtDNA function.

Lon is important under normal survival and proliferation as well as for the defence against oxidative stress induced protein damage. For instance, Lon is significantly upregulated under different stress conditions, like heat stress.[25] The importance of Lon for cellular and mitochondrial viability is supported by dramatically disordered phenotypes within organisms being deficient for Lon protease. Yeast cells lacking Pim1, the *S. cerevisiae* homologue of Lon, exhibit large mtDNA deletions and accumulation of electron dense aggregates.[25,88] As mtDNA codes for respiratory chain subunits, Pim1 deletion mutants are unable to grow on nonfermentable carbon sources, leading to a *petite* phenotype.[88] Deletion of Lon in higher eukaryotes seems to be lethal. Even RNAi-treated cell cultures offering decreased Lon protein amounts are characterized by broad mitochondrial dysfunctions. In human lung fibroblasts down-regulation of Lon was reported to impair mitochondrial structure, function and causes cell death by activation of apoptotic pathways.[89] Also in plants mitochondrial Lon protease is required to maintain mitochondrial function. Mitochondria isolated from *Arabidopsis thaliana* Lon deletion mutants displayed lower capacity for respiration, the activity of key enzymes of the tricarboxylic acid cycle was significantly reduced and mitochondria had an aberrant morphology, respectively.[90] These severe phenotypes of Lon deficiency in different eukaryotes demonstrate the importance of this protease to maintain proper mitochondrial and cellular function by selectively eliminating misfolded, damaged and certain short-lived regulatory proteins.

Lon protease, as well as the membrane-bound i-AAA protease, is also involved in coping ER stress induced mitochondrial dysfunctions. In rats, it was found that expression and assembly of Cox are disturbed under ER stress. The signalling pathway improves assembly of respiratory chain complexes and sustains mitochondrial function by induction of Lon and i-AAA proteases.[85]

So far, the signal targeting a protein to be degraded by Lon is unidentified. *In organello* degradation of imported reporter proteins containing different structural properties in *S. cerevisiae* revealed that a nonnative amino-terminal exposed protein segment in combination with a limited unfolding capacity mediated by chaperones (mtHsp70) accounts for the selectivity of Lon protease for damaged or misfolded proteins.[91] Protein turnover mediated by Lon is likely modulated or regulated by factors that affect the enzymatic activity of Lon and/or the conformational state.[92] For example, mitochondrial processing peptidase subunit a (MPPa) is degraded by Lon only in its non-assembled form since active heterodimeric MPPa complexed with MPPb is stable.[93] Importantly, the proteolytic activity of Lon is dependent on the availability of ATP and thus on the energy status of mitochondria. ATP binding stimulates proteolysis, while ADP inhibits protease function.[94]

Up to now, only a limited number of endogenous substrates of eukaryotic Lon protease have been identified (Table 2). Most substrate proteins are components of the respiratory chain, the tricarboxylic acid cycle (TCA) and amino acid biosynthesis pathway. It has been proposed that Lon protease is the key proteolytic machinery in the mitochondrial matrix mainly degrading oxidatively damaged proteins. For example mitochondrial aconitase, an iron-sulfur cluster containing protein, has been identified as a substrate of Lon in various model systems.[95-97] Aconitase acts as a key enzyme of the TCA cycle as well as in the maintenance of mtDNA.[98] The enzyme is highly vulnerable to oxidative damage, suffering a loss of activity and consequently mitochondrial damage and dysfunction. However, comparing the number of known substrates of Lon protease to the assumed number of mitochondrial matrix proteins, the substrate knowledge is still a puzzle. Probably Lon degrades not a restricted set of proteins, but rather a wide range of unfolded proteins.

Role of the Membrane-Bound AAA Proteases on Diseases, Apoptosis and Aging

During the years of investigations of mitochondrial PQC it turned out that proteases are highly important to sustain mitochondrial function. Loss of function or down-regulation of mitochondrial proteases has been reported to be linked to neurodegenerative diseases (ND) and aging. Parkinson's disease, autosomal dominant optic atrophy (ADOA) and hereditary spastic paraplegia (HSP) are prominent examples.[99]

For instance, autosomal dominant optic atrophy strongly depends on the activity of mitochondrial membrane-bound proteases regulating mitochondrial morphology. The mitochondrial morphology is controlled by several proteins involved in fusion and fission.[100] One of them is Opa1 (Mgm1 in yeast), a dynamin-like GTPase in the inner mitochondrial membrane, as described above. Mutation of Opa1 leads to ADOA.[101,102] The balanced formation of long membrane-bound and shorter soluble isoforms appears to be crucial for mitochondrial morphology, respiratory competence and the maintenance of mtDNA.[103] This balance strongly depends on the activity of membrane-bound proteases, demonstrating a prominent example for mitochondrial PQC mediated by mitochondrial proteases. As the downregulation of Opa1 causes mitochondrial fragmentation and alters the shape of the cristae, Opa1 has been linked to apoptosis and aging.[104] It has been suggested that during apoptosis the protease-mediated balance between long and shorter Opa1 isoforms is disrupted.[105,106] This leads to the unravelling of the cristae junctions allowing pro-apoptotic factors like cytochrome c and HtrA2/Omi to diffuse into the intermembrane space and finally into the cytosol. So far three proteases have been linked to the processing of Opa1. In yeast, the rhomboid protease Pcp1 cleaves a subset of the long Mgm1 isoform at their transmembrane domain to produce the shorter Mgm1 isoform which is released into the intermembrane space.[71,73] Similarly, the mammalian rhomboid PARL seems to be involved in the generation of the soluble Opa1 isoforms, although the action of PARL on Opa1 could be indirect.[107] In yeast, PARL is able to functionally replace Pcp1 and process Mgm1.[73] In contrast, the m-AAA protease Paraplegin has also been proposed to stimulate Opa1 processing.[108] Surprisingly, neither PARL-deficient cells nor Paraplegin-deficient cells show impaired Opa1 processing.[109] To identify proteases which are capable of cleaving Opa1, Opa1 was expressed in yeast cells where either yeast or mammalian m-AAA proteases mediate Opa1 cleavage. Finally, knockdown of human Yme1L has also been shown to constitutively regulate a subset of Opa1 isoforms.[110,111] Therefore, the question which mitochondrial protease(s) mediates Opa1 cleavage seems to remain unsolved.

In mammals, nonsense mutations in Paraplegin, a subunit of the m-AAA protease, were found to cause an autosomal recessive form of HSP, a genetically heterogeneous disease that is characterized by progressive and cell-specific axonal degeneration.[22,112] Paraplegin and Afg3L2 assemble into a high molecular mass complex in the inner mitochondrial membrane.[113] This complex formation is impaired in fibroblasts from HSP patients leading to a decreased activity of the respiratory complexes, decreased ATP levels and an increased sensitivity towards oxidative stress.[22,113] Likewise Paraplegin-deficient mice show a slow progressive and cell-specific axonal degeneration.[74] The lack of Paraplegin leads to enlarged and morphologically abnormal mitochondria. In contrast, mice lacking functional Afg3L2 are affected by a severe neuromuscular syndrome.[66] They contain swollen giant mitochondria with damaged cristae and the activity of the respiratory complexes I and III is strongly reduced resulting in reduced ATP synthesis. Only recently, mice carrying different dosages of Afg3L2 and Paraplegin have been generated showing that the homo- and hetero-oligomeric complexes are largely redundant.[114]

These tissue-specific consequences of the loss of Paraplegin or Afg3L2 are somehow puzzling since both isoenzymes should be widely expressed in mitochondria from all tissues. A possible explanation could be the variation of the relative abundance of the subunits between different tissues and their versatile assembly as homo- or hetero-oligomeric complexes.[44,114] Additionally, homo-oligomeric complexes composed of Afg3L2 seem to have different substrate specificities than hetero-oligomeric complexes containing Afg3L2 and Paraplegin.[109] It cannot be excluded that other yet unknown substrates exist which contribute to the tissue-specificity of the disease even though this is somehow contradictory to the observed redundancy of the homo- and hetero-oligomeric complexes. However, initial efforts towards a gene-replacement therapy have successfully halted the progression of the degeneration in the peripheral nerves of Paraplegin-deficient mice pointing to a promising path to pursue.[115]

Mitochondrial Lon Protease Activity and Aging

Chaperones and molecular proteases control the balance of functional, correctly folded and misfolded proteins by removing damaged polypeptides and toxic protein aggregates from cells, delaying the onset and the outcome of disease and aging. While mutation or loss of mitochondrial membrane-bound proteases very often leads to neurodegenerative disease, soluble matrix proteases are more responsible for preventing protein aggregation and, thus, avoiding premature aging. However, a few neurodegenerative diseases have been described so far which negatively affect the activity of matrix proteases. For instance patients with SPG13 display reduced transcript and protein levels of the mitochondrial matrix PQC proteases Lon and Clp.[24]

Numerous investigations describe the accumulation of aggregation-prone misfolded proteins during lifetime and a misbalance between functional and nonfunctional proteins. For instance, proteasome activity in the cytoplasm is known to decline during aging, amplifying the misbalance and, additionally, decreasing the pool of free accessible amino acids for protein synthesis.[116] Significantly this observation is linked to an age-related increased ROS production and thus to the free radical theory of aging, suggesting that aging is the result of an accumulation of molecular damage via ROS.[117]

Lon protease is known to play an essential role in the degradation of damaged proteins in mitochondria. Different age-related declines in transcript and protein levels and activity of Lon have been reported so far, demonstrating an important function of this protease in the aging process. In skeletal muscle samples transcript levels of aged mice were four-fold decreased compared to samples from young individuals.[118] Interestingly, this observation was completely prevented when animals were fed a calorie restricted food, an intervention known to slow down protein damage. In a comparable study an approximate five-fold reduction was observed in mitochondrial Lon protein levels and activity in old mice compared to young animals.[119] Additionally, old mice exhibit higher levels of oxidatively damaged proteins, particularly oxidized aconitase, probably due to the diminished protease activity of Lon to degrade misfolded proteins.[119] Selective down-regulation of Lon in human lung fibroblasts by antisense RNA was reported to lead to the formation of abnormal mitochondria and induce apoptosis. Giant mitochondria with large empty vacuoles, mitochondria filled with electron dense inclusion bodies and miniature mitochondria were found in these cells.[89] These pathological phenotypes are similar to what is observed in aging tissues, suggesting that Lon downregulation may contribute to the aging effect. Studies on isolated mitochondria of rat liver samples demonstrated a decline in Lon protease activity in senescent samples as well.[120] This observation is supported by the observation that oxidative protein modifications increase in old mitochondria. Interestingly, in these studies overall Lon protein amounts did not change during aging, indicating that Lon activity might be abolished in aged cells either by reduced disposability of ATP or by oxidative protein damage of Lon itself and consequently loss of proteolytic activity.

Investigations studying Lon protease activity in different mammalian tissues and in different organisms suggest a tissue- and organism-specific regulation during aging. In rat heart an age-related increase in mitochondrial transcript levels was found, while protease activity remained unchanged.[121] Similar observations were reported for the filamentous fungus *Podospora anserina*, an experimentally tractable short lived organism with a well defined mitochondrial etiology of aging.[122-124] Like in rats, transcript levels of lon were found to be increased in aged *P. anserina* cultures. However, protein abundance and activity of Lon protease did not change.[96] These discrepancies may be explained by the fact that (1) Lon protein synthesis in the cytoplasm is limited in aged samples and that (2) Lon protein import into aged mitochondria is disturbed. The first explanation is supported by different studies describing a decline in the general protein synthesis during aging in different model systems.[125] Global control of protein synthesis is generally achieved by changes in the phosphorylation state of initiation factors of their regulators. This phosphorylation state is known to change during aging, suggesting that protein synthesis constitutes a basic cellular process targeted by signalling pathways that modulate aging. The second explanation is supported by observations, that mitochondrial membrane potential and ATP amounts decrease during aging.[126] A high membrane potential is required for efficient import of nuclear encoded mitochondrial proteins by the import machinery complex TOM/TIM into the mitochondrial matrix.[127] Either reduced translation and/or inefficient transport into mitochondria might explain the unchanged Lon protease activity in aged rat heart tissue and *P. anserina* cultures while transcript is increased.

Recent experimental interventions into Lon expression in *P. anserina* support the importance of a functional Lon protease in aging. Constitutive over-expression of lon resulted in transgenic strains with increased ATP-dependent serine protease activity.[96] These strains display lower levels of carbonylated and carboxymethylated proteins, reduced secretion of hydrogen peroxide and a higher resistance against exogenous oxidative stress. Moreover, they are characterized by an extended health span (i.e., lifespan without impairments of vital functions like growth and fertility). Also mitochondrial morphology and the respiratory chain composition are not abolished. Additionally, mitochondrial respiration is increased in old cultures over-expressing lon in comparison to wild-type samples. Collectively, these data demonstrate a beneficial effect of increasing Lon protease abundance on stress resistance and organism aging and provide mechanistic clues suggesting that increased degradation of damaged mitochondrial proteins leads to a reduction of endogenous oxidative stress, an increased tolerance against exogenous oxidative stress and an increased life span. Thus, overexpression of lon appears to be a successful intervention and supports the important role of the mitochondrial PQC into the aging process.

Conclusion

Mitochondrial protein quality control processes have now been recognized to play crucial roles in maintaining mitochondrial functions. In particular chaperones, proteases and their regulation and interactions appear to play a fundamental role in preventing protein aggregation and maintaining mitochondrial requirement for cellular survival. Novel functional links between these systems are indicating that the understanding of how mitochondrial PQC works is still at the beginning. The highly important regulation and flawless function of all different components of this system is of fundamental importance, as decrease or inactivation of one of the proteases or chaperones leads to severe phenotypes, diseases and finally in cell death. However, details about their regulation and interaction remain to be investigated in more detail. Further analysis of proteases and chaperones involved in mitochondrial PQC, e.g., through the identification of their targets and substrates will almost certainly provide new insights into the regulation of this system. Importantly, the role of PQC systems on aging and disease needs to be thoroughly investigated, given that they are characterized by the loss or change of mitochondrial protease and chaperone activities, leading to increased damage and loss of mitochondrial function. Other pathways (e.g., mtDNA repair, mitochondrial dynamics) also need to be explored more carefully. Different tasks of mitochondrial PQC should be accomplished in different model systems to obtain a detailed overview about PQC mechanisms and interactions of the individual components of this hierarchical network. Such knowledge will be a good starting point for translational research with the ultimate goal to hold up degenerative processes leading to severe impairments and diseases and gain a basic understanding of the mechanistic processes leading to biological aging.

References

1. Mogk A, Haslberger T, Tessarz P et al. Common and specific mechanisms of AAA+ proteins involved in protein quality control. Biochem Soc Trans 2008; 36(Pt 1):120-125.
2. Pickart CM, Cohen RE. Proteasomes and their kin: proteases in the machine age. Nat Rev Mol Cell Biol 2004; 5(3):177-187.
3. Bukau B, Weissman J, Horwich A. Molecular chaperones and protein quality control. Cell 2006; 125(3):443-451.
4. Mogk A, Schmidt R, Bukau B. The N-end rule pathway for regulated proteolysis: prokaryotic and eukaryotic strategies. Trends Cell Biol 2007; 17(4):165-172.
5. Tolkovsky AM. Mitophagy. Biochim Biophys Acta 2009; 1793(9):1508-1515.
6. Okamoto K, Shaw JM. Mitochondrial morphology and dynamics in yeast and multicellular eukaryotes. Annu Rev Genet 2005; 39:503-536.
7. Hoppins S, Lackner L, Nunnari J. The machines that divide and fuse mitochondria. Annu Rev Biochem 2007; 76:751-780.
8. Kim I, Rodriguez-Enriquez S, Lemasters JJ. Selective degradation of mitochondria by mitophagy. Arch Biochem Biophys 2007; 462(2):245-253.
9. Neupert W, Herrmann JM. Translocation of proteins into mitochondria. Annu Rev Biochem 2007; 76:723-749.
10. Craig EA, Kramer J, Kosic-Smithers J. SSC1, a member of the 70-kDa heat shock protein multigene family of Saccharomyces cerevisiae, is essential for growth. Proc Natl Acad Sci USA 1987; 84(12):4156-4160.

11. Wagner I, Arlt H, Van Dyck L et al. Molecular chaperones cooperate with PIM1 protease in the degradation of misfolded proteins in mitochondria. EMBO J 1994; 13(21):5135-5145.
12. Savel'ev AS, Novikova LA, Kovaleva IE et al. ATP-dependent proteolysis in mitochondria. m-AAA protease and PIM1 protease exert overlapping substrate specificities and cooperate with the mtHsp70 system. J Biol Chem 1998; 273(32):20596-20602.
13. Van Dyck L, Dembowski M, Neupert W et al. Mcx1p, a ClpX homologue in mitochondria of Saccharomyces cerevisiae. FEBS Lett 1998; 438(3):250-254.
14. Hess DC, Myers CL, Huttenhower C et al. Computationally driven, quantitative experiments discover genes required for mitochondrial biogenesis. PLoS Genet 2009; 5(3):e1000407.
15. Gottesman S, Maurizi MR, Wickner S. Regulatory subunits of energy-dependent proteases. Cell 1997; 91(4):435-438.
16. Gottesman S, Wickner S, Maurizi MR. Protein quality control: triage by chaperones and proteases. Genes Dev 1997; 11(7):815-823.
17. Schmitt M, Neupert W, Langer T. Hsp78, a Clp homologue within mitochondria, can substitute for chaperone functions of mt-hsp70. EMBO J 1995; 14(14):3434-3444.
18. Schmitt M, Neupert W, Langer T. The molecular chaperone Hsp78 confers compartment-specific thermotolerance to mitochondria. J Cell Biol 1996; 134(6):1375-1386.
19. von Janowsky B, Major T, Knapp K et al. The disaggregation activity of the mitochondrial ClpB homolog Hsp78 maintains Hsp70 function during heat stress. J Mol Biol 2006; 357(3):793-807.
20. Yoneda T, Benedetti C, Urano F et al. Compartment-specific perturbation of protein handling activates genes encoding mitochondrial chaperones. J Cell Sci 2004; 117(Pt 18):4055-4066.
21. Reading DS, Hallberg RL, Myers AM. Characterization of the yeast HSP60 gene coding for a mitochondrial assembly factor. Nature 1989; 337(6208):655-659.
22. Casari G, De Fusco M, Ciarmatori S et al. Spastic paraplegia and OXPHOS impairment caused by mutations in paraplegin, a nuclear-encoded mitochondrial metalloprotease. Cell 1998; 93(6):973-983.
23. Bross P, Naundrup S, Hansen J et al. The Hsp60-(p.V98I) mutation associated with hereditary spastic paraplegia SPG13 compromises chaperonin function both in vitro and in vivo. J Biol Chem 2008; 283(23):15694-15700.
24. Hansen J, Corydon TJ, Palmfeldt J et al. Decreased expression of the mitochondrial matrix proteases Lon and ClpP in cells from a patient with hereditary spastic paraplegia (SPG13). Neuroscience 2008; 153(2):474-482.
25. Van Dyck L, Pearce DA, Sherman F. PIM1 encodes a mitochondrial ATP-dependent protease that is required for mitochondrial function in the yeast Saccharomyces cerevisiae. J Biol Chem 1994; 269(1):238-242.
26. Corydon TJ, Bross P, Holst HU et al. A human homologue of Escherichia coli ClpP caseinolytic protease: recombinant expression, intracellular processing and subcellular localization. Biochem J 1998; 331(Pt 1):309-316.
27. Thorsness PE, White KH, Fox TD. Inactivation of YME1, a member of the ftsH-SEC18-PAS1-CDC48 family of putative ATPase-encoding genes, causes increased escape of DNA from mitochondria in Saccharomyces cerevisiae. Mol Cell Biol 1993; 13(9):5418-5426.
28. Lemberg MK, Menendez J, Misik A et al. Mechanism of intramembrane proteolysis investigated with purified rhomboid proteases. EMBO J 2005; 24(3):464-472.
29. Arlt H, Tauer R, Feldmann H et al. The YTA10-12 complex, an AAA protease with chaperone-like activity in the inner membrane of mitochondria. Cell 1996; 85(6):875-885.
30. Vande Walle L, Lamkanfi M, Vandenabeele P. The mitochondrial serine protease HtrA2/Omi: an overview. Cell Death Differ 2008; 15(3):453-460.
31. Kambacheld M, Augustin S, Tatsuta T et al. Role of the novel metallopeptidase Mop112 and saccharolysin for the complete degradation of proteins residing in different subcompartments of mitochondria. J Biol Chem 2005; 280(20):20132-20139.
32. Buchler M, Tisljar U, Wolf DH. Proteinase yscD (oligopeptidase yscD). Structure, function and relationship of the yeast enzyme with mammalian thimet oligopeptidase (metalloendopeptidase, EP 24.15). Eur J Biochem 1994; 219(1-2):627-639.
33. Margineantu DH, Emerson CB, Diaz D et al. Hsp90 inhibition decreases mitochondrial protein turnover. PLoS ONE 2007; 2(10):e1066.
34. Radke S, Chander H, Schafer P et al. Mitochondrial protein quality control by the proteasome involves ubiquitination and the protease Omi. J Biol Chem 2008; 283(19):12681-12685.
35. Graef M, Seewald G, Langer T. Substrate recognition by AAA+ ATPases: distinct substrate binding modes in ATP-dependent protease Yme1 of the mitochondrial intermembrane space. Mol Cell Biol 2007; 27(7):2476-2485.
36. Poyton RO, McEwen JE. Crosstalk between nuclear and mitochondrial genomes. Annu Rev Biochem 1996; 65:563-607.

37. Leonhard K, Herrmann JM, Stuart RA et al. AAA proteases with catalytic sites on opposite membrane surfaces comprise a proteolytic system for the ATP-dependent degradation of inner membrane proteins in mitochondria. EMBO J 1996; 15(16):4218-4229.
38. Leonhard K, Guiard B, Pellecchia G et al. Membrane protein degradation by AAA proteases in mitochondria: extraction of substrates from either membrane surface. Mol Cell 2000; 5(4):629-638.
39. Leonhard K, Stiegler A, Neupert W et al. Chaperone-like activity of the AAA domain of the yeast Yme1 AAA protease. Nature 1999; 398(6725):348-351.
40. Hanson PI, Whiteheart SW. AAA+ proteins: have engine, will work. Nat Rev Mol Cell Biol 2005; 6(7):519-529.
41. Weber ER, Hanekamp T, Thorsness PE. Biochemical and functional analysis of the YME1 gene product, an ATP and zinc-dependent mitochondrial protease from S. cerevisiae. Mol Biol Cell 1996; 7(2):307-317.
42. Pajic A, Tauer R, Feldmann H et al. Yta10p is required for the ATP-dependent degradation of polypeptides in the inner membrane of mitochondria. FEBS Lett 1994; 353(2):201-206.
43. Kremmidiotis G, Gardner AE, Settasatian C et al. Molecular and functional analyses of the human and mouse genes encoding AFG3L1, a mitochondrial metalloprotease homologous to the human spastic paraplegia protein. Genomics 2001; 76(1-3):58-65.
44. Koppen M, Metodiev MD, Casari G et al. Variable and tissue-specific subunit composition of mitochondrial m-AAA protease complexes linked to hereditary spastic paraplegia. Mol Cell Biol 2007; 27(2):758-767.
45. Steglich G, Neupert W, Langer T. Prohibitins regulate membrane protein degradation by the m-AAA protease in mitochondria. Mol Cell Biol 1999; 19(5):3435-3442.
46. Campbell CL, Tanaka N, White KH et al. Mitochondrial morphological and functional defects in yeast caused by yme1 are suppressed by mutation of a 26S protease subunit homologue. Mol Biol Cell 1994; 5(8):899-905.
47. Nebauer R, Schuiki I, Kulterer B et al. The phosphatidylethanolamine level of yeast mitochondria is affected by the mitochondrial components Oxa1p and Yme1p. FEBS J 2007; 274(23):6180-6190.
48. Campbell CL, Thorsness PE. Escape of mitochondrial DNA to the nucleus in yme1 yeast is mediated by vacuolar-dependent turnover of abnormal mitochondrial compartments. J Cell Sci 1998; 111(Pt 16):2455-2464.
49. Pearce DA, Sherman F. Degradation of cytochrome oxidase subunits in mutants of yeast lacking cytochrome c and suppression of the degradation by mutation of yme1. J Biol Chem 1995; 270(36):20879-20882.
50. Nakai T, Yasuhara T, Fujiki Y et al. Multiple genes, including a member of the AAA family, are essential for degradation of unassembled subunit 2 of cytochrome c oxidase in yeast mitochondria. Mol Cell Biol 1995; 15(8):4441-4452.
51. Augustin S, Nolden M, Muller S et al. Characterization of peptides released from mitochondria: evidence for constant proteolysis and peptide efflux. J Biol Chem 2005; 280(4):2691-2699.
52. Arnold I, Wagner-Ecker M, Ansorge W et al. Evidence for a novel mitochondria-to-nucleus signalling pathway in respiring cells lacking i-AAA protease and the ABC-transporter Mdl1. Gene 2006; 367:74-88.
53. Kominsky DJ, Brownson MP, Updike DL et al. Genetic and biochemical basis for viability of yeast lacking mitochondrial genomes. Genetics 2002; 162(4):1595-1604.
54. Thorsness PE, Fox TD. Nuclear mutations in Saccharomyces cerevisiae that affect the escape of DNA from mitochondria to the nucleus. Genetics 1993; 134(1):21-28.
55. Kolodziejczak M, Gibala M, Urantowka A et al. The significance of Arabidopsis AAA proteases for activity and assembly/stability of mitochondrial OXPHOS complexes. Physiologia Plantarum 2007; 129:135-142.
56. Coppola M, Pizzigoni A, Banfi S et al. Identification and characterization of YME1L1, a novel paraplegin-related gene. Genomics 2000; 66(1):48-54.
57. Shah ZH, Hakkaart GA, Arku B et al. The human homologue of the yeast mitochondrial AAA metalloprotease Yme1p complements a yeast yme1 disruptant. FEBS Lett 2000; 478(3):267-270.
58. Tzagoloff A, Yue J, Jang J et al. A new member of a family of ATPases is essential for assembly of mitochondrial respiratory chain and ATP synthetase complexes in Saccharomyces cerevisiae. J Biol Chem 1994; 269(42):26144-26151.
59. Tauer R, Mannhaupt G, Schnall R et al. Yta10p, a member of a novel ATPase family in yeast, is essential for mitochondrial function. FEBS Lett 1994; 353(2):197-200.
60. Guélin E, Rep M, Grivell LA. Sequence of the AFG3 gene encoding a new member of the FtsH/Yme1/Tma subfamily of the AAA-protein family. Yeast 1994; 10(10):1389-1394.
61. Arlt H, Steglich G, Perryman R et al. The formation of respiratory chain complexes in mitochondria is under the proteolytic control of the m-AAA protease. EMBO J 1998; 17(16):4837-4847.

62. Guélin E, Rep M, Grivell LA. Afg3p, a mitochondrial ATP-dependent metalloprotease, is involved in degradation of mitochondrially-encoded Coxl, Cox3, Cob, Su6, Su8 and Su9 subunits of the inner membrane complexes III, IV and V. FEBS Letters 1996; 381:42-46.
63. Van Dyck L, Neupert W, Langer T. The ATP-dependent PIM1 protease is required for the expression of intron-containing genes in mitochondria. Genes Dev 1998; 12(10):1515-1524.
64. Paul MF, Tzagoloff A. Mutations in RCA1 and AFG3 inhibit F1-ATPase assembly in Saccharomyces cerevisiae. FEBS Lett 1995; 373(1):66-70.
65. Nolden M, Ehses S, Koppen M et al. The m-AAA protease defective in hereditary spastic paraplegia controls ribosome assembly in mitochondria. Cell 2005; 123(2):277-289.
66. Maltecca F, Aghaie A, Schroeder DG et al. The mitochondrial protease AFG3L2 is essential for axonal development. J Neurosci 2008; 28(11):2827-2836.
67. Rainey RN, Glavin JD, Chen HW et al. A new function in translocation for the mitochondrial i-AAA protease Yme1: import of polynucleotide phosphorylase into the intermembrane space. Mol Cell Biol 2006; 26(22):8488-8497.
68. Tatsuta T, Augustin S, Nolden M et al. m-AAA protease-driven membrane dislocation allows intramembrane cleavage by rhomboid in mitochondria. EMBO J 2007; 26(2):325-335.
69. Lemaire C, Hamel P, Velours J et al. Absence of the mitochondrial AAA protease Yme1p restores F_0-ATPase subunit accumulation in an oxa1 deletion mutant of Saccharomyces cerevisiae. J Biol Chem 2000; 275(31):23471-23475.
70. Bier E, Jan LY, Jan YN. Rhomboid, a gene required for dorsoventral axis establishment and peripheral nervous system development in Drosophila melanogaster. Genes Dev 1990; 4(2):190-203.
71. Herlan M, Vogel F, Bornhovd C et al. Processing of Mgm1 by the rhomboid-type protease Pcp1 is required for maintenance of mitochondrial morphology and of mitochondrial DNA. J Biol Chem 2003; 278(30):27781-27788.
72. Esser K, Tursun B, Ingenhoven M et al. A novel two-step mechanism for removal of a mitochondrial signal sequence involves the mAAA complex and the putative rhomboid protease Pcp1. J Mol Biol 2002; 323(5):835-843.
73. McQuibban GA, Saurya S, Freeman M. Mitochondrial membrane remodelling regulated by a conserved rhomboid protease. Nature 2003; 423(6939):537-541.
74. Ferreirinha F, Quattrini A, Pirozzi M et al. Axonal degeneration in paraplegin-deficient mice is associated with abnormal mitochondria and impairment of axonal transport. J Clin Invest 2004; 113(2):231-242.
75. Gottesman S. Proteolysis in bacterial regulatory circuits. Annu Rev Cell Dev Biol 2003; 19:565-587.
76. Kang SG, Ortega J, Singh SK et al. Functional proteolytic complexes of the human mitochondrial ATP-dependent protease, hClpXP. J Biol Chem 2002; 277(23):21095-21102.
77. Yu AY, Houry WA. ClpP: a distinctive family of cylindrical energy-dependent serine proteases. FEBS Lett 2007; 581(19):3749-3757.
78. Keiler KC, Waller PR, Sauer RT. Role of a peptide tagging system in degradation of proteins synthesized from damaged messenger RNA. Science 1996; 271(5251):990-993.
79. Haynes CM, Petrova K, Benedetti C et al. ClpP mediates activation of a mitochondrial unfolded protein response in C. elegans. Dev Cell 2007; 13(4):467-480.
80. Aldridge JE, Horibe T, Hoogenraad NJ. Discovery of genes activated by the mitochondrial unfolded protein response (mtUPR) and cognate promoter elements. PLoS ONE 2007; 2(9):e874.
81. Wang N, Gottesman S, Willingham MC et al. A human mitochondrial ATP-dependent protease that is highly homologous to bacterial Lon protease. Proc Natl Acad Sci USA 1993; 90(23):11247-11251.
82. Stahlberg H, Kutejova E, Suda K et al. Mitochondrial Lon of Saccharomyces cerevisiae is a ring-shaped protease with seven flexible subunits. Proc Natl Acad Sci USA 1999; 96(12):6787-6790.
83. Rep M, van Dijl JM, Suda K et al. Promotion of mitochondrial membrane complex assembly by a proteolytically inactive yeast Lon. Science 1996; 274(5284):103-106.
84. Liu T, Lu B, Lee I et al. DNA and RNA binding by the mitochondrial lon protease is regulated by nucleotide and protein substrate. J Biol Chem 2004; 279(14):13902-13910.
85. Hori O, Ichinoda F, Tamatani T et al. Transmission of cell stress from endoplasmic reticulum to mitochondria: enhanced expression of Lon protease. J Cell Biol 2002; 157(7):1151-1160.
86. Bogenhagen DF, Rousseau D, Burke S. The layered structure of human mitochondrial DNA nucleoids. J Biol Chem 2008; 283(6):3665-3675.
87. Fu GK, Markovitz DM. The human LON protease binds to mitochondrial promoters in a single-stranded, site-specific, strand-specific manner. Biochemistry 1998; 37(7):1905-1909.
88. Suzuki CK, Suda K, Wang N et al. Requirement for the yeast gene LON in intramitochondrial proteolysis and maintenance of respiration. Science 1994; 264(5161):891.
89. Bota DA, Ngo JK, Davies KJ. Downregulation of the human Lon protease impairs mitochondrial structure and function and causes cell death. Free Radic Biol Med 2005; 38(5):665-677.

90. Rigas S, Daras G, Laxa M et al. Role of Lon1 protease in postgerminative growth and maintenance of mitochondrial function in Arabidopsis thaliana. New Phytol 2009; 181(3):588-600.
91. von Janowsky B, Knapp K, Major T et al. Structural properties of substrate proteins determine their proteolysis by the mitochondrial AAA+ protease Pim1. Biol Chem 2005; 386(12):1307-1317.
92. Lee I, Berdis AJ, Suzuki CK. Recent developments in the mechanistic enzymology of the ATP-dependent Lon protease from Escherichia coli: highlights from kinetic studies. Mol Biosyst 2006; 2(10):477-483.
93. Ondrovicova G, Liu T, Singh K et al. Cleavage site selection within a folded substrate by the ATP-dependent lon protease. J Biol Chem 2005; 280(26):25103-25110.
94. Bulteau AL, Szweda LI, Friguet B. Mitochondrial protein oxidation and degradation in response to oxidative stress and aging. Exp Gerontol 2006; 41(7):653-657.
95. Bota DA, Davies KJ. Lon protease preferentially degrades oxidized mitochondrial aconitase by an ATP-stimulated mechanism. Nat Cell Biol 2002; 4(9):674-680.
96. Luce K, Osiewacz HD. Increasing organism health span by enhancing mitochondrial protein quality control. Nat Cell Biol 2009; 11(7):852-858.
97. Major T, von Janowsky B, Ruppert T et al. Proteomic analysis of mitochondrial protein turnover: identification of novel substrate proteins of the matrix protease pim1. Mol Cell Biol 2006; 26(3):762-776.
98. Chen XJ, Wang X, Butow RA. Yeast aconitase binds and provides metabolically coupled protection to mitochondrial DNA. Proc Natl Acad Sci USA 2007; 104(34):13738-13743.
99. Tatsuta T, Langer T. Quality control of mitochondria: protection against neurodegeneration and ageing. EMBO J 2008; 27(2):306-314.
100. Griparic L, van der Bliek AM. The many shapes of mitochondrial membranes. Traffic 2001; 2(4):235-244.
101. Alexander C, Votruba M, Pesch UE et al. OPA1, encoding a dynamin-related GTPase, is mutated in autosomal dominant optic atrophy linked to chromosome 3q28. Nat Genet 2000; 26(2):211-215.
102. Delettre C, Lenaers G, Griffoin JM et al. Nuclear gene OPA1, encoding a mitochondrial dynamin-related protein, is mutated in dominant optic atrophy. Nat Genet 2000; 26(2):207-210.
103. Wong ED, Wagner JA, Gorsich SW et al. The dynamin-related GTPase, Mgm1p, is an intermembrane space protein required for maintenance of fusion competent mitochondria. J Cell Biol 2000; 151(2):341-352.
104. Olichon A, Baricault L, Gas N et al. Loss of OPA1 perturbates the mitochondrial inner membrane structure and integrity, leading to cytochrome c release and apoptosis. J Biol Chem 2003; 278(10):7743-7746.
105. Gottlieb E. OPA1 and PARL keep a lid on apoptosis. Cell 2006; 126(1):27-29.
106. Frezza C, Cipolat S, Martins de Brito O et al. OPA1 controls apoptotic cristae remodeling independently from mitochondrial fusion. Cell 2006; 126(1):177-189.
107. Cipolat S, Rudka T, Hartmann D et al. Mitochondrial rhomboid PARL regulates cytochrome c release during apoptosis via OPA1-dependent cristae remodeling. Cell 2006; 126(1):163-175.
108. Ishihara N, Fujita Y, Oka T et al. Regulation of mitochondrial morphology through proteolytic cleavage of OPA1. EMBO J 2006; 25(13):2966-2977.
109. Duvezin-Caubet S, Koppen M, Wagener J et al. OPA1 processing reconstituted in yeast depends on the subunit composition of the m-AAA protease in mitochondria. Mol Biol Cell 2007; 18(9):3582-3590.
110. Song Z, Chen H, Fiket M et al. OPA1 processing controls mitochondrial fusion and is regulated by mRNA splicing, membrane potential and Yme1L. J Cell Biol 2007; 178(5):749-755.
111. Griparic L, Kanazawa T, van der Bliek AM. Regulation of the mitochondrial dynamin-like protein Opa1 by proteolytic cleavage. J Cell Biol 2007; 178(5):757-764.
112. Casari G, Rugarli E. Molecular basis of inherited spastic paraplegias. Curr Opin Genet Dev 2001; 11(3):336-342.
113. Atorino L, Silvestri L, Koppen M et al. Loss of m-AAA protease in mitochondria causes complex I deficiency and increased sensitivity to oxidative stress in hereditary spastic paraplegia. J Cell Biol 2003; 163(4):777-787.
114. Martinelli P, La Mattina V, Bernacchia A et al. Genetic interaction between the m-AAA protease isoenzymes reveals novel roles in cerebellar degeneration. Hum Mol Genet 2009; 18(11):2001-2013.
115. Pirozzi M, Quattrini A, Andolfi G et al. Intramuscular viral delivery of paraplegin rescues peripheral axonopathy in a model of hereditary spastic paraplegia. J Clin Invest 2006; 116(1):202-208.
116. Chondrogianni N, Stratford FL, Trougakos IP et al. Central role of the proteasome in senescence and survival of human fibroblasts: induction of a senescence-like phenotype upon its inhibition and resistance to stress upon its activation. J Biol Chem 2003; 278(30):28026-28037.
117. Harman D. Aging: a theory based on free radical and radiation chemistry. J Gerontol 1956; 11(3):298-300.

118. Lee CK, Klopp RG, Weindruch R et al. Gene expression profile of aging and its retardation by caloric restriction. Science 1999; 285(5432):1390-1393.
119. Bota DA, Van Remmen H, Davies KJ. Modulation of Lon protease activity and aconitase turnover during aging and oxidative stress. FEBS Lett 2002; 532(1-2):103-106.
120. Bakala H, Delaval E, Hamelin M et al. Changes in rat liver mitochondria with aging. Lon protease-like reactivity and N(epsilon)-carboxymethyllysine accumulation in the matrix. Eur J Biochem 2003; 270(10):2295-2302.
121. Delaval E, Perichon M, Friguet B. Age-related impairment of mitochondrial matrix aconitase and ATP-stimulated protease in rat liver and heart. Eur J Biochem 2004; 271(22):4559-4564.
122. Scheckhuber CQ, Osiewacz HD. Podospora anserina: a model organism to study mechanisms of healthy ageing. Mol Genet Genomics 2008; 280(5):365-374.
123. Osiewacz HD. Molecular analysis of aging processes in fungi. Mutat Res 1990; 237(1):1-8.
124. Osiewacz HD. Aging in fungi: role of mitochondria in Podospora anserina. Mech Ageing Dev 2002; 123(7):755-764.
125. Tavernarakis N. Ageing and the regulation of protein synthesis: a balancing act? Trends Cell Biol 2008; 18(5):228-235.
126. Nicholls DG. Mitochondrial membrane potential and aging. Aging Cell 2004; 3(1):35-40.
127. Kutik S, Guiard B, Meyer HE et al. Cooperation of translocase complexes in mitochondrial protein import. J Cell Biol 2007; 179(4):585-591.
128. Granot Z, Kobiler O, Melamed-Book N et al. Turnover of mitochondrial steroidogenic acute regulatory (StAR) protein by Lon protease: the unexpected effect of proteasome inhibitors. Mol Endocrinol 2007; 21(9):2164-2177.

Chapter 10

$p38^{MAPK}$ in the Senescence of Human and Murine Fibroblasts

Florence Debacq-Chainiaux,* Emmanuelle Boilan, Jérémie Dedessus Le Moutier, Geoffroy Weemaels and Olivier Toussaint

Abstract

Oncogenic and environmental stresses, such as reactive oxygen species, UV radiation etc, can induce premature cellular senescence without critical telomere shortening. The role of the Ras/Raf/ERK signal transduction cascade in this process has been previously established, but recent evidence also indicates a critical role of the p38 MAP kinases pathway. Oncogenic and environmental stresses impinge upon the $p38^{MAPK}$ pathway, suggesting a major role of this pathway in senescence induced by stresses. Prematurely senescent cells are most likely to appear in several age-related pathologies associated with a stressful environment and/or the release of pro-inflammatory cytokines.

Introduction

Replicative senescence has been widely demonstrated for many non-immortalized mammalian cell types capable of proliferating in the adult.[1] Notable exceptions are embryonic stem cell lines and most tumor-derived cells in which telomerase is constitutively activated.[2] Since features of replicative senescence can appear after exposure of both non-immortalized cells and telomerase-immortalized cells to oncogenic stimuli, oxidative stress and/or DNA damaging agents, preliminary definitions of the various phenotypes arising in such conditions must be given, in order to reflect the different trends of the current literature.

Senescence Is the Hardest Word to Say

The narrowest definition of senescence is the irreversible growth arrest that occurs when non-immortalized, nontransformed cells have exausted their replicative potential.[1] This is triggered by critical telomere shortening and rearrangements of the telomere structure, due to the inability to replicate the chromosomal ends. This type of senescence, termed replicative senescence, counts cell generations (definition 1).[3] However most in vitro studies on replicative senescence were performed with cells cultivated at 20% O_2, which accelerates telomere shortening by generating a more severe oxidative stress when compared to the much lower in vivo physiological pressure of O_2.[4]

Other authors expanded the definition of senescence to a functional definition encompassing all kinds of irreversible arrests of nontransformed proliferative cell types induced by cellular stresses such as subcytotoxic oxidative stresses and DNA damaging agents (definition 2).[5-7] Notably most features of replicative senescence including irreversible growth arrest, also appear after exposure of proliferating nonimmortalized, nontransformed cells to cellular stresses. Irreversible growth

*Corresponding Author: Florence Debacq-Chainiaux—University of Namur, Research Unit on Cellular Biology, Rue de Bruxelles, 61, Namur B-5000, Belgium.
Email: florence.chainiaux@fundp.ac.be

Protein Metabolism and Homeostasis in Aging, edited by Nektarios Tavernarakis.

arrest of proliferative cell types induced by these cellular stresses may be called stress-induced senescence-like phenotype, according to definition 1, or stress-induced premature senescence, according to definition 2. Both phrases can be abbreviated as SIPS.[8] In addition, the induction of features of replicative senescence after exposure of cells to oncogenic stimuli, has been called oncongene-induced premature senescence (OIS).[5,9]

SIPS differs from replicative senescence namely it can be telomere-independent. While the ectopic expression of the catalytic subunit of telomerase (hTERT) rescues replicative senescence, oncogenic ras, H_2O_2, ultraviolet light and x-radiation induced similar levels of senescence features in the parental and hTERT-expressing human diploid fibroblasts (HDFs) and none or very limited telomere shortening was detected in the hTERT-expressing cells undergoing SIPS.[10] These data suggest that replicative senescence induced by telomere shortening and SIPS are mediated by at least partially different mechanisms.

The Role of DNA Damage Checkpoint Genes in Senescence

The mechanisms by which SIPS occurs are not yet fully understood. However, results from fibroblasts bearing specific mutations responsible for Ataxia teleangiectasia (AT) and Nijmegen breakage syndrome (NBS) have indicated a biologic relevance or the biological reference of these genes in senescence. AT and NBS patients have a high incidence of cancer and cellular defects including radiosensitivity, chromosome instability and premature senescence.[11] ATM and NBS1, the respective products of the gene mutated in AT and NBS patients, cooperate to activate cell cycle checkpoints triggered by DNA damage. ATM phosphorylates many proteins involved in cell cycle checkpoints and DNA repair. For example, ATM phosphorylates and activates Chk2 in response to ionizing radiation, resulting in the stabilization and activation of p53, induction of $p21^{WAF1}$ expression and cell cycle arrest at G_1 phase. ATM participates in telomere maintenance and senescence signaling originating from telomeres.[12,13] Deficiency of ATM or NBS1 leads to telomere shortening, suggesting that ATM and NBS1 proteins may be involved in regulation of telomere length and/or structure. Although the precise mechanism by which ATM and NBS1 regulate the structure and function of telomeres is currently not understood, the role of telomeres in the replicative senescence of ATM and NBS1 cells is demonstrated by the rescue of ATM and NBS1 cells through telomere elongation induced by ectopic expression of hTERT. However, like the parental cells, the hTERT-expressing AT and NBS fibroblasts remain sensitive to SIPS. Despite the defective response to DNA damage in hTERT-expressing AT fibroblasts, H_2O_2 and ionizing radiation induced SIPS in these cells, concurrently with the induction of p53, $p21^{WAF1}$, $p16^{INK\text{-}4a}$ protein and phosphorylation of the stress-induced $p38^{MAPK}$.[14-16] These observations suggest that the ATM-dependent signaling pathway is involved in replicative senescence, but is dispensable for inducing SIPS in response to ionizing radiation or oxidative stress and associate SIPS with $p38^{MAPK}$, a central mediator of SIPS.

Signal Transduction and Gene Expression in SIPS: Central Role of $p38^{MAPK}$

Depending on the cell type, stress can activate three major MAPK pathways, ERKs (extracellular-regulated kinases), JNKs (c-JUN N-terminal kinases) and $p38^{MAPKs}$.

$p38^{MAPK}$, but not JNK, has been shown to be activated in replicative senescence and SIPS induced by oncogenic ras and culture shock.[17,18] Culture shock has been defined after the observation that mouse fibroblasts implanted from in vivo to in vitro become senescent after a small number of cell divisions: approximately 10 population doublings, an estimate of the average number of cell divisions of a given population. The telomeres do not significantly shorten during this period given murine cells have extremely long telomeres. Rather, it is believed that senescence is induced by the environmental changes imposed on the mouse cells, such as oxidative environment due to culture conditions[19] and thus called "culture shock".[20]

As reviewed below, the activation of $p38^{MAPK}$ is delayed after the exposure of cells to subcytotoxic senescence-inducing stress. A working hypothesis could be that the activation of $p38^{MAPK}$

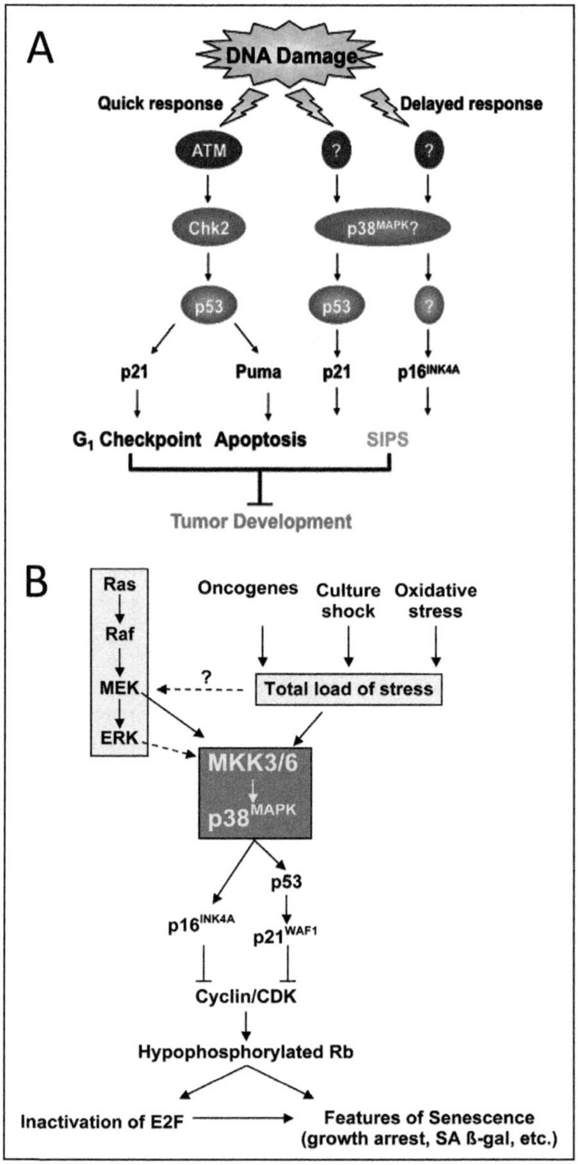

Figure 1. A) Working hypothesis. SIPS is not triggered by the immediate cascade of events activated by the initial DNA damage and involving proteins such as ATM, Chk2 and p53. In this hypothesis, the initial quick response triggered by DNA damage is responsible for transient growth arrest namely through the expression of $p21^{WAF-1}$, possibly via a first phase of activation of $p38^{MAPK}$. Then indirect mechanisms allow a delayed and sustained activation of $p38^{MAPK}$ responsible for the establishment of the phenotype of SIPS. B) Hypothesis of the stress load. Oncogenes, culture shock, oxidative stress and oncogenic activation can be represented as a total load of stress activating $p38^{MAPK}$ and triggering signaling cascades involving namely Rb as well as other signaling pathways. The possibility of activation of a cross-talk between Ras/Raf-1/MEK1/ERK pathway and MKK3-MKK6/$p38^{MAPK}$ pathway in Ras-induced cellular senescence is also taken into consideration.

that mediates the induction of SIPS is not triggered immediately by the initial stress but rather indirectly (Fig. 1A).

Cells lacking c-Jun undergo prolonged cell-cycle arrest, but resist apoptosis, whereas cells that ectopically express c-Jun do not arrest but undergo apoptosis.[21] A lack of activation of the JNKs after subcytotoxic stress could explain why fibroblasts become prematurely senescent instead of apoptotic when exposed to high doses of oxidants. Furthermore, activation of JNK by ectopically expressed active MKK7 does not induce premature senescence.[18] Thus, the JNK pathway does not seem to contribute to the induction of SIPS.

Ectopic activation of p38^{MAPK} by the constitutive active form of the upstream kinase, MKK3 and MKK6, induces cellular senescence.[17,18,22] Moreover, pharmacological or genetic inactivation of p38^{MAPK} bypassed or delayed the onset of various types of cellular senescence.[17] These results indicate that p38^{MAPK} plays a central role in the senescence-inducing pathway downstream of various stimuli. p38^{MAPK} activation is delayed after the exposure of cells to senescence-inducing stimuli. This p38^{MAPK}-activating condition appears to be quantitative and irreversible and p38^{MAPK} is activated only when the total level of the condition (stress, telomere shortening, oncogene overexpression and culture shock) reaches a critical threshold[23] (Fig. 1B). Specific activation of p38^{MAPK} by overexpression of upstream activators in primary or tumor cells triggers irreversible cell cycle arrest and biochemical and morphologic features of SIPS. This effect is mitigated by concomitant addition of a chemical inhibitor of the p38^{MAPK} alpha and beta isoforms. Blocking p38^{MAPK} activity after 72 h does not reverse senescence.

TGF-β1 and p38^{MAPK} in H_2O_2- and UVB-Induced SIPS

The abundance of *Transforming growth factor-β1 (TGF-β1)* mRNA increases in replicative senescence of WI-38 HDFs, in premature senescence of WI-38 HDFs as induced by *tert*-butylhydroperoxide (*t*-BHP) and ethanol, in premature senescence of IMR-90 HDFs as induced by H_2O_2, in premature senescence of skin HDFs as induced by UVB or by psoralen + UVA treatment, in premature senescence of human retina epithelial cells as induced by *t*-BHP and in the dermis of aged versus middle-aged probands.[24-30] Immunofluorescence allowed detection of both the latent and the active form of TGF-β1 in UVB-, *t*-BHP- and ethanol-induced senescence.[27,31]

It has been found that induction of SIPS by H_2O_2 in fetal lung IMR-90 HDFs requires a positive feedback loop engaged between the activation of p38^{MAPK} and the overexpression of TGF-β1. Similar results were obtained with AG04431 skin HDFs in SIPS induced by repeated mild exposures to UVB.[27] H_2O_2 triggers the phosphorylation of p38^{MAPK} within 10 minutes after the onset of exposure. Then p38^{MAPK} phosphorylates and activates ATF-2, which in turn participates in the short-term growth arrest and triggers the overexpression of TGF-β1. Phosphorylated ATF-2 and hypophosphorylated retinoblastoma protein (Rb) physically interact. Incubation of IMR-90 HDFs with TGF-β1 also induces the appearance of the features of senescence (senescence-like morphology, SA β-gal activity and overexpression of the senescence-associated genes *fibronectin, apolipoprotein J/clusterin, osteonectin, matrix metalloproteinase-2 and Smooth Muscle 22 (SM22 or transgelin)*.[28] Fibronectin is an essential component of the extracellular matrix and may contribute to the morphological changes observed in senescent HDFs as well as anchorage of cells to their substrate. Osteonectin is a calcium binding protein able to inhibit the cell entry into the S phase through selective binding of platelet-derived growth factor. SM22 encodes for a putative calcium binding protein involved in senescence-induced morphological changes. Apolipoprotein J is an extracellular protein with chaperone-like activity similar to small heat shock proteins. The secretion of apolipoprotein J is sharply increased in UVB-induced SIPS of skin AG04431 HDFs.[27]

Treatment of IMR-90 and AG04431 HDFs with antibodies against TGF-β1 (or TGF-β1 receptor II) for 72 h after respectively H_2O_2 and UVB subcytotoxic stress, abolishes the induction of these biomarkers. p38^{MAPK} and ATF-2 remain phosphorylated for at least 72 h after the initial stress. Neutralising antibodies against TGF-β1 prevent the sustained phosphorylation of p38^{MAPK} and ATF-2. This indicates that TGF-β1 sustains the long-term activation of p38^{MAPK}. TGF-β1 antibody does not prevent the phosphorylation of p38^{MAPK} taking place just after the onset of stress, when TGF-β1 is not yet overexpressed. It is known that TGF-β1 activates the

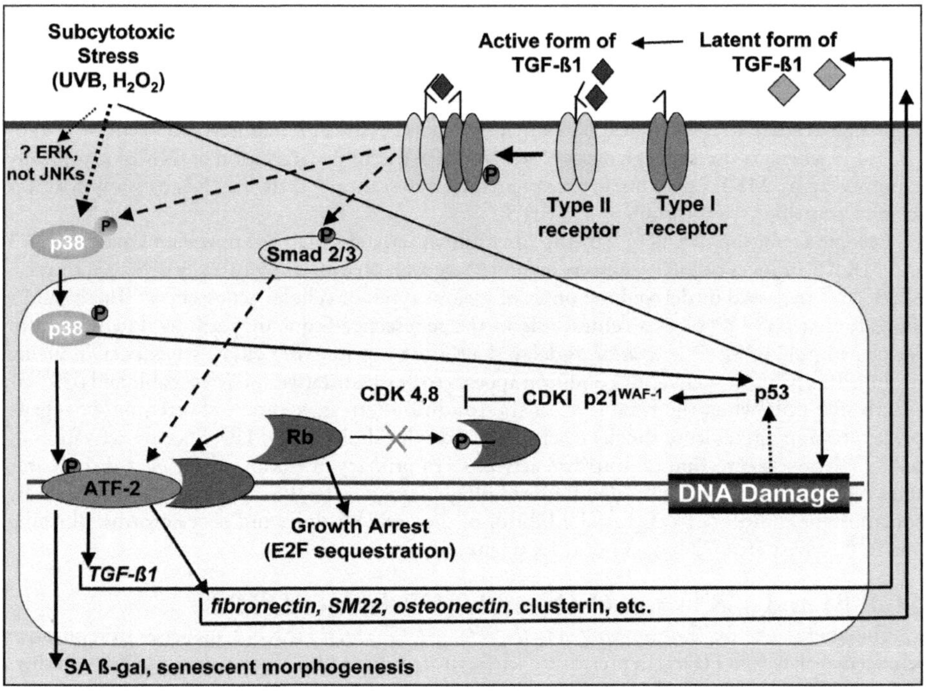

Figure 2. Model for the establishment of H_2O_2- and UVB-induced premature senescence in fetal lung and skin HDFs, based on a positive feed-back loop between $p38^{MAPK}$ and Transforming Growth Factor (TGF-β1) that is responsible for a sustained activation of $p38^{MAPK}$ for days after the initial stress has vanished. Upon DNA damage p53 becomes activated and induces the expression of the cyclin-dependent kinase inhibitor (CdKI) $p21^{WAF-1}$, thereby inhibiting the activity of the cyclin D/cyclin-dependent kinase 4 and 6 complexes activities, this could explain the hypophosphorylation of the retinoblastoma protein, which blocks the cell cycle through the sequestration of the E2F transcription factor. H_2O_2 and UVB phosphorylate $p38^{MAPK}$ within minutes after the onset of stress,[33] namely through the activation of the latent form of TGF-β1.[27] Once activated, $p38^{MAPK}$ phosphorylates the ATF-2 transcription factor. This triggers the overexpression of the inactive form of TGF-β1 increasing its secretion and cleavage into its active form are increased.[27] In a second phase active TGF-β1 binds its Type II receptors, which activates the Type I receptors and reactivates $p38^{MAPK}$ and p53, keeping Rb hypophosphorylated via the induction of the CdKIs. 24 h after stress, phosphorylated ATF-2 and hypophosphorylated retinoblastoma protein start to interact. ATF-2 and Rb are necessary for the appearance of biomarkers of senescence: senescence-like morphology, SA β-gal activity and overexpression of the senescence-associated genes *fibronectin, apolipoprotein J, osteonectin and SM22.*[27,28,33,77] The Smad proteins and the ERK proteins could also finely tune this process.

$p38^{MAPK}$ pathway.[32] Along the same line of evidence, inhibition of $p38^{MAPK}$ or ATF-2 by chemical inhibitor or by antisense oligonucleotides prevents the overexpression of TGF-β1 and blocks the appearance of H_2O_2-induced SIPS[33] (Fig. 2).

In addition, TGF-β1 triggers an increase in the concentration of H_2O_2 in IMR-90 HDFs,[34] generating a constant oxidative stress. Superoxide dismutase 2 dismutates superoxide anions into H_2O_2 in mitochondria and is overexpressed in cells in which $p38^{MAPK}$ is activated.[22] Cyclooxygenase-2 is overexpressed in replicative senescence, in *c-rel*-induced senescence and in H_2O_2-induced premature senescence.[35] $p38^{MAPK}$ and ATF-2 are partly responsible for the overexpression of cyclooxygenase-2 observed in H_2O_2-induced premature senescence.[36] $p21^{WAF1}$, a CDK inhibitor that blocks the kinase activity of cyclinD-CDK4/6 complexes, is found upregulated in H_2O_2- and *t*-BHP-induced

SIPS, explaining in part why Rb becomes hypophosphorylated.[28,33] The upregulation of $p21^{WAF1}$ expression is due to increases in both mRNA and protein levels in response to $p38^{MAPK}$.[22]

Immunofluorescence showed that the protein level of insulin-like growth factor binding protein-3 (IGFBP-3) was increased in replicative senescence and premature senescence induced by *t*-BHP and ethanol. The overexpression of IGFBP-3 is well documented in replicatively senescent HDFs.

Endogenous expression of IGFBP-3 by HDFs was knocked down using siRNA specifically targeting IGFBP-3 when the stress protocols were applied to induced SIPS with *t*-BHP and ethanol. In these conditions, the increase of apolipoprotein J, fibronectin and osteonectin mRNA abundance was abolished. IGFBP-3 siRNA did not alter the stress-induced variation of the RNA level of *TGF-β1*. This is interesting because TGF-β1 has been described as regulating IGFPB-3 expression in many biological models but not vice versa. After 72 h of stimulation with TGF-β1, there was an increase of more than 20-fold in the relative transcript level of *IGFBP-3* as well as an increase in IGFBP-3 protein level as seen by immunofluorescence. Moreover neutralizing antibody against TGF-β1 added in the medium after the last *t*-BHP and ethanol stress, in conditions inducing premature senescence, diminished the increased transcript and protein levels of IGFBP-3.[31]

$p38^{MAPK}$, p53 and Rb

p53 and Rb and are intimately involved in cellular senescence. Given that $p38^{MAPK}$ phosphorylates p53,[37] it is important to determine whether Rb and p53 are upstream or downstream of $p38^{MAPK}$. This question was addressed by assessing the consequence of $p38^{MAPK}$ activation in HDFs expressing the human papilloma virus (HPV) E6 that inactivates p53, E7 that inactivates Rb family proteins, or both E6 and E7 proteins. It was found that $p38^{MAPK}$ activation led to growth arrest in E6-expressing HDFs and not E7-expressing HDFs, indicating that Rb, but not p53, is an essential downstream molecule of $p38^{MAPK}$ for inducing cell cycle arrest in this system.[17] It was also found that, upon subcytotoxic acute exposure of E7-expressing IMR-90 HDFs to H_2O_2, the presence of Rb was necessary for the appearance of senescence morphology, senescence-associated β-galactosidase activity and overexpression of several senescence-associated genes such a *fibronectin, osteonectin, apolipoprotein J*.[28,38] Supporting the role of the Rb pathway in $p38^{MAPK}$-mediated SIPS, activation of $p38^{MAPK}$ led to the accumulation of $p16^{INK4A}$ at both protein and mRNA levels.[18]

In contrast, other studies suggest that p53 may be a downstream effector of $p38^{MAPK}$ in the senescence pathway. $p38^{MAPK}$ can phosphorylate and activate p53 either directly[37,39] or through its downstream kinase casein kinase 2.[40,41] In IMR-90 HDFs, oncogenic ras induces the phosphorylation of p53 at Ser33 and Ser46, the sites that had been shown to be phosphorylated by $p38^{MAPK}$ in response to UV irradiation.[42] In the same study, elimination of $p38^{MAPK}$ activity either by knocking out the $p38^{MAPK}$ gene or overexpressing PPM1D $p38^{MAPK}$ phosphatase, mimicked the effect of p53 inactivation in promoting the transformation of MEF cells by oncogenic ras and E1A. These data have placed $p38^{MAPK}$ and p53 in the same pathway that mediates senescence and tumor suppression. It remains to be determined whether the epistatic relationship between $p38^{MAPK}$ and p53 can be reproduced in primary human cells.

Role of Caveolin-1 in Cellular Senescence and Interplay with $p38^{MAPK}$

A possible interplay between caveolin-1 and $p38^{MAPK}$ has been shown to contribute to replicative senescence of HDFs and premature senescence of transformed NIH-3T3 murine cells in a $p53/p21^{WAF1}$-dependent manner. However, evidence for this interaction in SIPS of HDFs is presently missing.

Caveolae are flask-shaped invaginations of the plasma membrane that mediate signal transduction events. Caveolin-1, the structural protein component of caveolar membranes, acts as a scaffolding protein to concentrate, organize and functionally modulate signaling molecules. Park et al reported higher levels of caveolin-1 in senescent HDFs, as compared to younger HDFs.[43] Up-regulation of caveolin-1 in senescent HDFs was associated to a significant inhibition of

EGF-stimulated ERK-1/2 phosphorylation and regulation of focal adhesion kinase activity.[44] MEFs transgenically overexpressing caveolin-1 show a reduced proliferative lifespan, senescence-like cell morphology and SA β-gal activity.[45] These results are consistent with the ability of caveolin-1 to arrest MEFs in the G_0/G_1 phase of the cell cycle. Interestingly, over-expression of caveolin-1 in MEFs resulted in the transactivation of a p53-responsive element and up-regulation of $p21^{WAF1}$ expression. Taken together, these data suggest that caveolin-1 may mediate premature senescence through a $p53/p21^{WAF1}$-dependent pathway.

Subcytotoxic concentration of H_2O_2 induces irreversible growth arrest and triggers an overexpression of caveolin-1 in NIH 3T3 cells, which is greatly reduced when the cells harbor antisense caveolin-1, whereas the induction of this growth arrest is recovered when caveolin-1 levels is restored.[45] A possible link between caveolin-1 and $p38^{MAPK}$ has been shown:[46] H_2O_2 and UV light stimulate the tyrosine phosphorylation of caveolin-1 in NIH 3T3 cells, while the $p38^{MAPK}$ inhibitor (SB203580) blocks this stress-induced process.

A similar role of $p38^{MAPK}$ has not been reported yet during senescence in non-immortalized MEFs nor in HDFs. However high density lipoproteins have recently been shown to induce cyclooxygenase-2 expression and prostacyclin release in human endothelial cells via a $p38^{MAPK}$- and caveolin-1-dependent mechanisms.[47] The activation of Cdc42 Rho GTPase plays a role in UV-induced activation of $p38^{MAPK}$[48] and activated Cdc42 directly interacts with caveolin-1 in senescent HDFs,[44] suggesting complex mechanisms involving Cdc42, caveolin-1 and $p38^{MAPK}$ activation in the senescence of normal human cells. In HDFs, RhoGTPase cdc42 was shown to be responsible for $p38^{MAPK}$ activation in turn triggering phosphorylation of L-caldesmon and HSP27. Cdc42 was also shown to be mainly responsible for the increase in *TGF-β1* mRNA level observed at 24 h after treatment with H_2O_2 onwards.[26] We reported the first evidence of increased nuclear and cytoplasmic localization of caveolin-1 during establishment of H_2O_2-induced premature senescence. Moreover, we demonstrated that phosphorylation of caveolin-1 during treatment with H_2O_2 is dependent on $p38\alpha^{MAPK}$. Caveolin1 down-regulation did not block this morphological change nor the actin and paxillin distribution normally seen after the treatment. Caveolin1 down-regulation with siRNA did not alter the sharp decrease in (^{3}H) thymidine incorporation into DNA, nor the increase in *p21*^{WAF1} and *TGF-β1* mRNA and in the proportion of SA β-gal positive HDFs, all normally observed after treatment with H_2O_2 at sublethal concentration.[49]

Premature Senescence as an Anti-Oncogenic Defense Mechanism

Premature senescence can be induced following the activation of proto-oncogenes in certain primary human and murine cells.[50-52] The ras proto-oncogenes encode small GTP binding proteins that transduce growth signals from cell surface receptors in response to extracellular stimuli.[53-55] In various cell types, ras-induced senescence is accompanied by the accumulation of different combinations of cell cycle inhibitors such as p53, $p21^{WAF1}$, $p16^{INK4A}$ and $p14/p19^{ARF}$, decreased expression of cyc A and reduced kinase activity of CDK2,[50] as also observed in H_2O_2-induced SIPS of IMR-90 HDFs.[56] All these events are consistent with a G1 growth arrest. It is conceivable that ras-induced premature senescence serves as a defensive mechanism against oncogenic transformation in mammalian cells, because it can effectively eliminate cells that have acquired oncogenic ras. Therefore, cellular transformation by oncogenic ras requires additional, cooperating genetic alterations that bypass ras-induced senescence. A number of genetic changes, including inhibition of p53 or $p14(p19)^{ARF}$ functions,[57,58] defects in TGF-β signaling[52] and acquisition of viral oncogenes such as adenoviral E1A,[50] HPV E6E7[59] and SV40 large T region,[60] have been reported to allow bypass of ras-induced senescence. These findings at least in part provide an explanation to the observation that ras needs to cooperate with other genetic alterations to transform primary cells.[61-65]

Signaling Pathway Mediating Ras-Induced Premature Senescence— The Tumor Suppressing Function of $p38^{MAPK}$

Oncogenic activation may be considered as a type of stress on primary cells, which leads to the withdrawal of transformed cells from the proliferating population by inducing premature senescence. A link between ras activation and oxidative stress was established.[66] In normal HDFs, oncogenic ras (Ha-rasV12) increased the levels of intracellular reactive oxygen species (ROS). The ability of ras to induce premature senescence seemed to depend on increased ROS levels, since low oxygen environments or scavengers of H_2O_2, conditions under which ROS production was inhibited, could rescue senescence. This finding, together with the observations that ROS induce premature senescence,[67,68] raised the possibility that increased oxidative stress may be a crucial step during the induction of senescence by oncogenic ras in HDFs.

Activation of the Raf-MEK-ERK cascade by ras has been shown to play an important role in ras-mediated transformation. Interestingly, the ability of oncogenic ras to induce premature senescence also depends on the same Raf-MEK-ERK pathway that mediates cell proliferation and transformation.[69,70] Constitutive activation of this pathway by overexpressing active forms of Raf-1 or MEK1 induces the overexpression of $p16^{INK4A}$ and/or $p21^{WAF1}$ and leads to premature senescence. In addition, ras fails to induce senescence when activation of the Raf-MEK-ERK pathway is pharmacologically inhibited.

Several recent studies have revealed the involvement of the $p38^{MAPK}$ pathway in premature senescence induced by active ras or raf-1.[17,18,42,71] Ha-rasV12 or active raf-1 stimulated $p38^{MAPK}$ activity in BJ foreskin,[18] IMR-90[17] and WI-38 fetal lung HDFs[42] and MEFs,[71] coinciding with the induction of premature senescence. The ras-activated $p38^{MAPK}$ activity appeared to be mediated by its upstream kinases MKK3 and MKK6, rather than by MKK4, because ras increased the activity of MKK3 and MKK6, but not that of MKK4 and a dominant negative mutant of MKK3 or MKK6 blocked the ability of ras to stimulate $p38^{MAPK}$.[18] Constitutive activation of $p38^{MAPK}$ by forced expression of active mutants of MKK3 and MKK6 was sufficient to cause premature senescence. Conversely, inhibition of $p38^{MAPK}$ activity by SB203580 or by stably expressed, dominant-negative mutants of MKK3 or MKK6 at least partially rescued the senescence phenotype induced by oncogenic ras.[17,18] Furthermore, ras-induced senescence can also be prevented by genetically modulating the activities of $p38^{MAPK}$ regulators. Ectopic expression of PPM1D, a phosphatase that inhibits the activation of $p38^{MAPK}$, rendered IMR-90 cells resistant to ras-induced senescence.[42] In addition, ras was unable to cause premature senescence in MEFs deficient in Gadd45a that could bind to and regulate the activity of $p38^{MAPK}$.[71] PPM1D not only rescued ras-induced senescence, but also facilitated the transformation of MEF cells by ras and E1A.[42] In the latter case, bypass of premature senescence by Gadd45a deficiency allowed ras to transform MEFs alone without any cooperating oncogenes.[72] These studies strongly indicate that $p38^{MAPK}$ plays a central role in mediating ras-induced premature senescence and argue that $p38^{MAPK}$ pathway has tumor suppressing functions in limiting the transformation potential of ras in primary cells.

Raf-MEK-ERK and MKK3/6-$p38^{MAPK}$ represent two independent MAPK pathways that mediate distinct cellular processes in response to different extracellular stimuli.[54] While the Raf-MEK-ERK cascade mainly transduces proliferative signals, the MKK3/6-$p38^{MAPK}$ cascade is a major mediator of stress responses that usually lead to growth arrest or apoptosis. It was initially puzzling how activation of either one of these two pathways with opposing functions was sufficient to cause senescence and why activation of both pathways was necessary for ras to induce senescence. Further investigation has revealed that ras-induced premature senescence is the result of a cross-talk between these two pathways and that the MKK3/6-$p38^{MAPK}$ cascade seems to act at least partially downstream of the Raf-MEK-ERK cascade to mediate senescence.[17,18] Constitutively active MEK induced MKK3/6 and $p38^{MAPK}$ activation and premature senescence in a similar fashion to oncogenic ras. The induction of $p38^{MAPK}$ activity by ras not only requires active MKK3 and MKK6, but also active MEK. More importantly, while the $p38^{MAPK}$ inhibitor SB203580 rescued premature senescence induced by oncogenic ras, active MEK1 and active MKK3 and MKK6, the MEK inhibitor U0126 rescued senescence induced by ras and active MEK1, but not that by active MKK3 and MKK6. These studies, conducted at either molecular or genetic levels, have all placed MKK3/6 and $p38^{MAPK}$ downstream of MEK in the pathway that induces SIPS.

Conclusion: The Next Steps...

Depending on the species, the length of telomeres of cells at early generations can greatly differ. Mice have much longer telomeres than humans. This greatly changes the biology of telomeres and the mechanisms of senescence.[73] However, $p38^{MAPK}$ seems to play a role in premature senescence in fibroblasts of both species. Recent studies have defined a pathway in which oncogenic ras induces premature senescence by sequentially activating Raf1, MEK, MKK3/6 and $p38^{MAPK}$. However, several issues remain unresolved concerning this senescence-inducing pathway involving the two MAPK cascades.

First, it remains to be determined whether the involvement of $p38^{MAPK}$ pathway in SIPS is universal. Most of the current studies focused on primary fibroblasts and relatively little is known about the role of $p38^{MAPK}$ in SIPS in other cells types. Despite the differences in the mechanism of senescence between mouse and human cells, $p38^{MAPK}$ seems to play a major role in SIPS in both species. However, mechanistic details regarding the role of $p38^{MAPK}$ pathway in SIPS may still differ in the 2 species. Furthermore, whether $p38^{MAPK}$ is essential for the response to all senescence inducers is unknown.

The second obvious uncertainty lies in the mechanism by which MKK3/6 is activated following MEK activation. While MEK and ERK are activated instantly following ras or raf1 activation, the accumulation of active MKK3/6 and $p38^{MAPK}$ kinase seemed to be delayed, lagging behind that of active MEK and ERK by hours or days depending on the experimental settings.[17,18,23] Thus, the $p38^{MAPK}$ pathway may be activated by ras as a result of the sustained presence of an active MEK-ERK cascade. This finding suggests a model in which MKK3/6 is not activated by MEK via direct signaling, but rather through an indirect, slow route. This model is consistent with the observation that senescence induced by active MKK3 and MKK6 occurred much more rapidly than that by oncogenic ras and active MEK1 in BJ cells.[18] The hypothesis of a slow route was also proposed above for H_2O_2-induced SIPS, where a sustained phosphorylation of $p38^{MAPK}$ is regulated by TGF-β1, which is overexpressed after 24 h or more after stress.[33]

Third, the signaling components that mediate SIPS need to be further defined. It is unclear whether ERK is required for the activation of MKK3/6. The roles of four $p38^{MAPK}$ isoforms, α, β, γ and δ, in the senescence pathway have yet to be determined. It was indicated that a dominant negative allele of p38β, γ or δ delayed but did not completely block ras-induced senescence. Thus in the case of ras-induced senescence, these 3 isoforms may all play a role. But it remains possible that these retrovirally transduced dominant negative alleles are not potent enough to completely inhibit the endogenous kinase and rescue senescence.[18] Further studies using siRNA may eventually solve this problem. A recent study reported that MEFs from MKK3/6 double knockout mice were not growth arrested by serum starvation. If we consider serum starvation as a form of stress, this finding fits the concept that these proteins are necessary for stress-induced growth arrest.[74] It is worth examining whether SIPS can be induced in these cells. Furthermore, it is unclear whether the recently reported MKK-independent activation of $p38^{MAPK}$ by an adaptor protein TAB1[75,76] contributes to SIPS.

Fourth, further studies are needed to define the $p38^{MAPK}$ downstream effectors that directly enforce the growth arrest and other senescence phenotypes. Previous studies have implicated either $p16^{INK4A}$/Rb or p53 as the downstream effector of $p38^{MAPK}$ in SIPS, as discussed above. It is important to further clarify the roles of $p16^{INK4A}$, Rb, p53 and other possible p38 downstram effectors in the SIPS pathway. It also remains to be determined whether $p38^{MAPK}$ activates p53 directly or via a slow, indirect mechanism. The latter could explain why senescence is favored over apoptosis, likely because the p53-dependent growth arresting mechanisms would need time to establish. It could also explain why HDFs that enter SIPS at some subcytotoxic concentration of oxidants, will start dying when a narrow threshold of oxidant is trespassed,[68] possibly due to an unbalance between the growth-arresting and the apoptosis-inducing signalling pathways.

Lastly, it has always been controversial whether it is senescence or premature senescence that occurs in vivo during age-related processes. The situation is not clear concerning normal aging. However, the physiopathological conditions linked with several age-related pathologies are associated with oxidative stress and/or release of pro-inflammatory cytokines. Senescent cells have been observed in such conditions, favoring the existence of SIPS in vivo, in addition to replicative senescence or not. Demonstrating the role of $p38^{MAPK}$ in these pathological conditions could represent a path to follow for new interventional strategies.

Acknowledgements

F.D.-C. and O.T. are respectively Research Associate and Research Assistant, Fonds National de la Recherche Scientifique, Belgium). E.B. is a recipient of a FRIA fellowship, Belgium. We acknowledge the Région Wallonne for the First-DEI '*Cosmet-X*' project, the Network II *Senegene* project and the Pole of Excellence *Nanotoxico* project. We thank the European Commission for the Integrated Projects *Proteomage* (LSHM-CT-2005-518230) and *GeHA* (LSHM-CT-2004-503270); for the Coordination and Support Actions *Link-Age* and *WhyWeAge* (LSHM-CT-2005-513866, HEALTH-F4-2008-200970); and the Marie Curie *Matiss* Project (MTKI-CT-2006–042768).

References

1. Campisi J, d'Adda di Fagagna F. Cellular senescence: when bad things happen to good cells. Nat Rev Mol Cell Biol 2007; 8:729-740.
2. Shay JW, Wright WE. Hayflick, his limit and cellular ageing. Nat Rev Mol Cell Biol 2000; 1:72-76.
3. Wright WE, Shay JW. Cellular senescence as a tumor-protection mechanism: the essential role of counting. Curr Opin Genet Dev 2001; 11:98-103.
4. von Zglinicki T. Oxidative stress shortens telomeres. Trends Biochem Sci 2002; 27:339-344.
5. Serrano M, Blasco MA. Putting the stress on senescence. Curr Opin Cell Biol 2001; 13:748-753.
6. Toussaint O, Remacle J, Dierick JF et al. From the Hayflick mosaic to the mosaics of ageing. Role of stress-induced premature senescence in human ageing. Int J Biochem Cell Biol 2002; 34:1415-1429.
7. Toussaint O, Salmon M, Royer V et al. In: Nyström T, Osiewacz H, eds. Topics in Current Genetics. Heidelberg: Springer-Verlag Berlin, 2003:3.
8. Toussaint O, Medrano EE, von Zglinicki. T-Cellular and molecular mechanisms of stress-induced premature senescence (SIPS) of human diploid fibroblasts and melanocytes. Experimental Gerontology 2000; 35:927-945.
9. Schmitt CA. Senescence, apoptosis and therapy—cutting the lifelines of cancer. Nat Rev Cancer 2003; 3:286-295.
10. de Magalhaes JP, Chainiaux F, de Longueville F et al. Gene expression and regulation in H_2O_2-induced premature senescence of human foreskin fibroblasts expressing or not telomerase. Experimental Gerontology 2004; 39:1379-1389.
11. Shiloh, Y. Ataxia-telangiectasia and the Nijmegen breakage syndrome: related disorders but genes apart. Annu Rev Genet 1997; 31:635-662.
12. d'Adda di Fagagna F, Reaper PM, Clay-Farrace L et al. A DNA damage checkpoint response in telomere-initiated senescence. Nature 2003; 426:194-198.
13. Wong KK, Maser RS, Bachoo RM et al. Telomere dysfunction and Atm deficiency compromises organ homeostasis and accelerates ageing. Nature 2003; 421:643-648.
14. de Magalhaes JP, Chainiaux F, Remacle J et al. Stress-induced premature senescence in BJ and hTERT-BJ1 human foreskin fibroblasts. FEBS letters 2002; 523:157-162.
15. Gorbunova V, Seluanov A, Pereira-Smith OM. Expression of human telomerase (hTERT) does not prevent stress-induced senescence in normal human fibroblasts but protects the cells from stress-induced apoptosis and necrosis. The Journal of Biological Chemistry 2002; 277:38540-38549.
16. Naka K, Tachibana A, Ikeda K et al. Stress-induced premature senescence in hTERT-expressing ataxia telangiectasia fibroblasts. The Journal of Biological Chemistry 2004; 279:2030-2037.
17. Iwasa H, Han J, Ishikawa F. Mitogen-activated protein kinase p38 defines the common senescence-signalling pathway. Genes Cells 2003; 8:131-144.
18. Wang W, Chen JX, Liao R et al. Sequential activation of the MEK-extracellular signal-regulated kinase and MKK3/6-p38 mitogen-activated protein kinase pathways mediates oncogenic ras-induced premature senescence. Mol Cell Biol 2002; 22:3389-3403.
19. Parrinello S, Samper E, Krtolica A et al. Oxygen sensitivity severely limits the replicative lifespan of murine fibroblasts. Nat Cell Biol 2003; 5:741-747.
20. Sherr CJ, DePinho RA. Cellular senescence: mitotic clock or culture shock? Cell 2000; 102:407-410.
21. Shaulian E, Schreiber M, Piu F et al. The mammalian UV response: c-Jun induction is required for exit from p53-imposed growth arrest. Cell 2000; 103:897-907.
22. Haq R, Brenton JD, Takahashi M et al. Constitutive p38HOG mitogen-activated protein kinase activation induces permanent cell cycle arrest and senescence. Cancer Res 2002; 62:5076-5082.
23. Deng Q, Liao R, Wu BL et al. High intensity ras signaling induces premature senescence by activating p38 pathway in primary human fibroblasts. The Journal of Biological Chemistry 2004; 279:1050-1059.
24. Borlon C, Debacq-Chainiaux F, Hinrichs C et al. The gene expression profile of psoralen plus UVA-induced premature senescence in skin fibroblasts resembles a combined DNA-damage and stress-induced cellular senescence response phenotype. Experimental Gerontology 2007; 42:911-923.
25. Borlon C, Weemaels G, Godard P et al. Expression profiling of senescent-associated genes in human dermis from young and old donors. Proof-of-concept study. Biogerontology 2008; 9:197-208.

26. Chretien A, Dierick JF, Delaive E et al. Role of TGF-beta1-independent changes in protein neosynthesis, p38alphaMAPK and cdc42 in hydrogen peroxide-induced senescence-like morphogenesis. Free Radical Biology & Medicine 2008; 44:1732-1751.
27. Debacq-Chainiaux F, Borlon C, Pascal T et al. Repeated exposure of human skin fibroblasts to UVB at subcytotoxic level triggers premature senescence through the TGF-beta1 signaling pathway. J Cell Sci 2005; 118:743-758.
28. Frippiat C, Chen QM, Zdanov S et al. Subcytotoxic H_2O_2 stress triggers a release of transforming growth factor-beta 1, which induces biomarkers of cellular senescence of human diploid fibroblasts. The Journal of Biological Chemistry 2001; 276:2531-2537.
29. Glotin AL, Debacq-Chainiaux F, Brossas JY et al. Prematurely senescent ARPE-19 cells display features of age-related macular degeneration. Free Radical Biology and Medicine 2008; 44:1348-1361.
30. Pascal T, Debacq-Chainiaux F, Chretien A et al. Comparison of replicative senescence and stress-induced premature senescence combining differential display and low-density DNA arrays. FEBS Letters 2005; 579:3651-3659.
31. Debacq-Chainiaux F, Pascal T, Boilan E et al. Screening of senescence-associated genes with specific DNA array reveals the role of IGFBP-3 in premature senescence of human diploid fibroblasts. Free Radical Biology and Medicine 2008; 44:1817-1832.
32. Hanafusa H, Ninomiya-Tsuji J, Masuyama N et al. Involvement of the p38 mitogen-activated protein kinase pathway in transforming growth factor-beta-induced gene expression. The Journal of Biological Chemistry 1999; 274:27161-27167.
33. Frippiat C, Dewelle J, Remacle J et al. Signal transduction in H_2O_2-induced senescence-like phenotype in human diploid fibroblasts. Free Radical Biology and Medicine 2002; 33:1334-1346.
34. Thannickal VJ, Fanburg BL. Activation of an H_2O_2-generating NADH oxidase in human lung fibroblasts by transforming growth factor beta 1. The Journal of Biological Chemistry 1995; 270:30334-30338.
35. Zdanov S, Bernard D, Debacq-Chainiaux F et al. Normal or stress-induced fibroblast senescence involves COX-2 activity. Experimental Cell Research 2007; 313:3046-3056.
36. Zdanov S, Toussaint O, Debacq-Chainiaux F. p53 and ATF-2 partly mediate the overexpression of COX-2 in H(2)O (2)-induced premature senescence of human fibroblasts. Biogerontology 2009; 10:291-298.
37. Bulavin DV, Saito S, Hollander MC et al. Phosphorylation of human p53 by p38 kinase coordinates N-terminal phosphorylation and apoptosis in response to UV radiation. EMBO J 1999; 18:6845-6854.
38. Chen QM, Bartholomew JC, Campisi J et al. Molecular analysis of H_2O_2-induced senescent-like growth arrest in normal human fibroblasts: p53 and Rb control G1 arrest but not cell replication. Biochem J 1998; 332 (Pt 1):43-50.
39. Huang C, Ma WY, Maxiner A et al. p38 kinase mediates UV-induced phosphorylation of p53 protein at serine 389. The Journal of Biological Chemistry 1999; 274:12229-12235.
40. Sayed M, Kim SO, Salh BS et al. Stress-induced activation of protein kinase CK2 by direct interaction with p38 mitogen-activated protein kinase. The Journal of Biological Chemistry 2000; 275:16569-16573.
41. Shi Y, Gaestel M. In the cellular garden of forking paths: how p38 MAPKs signal for downstream assistance. Biol Chem 2002; 383:1519-1536.
42. Bulavin DV, Demidov ON, Saito S et al. Amplification of PPM1D in human tumors abrogates p53 tumor-suppressor activity. Nat Genet 2002; 31:210-215.
43. Park JS, Kim HY, Kim HW et al. Increased caveolin-1, a cause for the declined adipogenic potential of senescent human mesenchymal stem cells. Mechanisms of Ageing and Development 2005; 126:551-559.
44. Cho KA, Ryu SJ, Oh YS et al. Morphological adjustment of senescent cells by modulating caveolin-1 status. The Journal of Biological Chemistry 2004; 279:42270-42278.
45. Volonte D, Zhang K, Lisanti MP et al. Expression of caveolin-1 induces premature cellular senescence in primary cultures of murine fibroblasts. Mol Biol Cell 2002; 13:2502-2517.
46. Volonte D, Galbiati F, Pestell RG et al. Cellular stress induces the tyrosine phosphorylation of caveolin-1 (Tyr(14)) via activation of p38 mitogen-activated protein kinase and c-Src kinase. Evidence for caveolae, the actin cytoskeleton and focal adhesions as mechanical sensors of osmotic stress. The Journal of Biological Chemistry 2001; 276:8094-8103.
47. Norata GD, Callegari E, Inoue H et al. HDL3 induces cyclooxygenase-2 expression and prostacyclin release in human endothelial cells via a p38 MAPK/CRE-dependent pathway: effects on COX-2/PGI-synthase coupling. Arterioscler Thromb Vasc Biol 2004; 24:871-877.
48. Seo M, Cho CH, Lee YI et al. Cdc42-dependent mediation of UV-induced p38 activation by G protein betagamma subunits. The Journal of Biological Chemistry 2004; 279:17366-17375.
49. Chretien A, Piront N, Delaive E et al. Increased abundance of cytoplasmic and nuclear caveolin 1 in human diploid fibroblasts in $H_{(2)}O_{(2)}$-induced premature senescence and interplay with p38alpha(MAPK). FEBS Letters 2008; 582:1685-1692.
50. Serrano M, Lin AW, McCurrach ME et al. Oncogenic ras provokes premature cell senescence associated with accumulation of p53 and p16INK4a. Cell 1997; 88:593-602.

51. Spyridopoulos I, Isner JM, Losordo DW. Oncogenic ras induces premature senescence in endothelial cells: role of p21(Cip1/Waf1). Basic Res Cardiol 2002; 97:117-124.
52. Tremain R, Marko M, Kinnimulki V et al. Defects in TGF-beta signaling overcome senescence of mouse keratinocytes expressing v-Ha-ras. Oncogene 2000; 19:1698-1709.
53. Barbacid M. ras genes. Annu Rev Biochem 1987; 56:779-827.
54. Cahill MA, Janknecht R, Nordheim A. Signalling pathways: jack of all cascades. Curr Biol 1996; 6:16-19.
55. Medema RH, Bos JL. The role of p21ras in receptor tyrosine kinase signaling. Crit Rev Oncog 1993; 4:615-661.
56. Frippiat C, Remacle J, Toussaint O. Down-regulation and decreased activity of cyclin-dependent kinase 2 in H_2O_2-induced premature senescence. Int J Biochem Cell Biol 2003; 35:246-254.
57. Peeper DS, Dannenberg JH, Douma S et al. Escape from premature senescence is not sufficient for oncogenic transformation by Ras. Nat Cell Biol 2001; 3:198-203.
58. Shvarts A, Brummelkamp TR, Scheeren F et al. A senescence rescue screen identifies BCL6 as an inhibitor of anti-proliferative p19(ARF)-p53 signaling. Genes Dev 2002; 16:681-686.
59. Morales CP, Holt SE, Ouellette M et al. Absence of cancer-associated changes in human fibroblasts immortalized with telomerase. Nat Genet 1999; 21:115-118.
60. Hahn WC, Dessain SK, Brooks MW et al. Enumeration of the simian virus 40 early region elements necessary for human cell transformation. Mol Cell Biol 2002; 22:2111-2123.
61. Franza BR Jr, Maruyama K, Garrels JI et al. In vitro establishment is not a sufficient prerequisite for transformation by activated ras oncogenes. Cell 1986; 44:409-418.
62. Hirakawa T, Ruley HE. Rescue of cells from ras oncogene-induced growth arrest by a second, complementing, oncogene. Proc Natl Acad Sci USA 1988; 85:1519-1523.
63. Hunter T. Cooperation between oncogenes. Cell 1991; 64:249-270.
64. Ruley HE. Transforming collaborations between ras and nuclear oncogenes. Cancer Cells 1990; 2:258-268.
65. Weinberg RA. Oncogenes, antioncogenes and the molecular bases of multistep carcinogenesis. Cancer Res 1989; 49:3713-3721.
66. Lee AC, Fenster BE, Ito H et al. Ras proteins induce senescence by altering the intracellular levels of reactive oxygen species. The Journal of Biological Chemistry 1999; 274:7936-7940.
67. Chen Q, Ames BN. Senescence-like growth arrest induced by hydrogen peroxide in human diploid fibroblast F65 cells. Proc Natl Acad Sci USA 1994; 91:4130-4134.
68. Dumont P, Burton M, Chen QM et al. Induction of replicative senescence biomarkers by sublethal oxidative stresses in normal human fibroblast. Free Radical Biology & Medicine 2000; 28:361-373.
69. Lin AW, Barradas M, Stone JC et al. Premature senescence involving p53 and p16 is activated in response to constitutive MEK/MAPK mitogenic signaling. Genes Dev 1998; 12:3008-3019.
70. Zhu J, Woods D, McMahon M et al. Senescence of human fibroblasts induced by oncogenic Raf. Genes Dev 1998; 12:2997-3007.
71. Bulavin DV, Kovalsky O, Hollander MC et al. Loss of oncogenic H-ras-induced cell cycle arrest and p38 mitogen-activated protein kinase activation by disruption of Gadd45a. Mol Cell Biol 2003; 23:3859-3871.
72. Hollander MC, Sheikh MS, Bulavin DV et al. Genomic instability in Gadd45a-deficient mice. Nat Genet 1999; 23:176-184.
73. Prowse KR, Greider CW. Developmental and tissue-specific regulation of mouse telomerase and telomere length. Proc Natl Acad Sci USA 1995; 92:4818-4822.
74. Brancho D, Tanaka N, Jaeschke A et al. Mechanism of p38 MAP kinase activation in vivo. Genes Dev 2003; 17:1969-1978.
75. Ge B, Gram H, Di Padova F et al. MAPKK-independent activation of p38alpha mediated by TAB1-dependent autophosphorylation of p38alpha. Science 2002; 295:1291-1294.
76. Tanno M, Bassi R, Gorog DA et al. Diverse mechanisms of myocardial p38 mitogen-activated protein kinase activation: evidence for MKK-independent activation by a TAB1-associated mechanism contributing to injury during myocardial ischemia. Circ Res 2003; 93:254-261.
77. Chen QM, Tu VC, Catania J et al. Involvement of Rb family proteins, focal adhesion proteins and protein synthesis in senescent morphogenesis induced by hydrogen peroxide. J Cell Sci 2000; 113(Pt 22):4087-4097.

Chapter 11

Protein Homeostasis in Models of Aging and Age-Related Conformational Disease

Elise A. Kikis[†], Tali Gidalevitz[†] and Richard I. Morimoto[*]

Abstract

The stability of the proteome is crucial to the health of the cell, and contributes significantly to the lifespan of the organism. Aging and many age-related diseases have in common the expression of misfolded and damaged proteins. The chronic expression of damaged proteins during disease can have devastating consequences on protein homeostasis (proteostasis), resulting in disruption of numerous biological processes. This chapter discusses our current understanding of the various contributors to protein misfolding, and the mechanisms by which misfolding, and accompanied aggregation/toxicity, is accelerated by stress and aging. Invertebrate models have been instrumental in studying the processes related to aggregation and toxicity of disease-associated proteins and how dysregulation of proteostasis leads to neurodegenerative diseases of aging.

Protein Folding Problem in Aging

Proteins are the predominant products of gene expression and contribute significantly to the shape and functionality of the cell. The status of the expressed proteome has an important role in the health of individual cells and the lifespan of the organism. In addition to cell type- and tissue-specific regulation of protein expression, maintenance of the proteome depends on efficient *protein folding homeostasis*, or *proteostasis*, that monitors and ensures folding, assembly and targeting of newly-synthesized proteins, repair of damaged proteins and clearance. Proteotoxic conditions arise by external stresses, or as a byproduct of normal cellular metabolic and signaling events during development and aging (Fig. 1). This is in addition to the intrinsic variation in the proteome due to polymorphisms, which is further exacerbated by random errors that can occur at every step of protein biogenesis. Together, these factors contribute to a flux of metastable proteins that are at risk for misfolding and aggregation.[1-4] In healthy young cells, these processes are balanced by the concerted action of molecular chaperones, detoxifying enzymes, degradation machinery and adaptive stress responses (Fig. 2).[5-7] In aging and disease, damaged proteins accumulate, leading to both loss-of-function and gain-of-function toxicity as these homeostatic mechanisms fail and contribute to pathology.

[†]These authors contributed equally

*Corresponding Author: Richard I. Morimoto—Department of Biochemistry, Molecular Biology and Cell Biology, Rice Institute for Biomedical Research, 2205 Tech Drive, Hogan 2-100, Northwestern University, Evanston, Illinois 60208. USA. Email: r-morimoto@northwestern.edu

Protein Metabolism and Homeostasis in Aging, edited by Nektarios Tavernarakis.

Factors detrimental to protein folding

Stressors
- pathophysiological states (fever, inflammation, infection, hypoxia, reperfusion, *etc.*)
- hormonal imbalance (thyroid and growth hormone, IGF-I, *etc.*), emotional/psychological stress
- nutrient intake and imbalance
- amino acid analogues (including from food sources - azetidine, canavanine)
- physical stress (temperature, irradiation, UV light, mechanical stress)
- environmental stressors (heavy/transition metals, alcohols, arsenates, pesticides, mutagens, irritants, *etc.*)
- aberrant physiological conditions (oxidation, osmolarity, pH)
- trauma, injury
- antibiotics and anti-inflammatory drugs (puromycin, tetracycline, NSAID, indomethacin)
- alcohols, nicotine
- anesthetics

"Activities of daily living"
- respiration, glycolysis
- neurotransmission
- bursts of protein synthesis (differentiation)
- signaling

Mutations, Polymorphisms, Mistranslation
- protein conformation
- subunit stoichiometry
- protein stability

} **protein misfolding → protein conformation diseases of aging**

Damaged/ mutated chaperones
- protein misfolding → **chaperonopathies**

Figure 1. Extrinsic and intrinsic factors detrimental to protein folding.

Sources of Protein Damage—Oxidative Modifications

The '*free radical theory*'[8] and the '*oxidative stress theory*'[9,10] of aging postulate that aging and many age-related diseases may be attributed to the generation of oxygen free radicals and reactive oxygen and nitrogen species (ROS and RNS), in excess of cellular antioxidants, resulting in oxidative damage to DNA, lipids and proteins. Early evidence connected many of the aging-associated diseases, including atherosclerosis, arthritis, muscular dystrophy, cataracts, pulmonary dysfunction, neurological disorders and cancer with oxidative damage.[11] Levels of oxidized proteins increase exponentially in regions of aging brains, concomitant with a decrease in the activities of the enzymes glutamine synthetase and alcohol dehydrogenase, which are susceptible to oxidative damage.[12] The decrease in glutamine synthetase activity was found to distinguish the brains of Alzheimer's disease (AD) patients from age-matched individuals, leading to the suggestion that AD may represent a specific brain vulnerability to age-related oxidation. Furthermore, a functional connection between the accumulation of oxidized proteins and the phenotypic manifestations of cellular disfunction in aging has been suggested.[13]

In addition to oxygen radicals, glucose, galactose, fructose and many glycolytic intermediates participate in non-enzymatic protein glycosylation (glycation) and glycoxidation, contributing to age-related protein modifications. Some glycolytic intermediates can generate methylglyoxal (MG), which is a highly reactive glycating agent,[14,15] leading to the formation of advanced glycation end products (AGEs), that have been implicated in age-related diseases, including Alzheimer's disease and complications associated with diabetes.[15-17] A decrease in circulating glucose and reduction of MG, resulting from dietary restriction or fasting, may explain some of the health-improving effects of these treatments.[18] Recently, an inactivating mutation in a glycolytic enzyme triosephosphate isomerase (Tpi) was shown to cause progressive motor impairment, severely reduced lifespan and neurodegeneration in *Drosophila*, presumably by causing an accumulation of methylglyoxal and subsequent increase in generation of AGEs.[19]

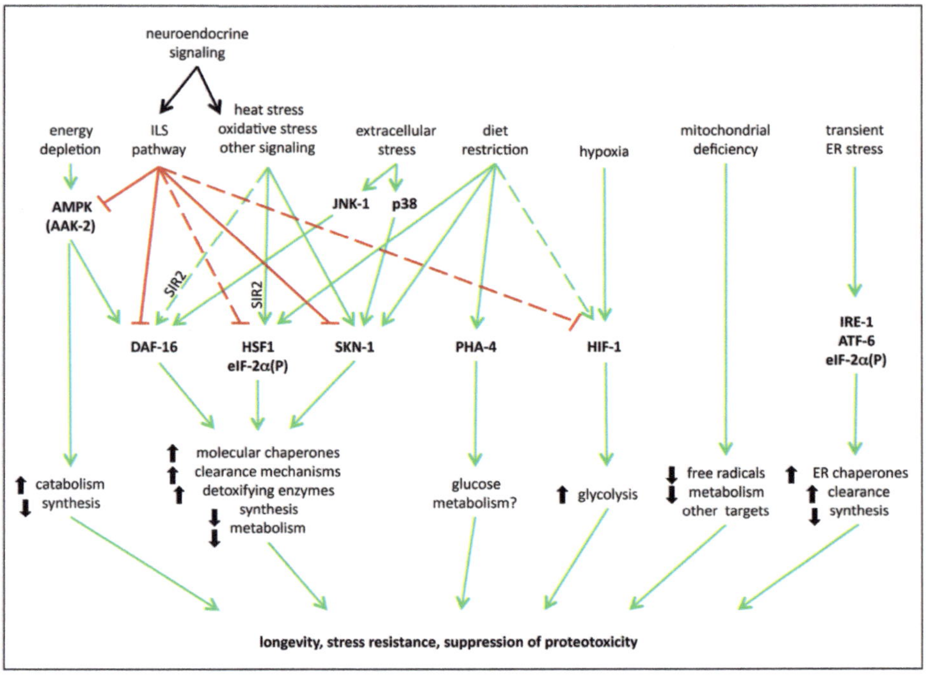

Figure 2. Proteostasis pathways. Multiple interconnected pathways regulate expression of genes that contribute to maintenance of protein folding homeostasis during growth, development, and under various stress conditions. The complex genetic interactions among these pathways necessitate precise control over their activities. Many of these pathways also possess regulatory feedback mechanisms, which are not well characterized/understood.

Loss of protein function due to oxidative modification can be due either to direct inhibition of activity of a protein (for example, covalent modification of a side chain in proximity of an active site), or to conformational changes or alteration of protein stability. Most modifications of the first type are irreversible, with the exception of cysteine and methionine modifications, which can be reversed by thiol transferases, thioredoxin, glutaredoxin and methionine sulfoxide reductases (MsrA and MsrB). The detoxification of methylglyoxal and glyoxal is achieved by the glutathione-dependent glyoxalase system. These and other enzymes are part of the cellular proteostasis machinery that is activated in response to various protein damaging conditions. For example, MrsA and B allow recovery of protein function in specific cases[20,21] and their overexpression confers protection against oxidative damage in yeast, *Drosophila* and human fibroblasts.[21]

In addition to functional inactivation of proteins, free radicals can have a local effect on charged side chains, or cause cleavage of the polypeptide backbone, leading to a spectrum of consequences, from complete unfolding, to local conformational changes,[22-24] to targeting for disposal. If not degraded, modified proteins often become cross-linked, leading to the accumulation of nonspecific aggregated material, such as lipofuscin, ceroids and AGE-containing pigments. Nuclear and cytosolic oxidized proteins have been shown to be degraded by an ATP- and ubiquitin-independent 20S proteasome[25,26] as well as by classical ubiquitin-26S proteasome system,[27] whereas mitochondrial oxidized proteins are degraded by the ATP-dependent Lon protease.[28] Interestingly, the 20S proteasome appears to be resistant to oxidative damage, while 26S proteasome is not.[29] Chaperone-mediated autophagy represents an alternative protein degradation pathway that is induced under oxidative stress conditions.[30] Under chronic oxidative stress conditions, there seems to be a damage cascade, whereby the initial appearance of modified proteins, which signifies the insufficiency of the repair and degradation arm of proteostasis machinery, results in further inhibition of both proteasomal and lysosomal functions (reviewed in refs. 21,24) and thus in further accumulation of protein damage.

Sources of Damaged Proteins—Misfolding

One source of damaged—often misfolded, mislocalized, or aggregated—proteins are errors inherent in protein biosynthesis and turnover. Variations in amino acid sequence, which may result from genetic mutations and polymorphisms, genomic instability, mistranslation, or incorporation of amino-acid analogues (such as certain antibiotics or plant metabolites, Fig. 1), have the potential to affect folding pathways and the stability of the native state.[2,31-35] For example, coding polymorphisms are not rare and are estimated to occur at an average of two per coding sequence,[36] while misincorporation during translation might cause up to 18% of expressed proteins to contain an amino acid substitution.[37] Our ability to understand, let alone predict, the consequences of missense mutations on the folding, stability or functionality of a given protein in vivo remains very limited, with data limited to a few model proteins studied in vitro, or to computational models. This is further complicated by the fact that the phenotypic expression of mildly destabilizing protein polymorphisms appears to depend on the robustness of the protein folding environment[1,4,38] and the capacity of the proteostasis network. Recent work on the evolution of protein sequences suggests that selection against the toxicity of misfolding due to mistranslation may represent an important evolutionary pressure in the case of highly expressed proteins.[37] From a physiological perspective, this may be interpreted to indicate that the flux of misfolded or destabilized proteins in a cell bears a significant fitness cost, not only through the loss-of-function of the misfolded proteins, but also because of the toxic effects of induced aggregation and the consequences of inappropriate intermolecular interactions,[2,7,39,40] as well as abnormal engagement of molecular chaperones and degradative machinery. Because of these gain-of-function properties of destabilized proteins, their chronic presence may exceed the capacity of the proteostasis machinery to cope, leading to the disruption of folding homeostasis and cellular dysfunction. Indeed, introduction into a cell of even one protein with a strong tendency to misfold has been shown to interfere with the folding and function of other cellular proteins harboring mildly destabilizing amino acid substitutions (Fig. 4).[3,38]

At the organismal level, consequences of misfolding due to biosynthetic errors can be gleaned from a *sti* mouse model, in which low levels of mischarged transfer RNAs, due to an editing defect in alanyl-tRNA synthetase, lead to neurodegeneration, with cerebellar Purkinje cell loss and ataxia.[41] Neurodegeneration is also a consequence of a mutation of the tyrosyl-tRNA synthetase in a subtype of Charcot-Marie-Tooth neuropathy.[42] The mouse *sti* mutation leads to the production of heterogeneous misfolded proteins, accompanied by increased expression of chaperones in the cytoplasm and the ER.[41] This increase in the components of the proteostasis machinery indicates that adaptive transcriptional responses—the heat shock response and the unfolded protein response—are activated in the face of protein misfolding, as cells attempt to maintain folding homeostasis.[7] However, the cellular dysfunction and degeneration in this mouse model together indicate that proteostasis is overwhelmed by the chronic elevation of misfolded proteins. A provocative question, then, is whether all misfolded and damaged proteins are recognized, refolded or cleared with equal efficiency, or whether certain proteins, such as those implicated in neurodegenerative diseases and other misfolding diseases of aging, are particularly challenging for the quality control machineries.

The lack of robustness of cellular folding homeostasis to chronic misfolding has significant implications for aging and age-related diseases. First, the aging of an organism is accompanied by an increased accumulation of damaged proteins, with contributions from both external stresses and physiological processes (Fig. 1). Second, the fidelity of biosynthetic processes decreases, exacerbated by a decline in the functionality of repair enzymes (reviewed in ref. 43) and the dampening of the heat shock response (reviewed in ref. 44). As accumulated misfolded proteins deplete essential components of the proteostasis machinery[45-47] and the quality control and repair mechanisms fail to respond and keep damaged proteins in check, the maintenance of the proteome becomes further compromised. With as much as 70% of rare missense alleles predicted to be mildly deleterious in humans[48] and approximately half of the genetic changes known to cause inherited disease (in OMIM and HGMD databases) being due to nonsynonymous changes,[33] such compromise should manifest in a gradual increase in cellular dysfunction and onset of disease. In this scenario, the strength and composition of the proteostatic buffer, the specific complement of mutations and polymorphisms in expressed genes and the accumulation of molecular damage, all serve to set a dynamic threshold for the onset of dysfunction, both in a cell-specific and an individual organism-specific manner.

Proteostasis Regulation in Aging and Late Onset Diseases

Adaptive transcriptional stress responses such as the heat shock response and the unfolded protein response (UPR) serve to protect cells from extreme flux of misfolded and damaged proteins when cells or organism are exposed to an acute proteotoxic condition (reviewed in refs. 49,50). Parallel to the sharply increased transcription/translation of molecular chaperones and components of degradative quality control machineries, mediated by the transcription factor HSF1, a generalized inhibition of protein synthesis allows the switch to refolding or degradation of existing proteins that become unstable during such proteotoxic shock.[51] However, the fact that HSF1,[52,53] the components of the UPR signaling,[54-56] as well as some of the stress-inducible molecular chaperones are essential for life or development suggests that these adaptive stress responses are also necessary for the maintenance of proteostasis under normal, nonstress conditions. On the other hand, forced overexpression of molecular chaperones is deleterious to growth of normal cells,[57,58] while the transformed phenotype of cancer cells depends on HSF1 and on an increased expression of chaperones.[59-61] It appears then that the precise regulation of proteostasis networks is essential for the health of the organism, while accumulating evidence points to its disregulation during aging as one of the main causes of cellular dysfunction and disease onset.

The association with aging is one of the most distinctive characteristics of protein conformation diseases. This connection is particularly striking in neurodegeneration where, for each specific disease, age is the strongest predictor of disease onset even for the familial variants. On the other hand, age appears to have a modifying, rather than causative, influence on disease onset, as each disease has its characteristic age of onset, with Alzheimer's disease and Parkinson's disease being late onset, ALS occurring most frequently in early to mid-life and Huntington's disease exhibiting a strong positive correlation between age of onset and polyQ length polymorphism.[62] Aging has been shown to be a potent enhancer of aggregation/toxicity in *C. elegans*, in which length-dependent polyQ aggregation and toxicity phenotypes are enhanced during aging, but suppressed in animals rendered long-lived by mutations in the lifespan regulating insulin-like signaling (ILS) pathway (Fig. 2).[63]

The modulatory effects of aging are most likely due to the functional and regulatory connections between aging and protein homeostasis. Activation of the kinase activity of IP3K (AGE-1), by downregulation of the insulin-like growth factor receptor DAF-2, in wild type *C. elegans*, initiates a signaling cascade that represses the fork head transcription factor DAF-16.[64,65] De-repression of DAF-16 in *age-1* or *daf-2* mutant animals extends lifespan, while inactivating *daf-16* mutations suppress longevity. The *daf-2/age-1* effects on polyQ aggregation toxicity also require DAF-16 activation, revealing that the dual effects of *daf-2/age-1* on longevity and polyglutamine toxicity share a common genetic pathway (ILS pathway, Fig. 2).[66,67] The modulation of protein misfolding and aggregation by the ILS pathway appears to be a general feature and has also been observed in *C. elegans* expressing Aβ as a model for Alzheimer's disease.[68]

How do genetic pathways that regulate lifespan suppress proteotoxicity? The molecular interactions between these pathways are mediated, in part, by factors that detect and respond to misfolded proteins—molecular chaperones, HSF1, DAF-16 and other transcription factors

(Fig. 2), and molecular chaperones have been shown to accumulate in the long-lived mutants of *C. elegans*. Downregulation of *hsf-1* suppresses both the ILS-mediated lifespan extension and the protection against proteotoxicity. Moreover, *hsf-1* downregulation leads to a decrease in normal lifespan and an accelerated aging phenotype in *C. elegans*, while overexpression of HSF1 extends lifespan.[66,67] The functional relationship between ILS and protein folding homeostasis can be demonstrated by the induction of both thermotolerance and life span extension not only by mutations in the ILS pathway, but also by a sublethal heat stress.[69] Furthermore, cells from naturally long-lived[70] or lifespan mutant[71] rodents appear to be resistant to multiple proteotoxic stresses. The mechanistic relationship can be shown by the induction of stress resistance by ablation of cells making insulin-like ligands in wild type *Drosophila*,[72] thus excluding indirect effects and adaptation to mutations. In *C. elegans*, inactivation of *daf-16, hsf-1, hsp-1* (the major cytosolic Hsp70 chaperone) and small heat shock proteins accelerates polyQ protein aggregation (Fig. 3C), supporting the proposition that the ILS could coordinately influence aging and proteostasis through the action of DAF-16 and HSF1.[66,67] The regulation of ILS pathway and of HSF1 and other transcription factors (such as SKN-1,[73] the *C. elegans* Nrf2 homologue) in the context of adaptive stress responses must serve to integrate organismal growth and development with multiple networks that regulate different aspects of protein homeostasis.[7,66-68] We are only now beginning to understand how such integration could be achieved, with recent evidence showing that the ability of individual cells in an organism to respond to proteotoxic conditions is controlled by the activity of a subset of neurons.[74]

In addition to the control of the heat shock response by neuronal signaling, there have been numerous observations in which the heat shock response is poorly or incompletely activated, including early in development.[75] Of particular interest are studies of the deficiency of the heat shock response in the brain and during aging.[76,77] Restricted expression of heat shock genes has been observed in different regions of the brain, while neuronal cells in culture can exhibit selective induction of chaperone genes. Human neuroblastoma Y79 cells, for example, respond to heat shock by induction of many chaperones including Hsp90, but not Hsp70, despite activation of HSF1.[78] Intact primary hippocampal neurons from neonatal rat embryos express HSF2 but not HSF1 until later stages of development.[79] Consequently, the heat shock response of primary hippocampal neurons is deficient, while astrocytes have a robust stress response. Similar observations have been made in primary motor neurons that exhibit a deficient heat shock response thought to be due to a defect in activation of HSF1.[80]

The involvement of proteostasis networks in conformational diseases and aging is underscored by a decrease in toxicity when individual molecular chaperones are overexpressed in various cell-based and animal models (reviewed in ref. 81). Overexpression of HSF1 and certain molecular chaperones has been shown to extend lifespan.[66,67,82,83] Furthermore, the ability to induce the heat shock response under conditions of environmental proteotoxic stress is predictive of the remaining poststress lifespan.[84,85] Likewise, proteasomal adaptation (by modulation of substrate accessibility to the proteasome core) to environmental stress in *C. elegans* ensures both resistance to proteotoxic conditions and maintenance of lifespan under normal conditions, arguably through regulating the degradation of misfolded proteins.[86] Moreover, because both HSF-1 and DAF-16 are regulated by the NAD-dependent sirtuin, SIRT1,[87-89] in addition to the ILS-pathway, it is reasonable to suggest that regulation of proteostasis is intimately linked by multiple pathways to metabolic control and lifespan.

Late Onset Diseases are Mainly Protein Folding Diseases

Common features of protein conformation diseases are the accumulation of protein deposits (such as aggregates, inclusion bodies and plaques) and a presumably consequential "gain-of-function" proteotoxicity.[90,91] These features, which are characteristic of misfolded protein species, are present in neurodegenerative diseases of aging such as Parkinson's disease, amyotrophic lateral sclerosis (ALS), prion disease, Alzheimer's disease and the family of disorders generally referred to as the CAG-repeat/polyglutamine-expansion diseases (Huntington's disease), Kennedy's disease,

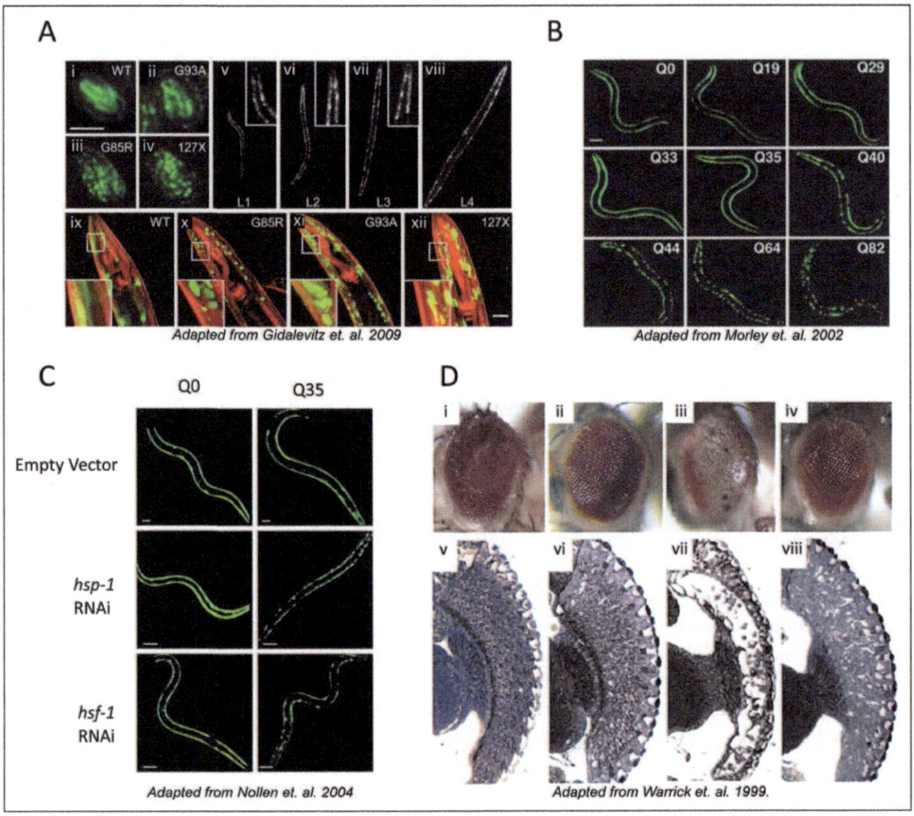

Figure 3. A) SOD1 mutant proteins aggregate in the body wall muscle cells of *C. elegans*. Fluorescent micrographs of SOD1transgenic embryos (i-iv) and adult animals (ix through xii) showing the distribution of the SOD-YFP fluorescence (green) and Rhodamine-phalloidin stained myofilaments (red). WT SOD1 protein exhibits diffuse, if patchy, fluorescence, while all mutant strains contain discrete fluorescent foci as well as some diffuse fluorescence. Panels v through viii show G85R animals in all larval stages (L1 through L4). Adapted from Gidalevitz et al, 2009 (ref. 38). B) Length-dependent aggregation of polyQ-YFP fusion proteins in *C. elegans*. Epifluoresence micrographs of 3- to 4-day-old *C. elegans* expressing different lengths of polyQ-YFP (Q0, Q19, Q29, Q33, Q35, Q40, Q44, Q64, Q82). Adapted from Bieschke et al, 2006 (ref. 63). C) RNAi for *hsp-1* and *hsf-1* induces aggregate formation of polyglutamine YFP proteins in a Q stretch length-dependent manner. Fluorescence microscopy pictures of 4-day-old *C. elegans* adults expressing Q0-YFP or Q35-YFP. RNAi of either *hsp-1* or *hsf-1* results in premature onset of aggregation in Q35 animals. Adapted from Nollen et al, 2004 (ref. 100). B,C) Copyright National Academy of Sciences, USA. Used with permission. D) HSP70 overexpression suppresses polyglutamine-induced neurodegeneration in *Drosophila*. Eyes (i-iv) and retinal sections (v,vi) of flies expressing expanded polyglutamine protein and human *HSPA1L* are shown. i,v, Control fly expressing only the promoter transgene. ii,vi, Flies expressing *HSPA1L*, coding for Hsp70 protein. Eye structure appears grossly normal. More detailed analysis revealed abnormalities in nuclear position and photoreceptor rhabdomere morphology when *HSPA1L* is highly expressed. iii,vii, Flies expressing the expanded polyglutamine protein MJDtr-Q78. These flies have degenerate eyes that lack pigment and show severe loss of retinal structure. iv,viii, Flies expressing both MJDtr-Q78 and *HSPA1L*. Co-expression of *HSPA1L* ameliorates the degenerative effects of MJDtr-Q78. The eye appears normal externally. Internally, eye structure is largely restored, although photoreceptor rhabdomere specializations are not made. Adapted from Warrick et al, 1999 (ref. 98).

spinocerebellar ataxias). Each disease has the distinctive characteristic of age-dependent onset, a progressive, usually fatal, clinical course and selective neuronal vulnerability despite broader expression of the causative protein.

Despite the fact that the proteins that are now known to "cause" neurodegenerative diseases lack similarities in primary sequence (other than the polyQ tract in CAG-repeat expansion diseases), they all share the ability to form alternate unfolded or misfolded states that in turn aggregate and/or are toxic. Thus, misfolding and aggregation has been proposed to be the molecular underpinning of disease. This is supported by evidence from transgenic model systems including *S. cerevisiae*, *C. elegans* and *D. melanogaster*,[92-97] in which it has been possible to recapitulate many molecular, cellular and behavioral phenotypes associated with neurodegenerative disease. The development of these nonmammalian models for protein conformational disease has been invaluable for the discovery of modifiers, pathways and underlying mechanisms of toxicity and for the testing of small molecules.[63,98-101] Additionally, these models have solidified our understanding of the link between protein conformational disorders, molecular chaperones and proteostasis regulators and aging.[66,68]

Invertebrate Models of Late Onset Conformational Diseases

The link between protein misfolding and human neurodegenerative disease and the fact that the machinery involved in proteostasis maintenance, is highly conserved amongst eukaryotes, has led to the widespread use of invertebrate models systems, such as *C. elegans* and *Drosophila*. Ultimately, modeling aspects of human neurodegenerative diseases in invertebrates allows for genetic manipulations, such as mutant screens/identification of modifiers and aging studies that would be prohibitively time consuming if carried out in vertebrate animals.

These studies in *C. elegans* and *D. melanogaster* benefit from the sequenced genomes, relatively short life cycles and abundant genetic tools of these well-characterized model organisms. *C. elegans* is particularly well suited for the study of neurodegenerative diseases of aging due to its a relatively simple (but still sufficiently complex) nervous system, its genetically defined aging pathway and the relative ease to perform live cell imaging of fluorescent proteins for studies of disease-causing protein aggregation dynamics. Likewise, in *D. melanogaster* the photoreceptor neurons have proven to be highly amenable to these types of studies, as human proteins can be ectopically expressed and the readout for proteotoxicity is revealed by effects on eye morphology.

To date, a relatively large number of invertebrate models of human neurodegenerative diseases have been generated. Importantly, these models have solidified our understanding that protein misfolding underlies the mechanism(s) of disease action. Furthermore, the invertebrate models have opened up new ways to study the molecular underpinnings of disease and to identify genes and gene networks whose activities are involved, positively or negatively, in modulating aggregation/toxicity. We begin with an overview of the existing models and discuss what has been learned from them generally and in what ways they recapitulate key aspects of human disease.

PolyQ

Huntington's disease (HD) and a number of related neurodegenerative diseases are caused genetically by expansions of polyQ-encoding CAG tracts in specific individual genes/proteins. Expressing these proteins, with wild-type or expanded polyQ tracts, in invertebrates has proven to be an effective way to study the molecular underpinnings of protein misfolding/toxicity associated with polyQ disorders. Specifically, SCA3/MJD (Machado-Joseph disease) (Fig. 3D)[92] and HD[102] were modeled in *Drosophila* and shown to recapitulate many important aspects of disease including cell-type specific sensitivity to the expressed protein, nuclear inclusion formation and late-onset cellular degeneration. These findings were furthered by the expression of polyQ proteins in *C. elegans* (Fig. 3B,C). HD models have been generated via the expression, in neuronal subsets, of an N-terminal fragment of the protein Huntingtin (Htt) with the wild type or a long polyglutamine (polyQ) tract.[93,97] Ultimately, it was shown that human Htt with an expanded polyglutamine tract led to aggregate formation in, and apoptotic cell death of, ASH sensory neurons, in a sensitized genetic background.[93] Additionally, Htt with 128 glutamines fused to YFP aggregated when expressed in PLM mechanosensory neurons and caused neuronal abnormalities but not death.[97] Having such fluorescently tagged variants of disease-causing proteins in *C. elegans* has proven highly advantageous by allowing for the real-time monitoring of aggregation and using dynamic imaging techniques in this transparent organism.[38,63,96,103]

The overall similarities between various polyQ disorders suggested that the polyQ expansions are predominantly, if not solely, responsible for the observed disease pathologies. To address this, general polyQ models, in which only the polyQ tract, without any of its normal flanking protein sequences (but fused in some cases to YFP, or to another protein not usually containing a polyQ tract) were developed (Fig. 3B).[63,96,104] The data from these models provided the first direct in vivo evidence that polyQ tracts themselves, independently of their protein context, aggregate and are toxic.

The significance of recapitulating key aspects of disease by expressing a single human protein in an invertebrate should not be underestimated. It confirmed that the mutant proteins are generally toxic even across species and in different cell types. This, in turn, lead to the hypothesis that the highly conserved cellular quality control system, involving protein folding/clearance, is inherently lacking the ability to prevent damage by these polyQ-containing disease-causing proteins.

Aβ

Alzheimer's disease (AD) is a relatively common neurodegenerative disorder affecting nearly 30% of all individuals older than 85 years of age. It is characterized by the accumulation of β-amyloid plaques and neurofibrillary tangles. The major proteinaceous components of β-amyloid plaques in AD patients are Aβ peptides; specifically, $A\beta_{1\text{-}40}$ and $A\beta_{1\text{-}42}$, which are produced via proteolysis of APP, the amyloid precursor protein. To model Aβ aggregation/toxicity in an invertebrate model system, $A\beta_{1\text{-}42}$ was expressed in *C. elegans* body wall muscle cells and was shown to form amyloid plaques that have biochemical characteristics similar to those found in the brains of AD patients.[105] Additionally, $A\beta_{1\text{-}42}$ and $A\beta_{1\text{-}40}$ were expressed in *Drosophila* photoreceptor neurons.[106] It was shown that $A\beta_{1\text{-}42}$ is substantially more toxic than $A\beta_{1\text{-}40}$,[106] consistent with the idea that $A\beta_{1\text{-}42}$ has a higher amyloidogenic propensity.

Tau

In addition to amyloid plaques, AD and other neurodegenerative diseases are characterized by the formation of neurofibrillary tangles comprised of the protein tau. Expression of mutant tau in specific neuronal subtypes or under pan-neuronal control in *Drosophila* resulted in neurodegeneration, but not neurofibrillary tangle/filamentous aggregate formation, as determined by electron microscopy, suggesting that the tangles themselves are not the source of mutant tau toxicity.[107] Expression of tau in all *C. elegans* neurons resulted in motility defects indicative of tau proteotoxicity, providing further support to the idea that low levels of tau protein, not necessarily neurofibrillarly tau tangles or otherwise insoluble tau, are required for toxicity.[108]

Alpha-Synuclein

Familial forms of Parkinson's disease (PD) are caused by mutations in the protein alpha-synuclein.[109,110] Expression of alpha-synuclein in *C. elegans* neurons was shown to cause neurodegeneration of neuronal subsets.[111] Expression of alpha-synuclein in body wall muscle cells resulted in age-dependent aggregation, a phenotype that was used to perform a genetic screen for suppressors of aggregation (see below).[112]

SOD1

Amyotrophic Lateral Sclerosis (ALS) is a neurodegenerative disease that, in its inherited form, is often associated with mutation of superoxide dismutase 1 (SOD1). A large number of mutations in SOD1 have been identified and shown to be linked to ALS.[113] Evidence suggests that mutations in SOD1 cause disease not by loss of SOD1 function, but instead by gain of toxic function by the mutant protein. Therefore, to model SOD1 protein misfolding/toxicity, these

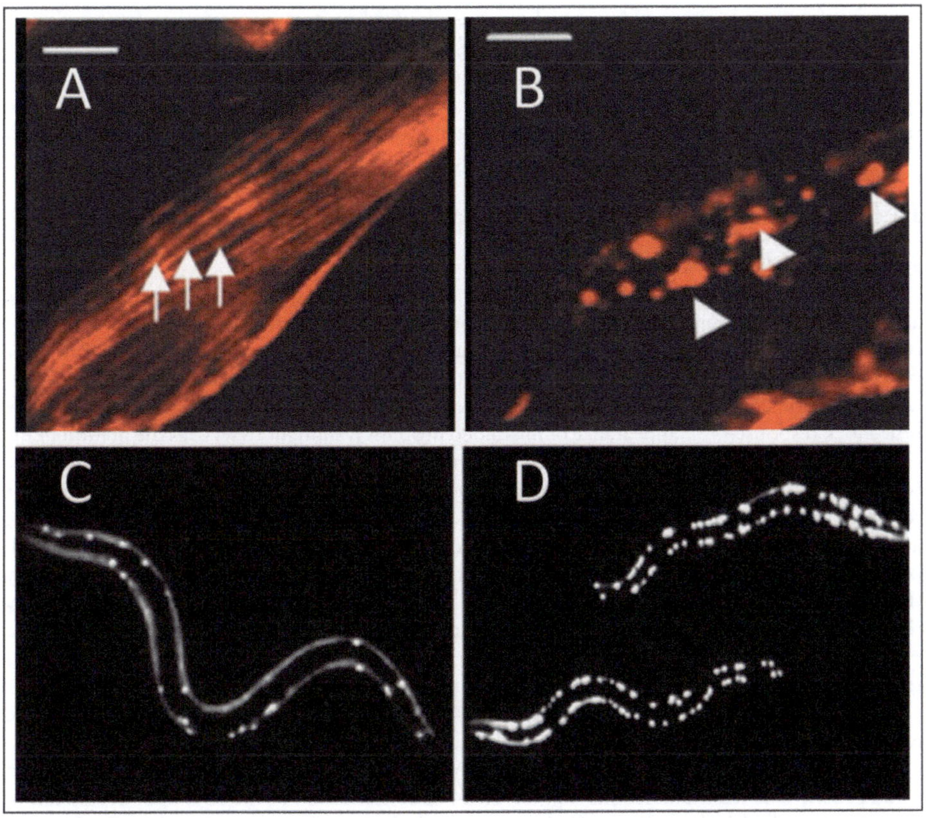

Figure 4. Progressive disruption of cellular folding capacity by misfolded proteins. A,B) Confocal images showing cellular localization of ts mutant paramyosin in the absence (A) or presence (B) of Q40-YFP. Arrows, normal muscle sarcomeres, arrowheads, abnormal paramyosin(ts) assemblies in the presence of Q40, both under permissive conditions. C,D) Images of Q40-YFP fluorescence in L2 larval stage animals expressing wild type (C) or ts mutant paramyosin (D) under permissive conditions. Adapted from: Gidalevitz T et al. Science 2006; 311:1471-1474;[3] ©2006 with permission from AAAS.

mutations have been expressed in a number of genetic systems including mouse, *Drosophila* and *C. elegans*. To model SOD1 protein misfolding in *Drosophila*, wild type human SOD1 (hSOD1), two mutant forms of SOD1 (G85R and A4V) and *Drosophila* SOD1 (dSOD1) were overexpressed in *Drosophila* motor neurons.[114] It was shown that unlike dSOD1, wild type and mutant hSOD1 are toxic, but non-aggregating in *Drosophila*.[114] This is in contrast to what was observed when hSOD1 was expressed in *C. elegans* body wall muscle cells,[38] or neurons.[115] Namely, expression of YFP-tagged SOD1 (G85R) under the control of a pan-neuronal promoter resulted in motility defects and distinct neuronal subtype-type-specific aggregation phenotypes that were confirmed by FRAP analysis.[115] Likewise, expression of a number of different disease-associated SOD1 mutants in *C. elegans* body wall muscle cells resulted in the formation of SOD1 protein aggregates (Fig. 3A).[38] Interestingly, the toxic effects of mutant forms of SOD1 were subtle in body wall muscle cells, but enhanced in the background of metastable ts proteins.[38] This is consistent both with the finding that temperature sensitive proteins can exacerbate the toxic effects of polyQ protein expression (Fig. 4)[3] and with the idea that disease phenotypes and/or proteotoxicity are highly sensitive to changes in the folding environment, presumably due to limited folding resources (see discussion below).

Why are these disease-causing proteins toxic to cells? In other words, why can't cells, or more specifically, affected neurons, combat the toxic effect of their expression either by activating the machinery responsible for assisting in the folding of misfolded protein species or in clearing them via autophagy and/or proteasome-mediated degradation?

Failure of Homeostasis in Conformational Disease of Aging

Failure of Adaptive Stress Responses

The cytoplasmic heat shock response and the ER and mitochondrial unfolded protein responses are the adaptive responses employed by cells to combat proteotoxic stress. The heat shock response is a multistep process that results in the immediate induction of transcription and translation of genes encoding molecular chaperones, proteases and other proteostasis regulators and in a generalized translational silencing. Although recent evidence indicates that additional signals may activate heat shock response,[116-118] appearance of misfolded proteins is thought to serve as the predominant signal for its induction.[119-123] Misfolded proteins are thought to titrate HSF1-associated chaperones, releasing the inhibition of HSF1 and enabling it to trimerize and translocate into the nucleus. This results in the activation of Hsp gene transcription, rebalancing of proteostasis and, finally, the attenuation of the heat shock response (reviewed in ref. 124). The main function of heat shock proteins during stress is to prevent inappropriate protein interactions and aggregation, mainly through binding to the exposed hydrophobic areas on a variety of cellular proteins, followed by the facilitation of refolding, or targeting to degradation during the recovery phase. Many of these proteins also function as molecular chaperones in the absence of stress, by guiding conformational transitions during synthesis, folding, translocation, assembly and degradation of proteins.[124-126] As chaperones affect other cellular functions, such as signaling,[127] the proper regulation of chaperone expression is critical to the health of the cell.

The ability of cells to manage general/widespread protein misfolding during heat shock and other proteotoxic stress conditions suggests that the combination of the abundant expression of chaperones under basal conditions,[50,128] together with the adaptive stress responses, provides sufficient 'folding capacity' to buffer unexpected folding requirements. Thus, the accumulation of misfolded and aggregated proteins associated with aging-related conformational diseases, and the misfolding-associated toxicity, indicate a failure of folding homeostasis.

Understanding the molecular mechanisms underlying this dysfunction is essential to both defining the mechanisms of toxicity and finding targets for corrective interventions in conformational disease and, perhaps, aging. Several possible, but not mutually exclusive, mechanisms are indicated by observations of dysregulation of proteostasis under these conditions. First, Hsp expression is often not induced in symptomatic cells, despite the accumulation of misfolded and aggregated proteins; moreover, decrease in specific chaperone expression has been noted.[129-134] In *C. elegans*, intracellular accumulation of misfolded and aggregated polyQ proteins only sporadically activated HSP expression[96] and in fact required the downregulation of ILS signaling for modulation of its toxicity.[66,68] This points to a potential override of the cellular stress response at the organismal level,[74] which traditionally has been considered to be cell-autonomous. Alternatively, it is possible that the accumulation of misfolded proteins in models of conformational diseases is either too gradual, or does not reach the threshold for heat shock activation. Second, molecular chaperones, components of degradative machinery and other proteins are often found trapped in aggregates, potentially mimicking hypomorphic phenotypes.[96,135,136] Third, there is evidence that the accumulation of misfolded proteins leads to the inactivation of components of proteostasis networks, in particular the proteasome[45,137,138] and of the cell's ability to induce a heat shock response[139] through the disease-associated misfolded proteins. The downregulation of specific chaperones and the inhibition of heat shock induction could potentially indicate that cells (or organisms) adapt to the chronic expression of misfolded proteins by actively preventing stress induction.

Whatever the specific mechanism of proteostasis dysfunction, it is clear that our ability to potentially adjust, if not correct, proteostatic networks to cope with the chronic protein misfolding may be essential to our ability to manage, or even prevent, conformational diseases of aging.[7]

Disruption of Proteostasis by Chronic Misfolding

Our knowledge of how protein misfolding and the disruption of proteostasis translate into cell-specific and disease-specific toxicity is still quite limited. Evidence from *C. elegans* suggests that the toxicity associated with expression of expanded polyQ and mutant SOD1 proteins is caused, in part, by the global disruption of the cellular protein folding homeostasis, resulting in a destabilization or misfolding of various metastable proteins.[3,38] The potential sources of metastable proteins, as described above, range from expressed protein polymorphisms, to protein damage from environmental stresses, to what has been termed 'activities of daily living', for example cell signaling, glycolysis, or respiration. When temperature-sensitive (ts) metastable proteins were used to mimic naturally occurring mildly destabilizing polymorphisms, they acted both as sensors of cellular protein folding capacity and contributors to cellular dysfunction. Expression of polyQ or mutant SOD1 in muscle or neuronal cells of *C. elegans* leads to the exposure of the temperature-sensitive phenotype at permissive conditions, mediated by misfolding and loss-of function of ts mutant protein present in the same cell (Fig. 4A).[3,38] Furthermore, the misfolding of ts proteins not only directed the specific phenotypes, but further increased aggregation of the polyQ proteins (Fig. 4B). The misfolding of ts proteins was most likely due to the depletion, by the polyQ or mutSOD1 proteins, of folding resources, necessary for maintaining these ts proteins in their folded and functional conformations.[140,141] A recent finding that many of the modifiers of toxicity of polyQ-expanded ataxin-3 in Drosophila also rescue the generic toxicity of protein misfolding due to the reduced function of Hsp70[142] strongly supports the disruption of proteostasis as a mechanism of toxicity. Furthermore, both the functionality of metastable proteins and the aggregation of polyglutamine expansions can be compromised by neuronally mediated overexcitation of the muscle cells in *C. elegans*.[101]

These findings parallel the evidence discussed above, that the selection against the toxicity of misfolding due to mistranslation exerts strong evolutionary pressure specifically on the highly expressed proteins.[37] This indicates that the flux of destabilized proteins in a cell bears a significant fitness cost and that folding homeostasis is indeed not robust, at least when it comes to chronically misfolded proteins. Importantly for this proposition, the toxic interaction between the destabilized protein polymorphisms (coding for ts proteins) and the disease-related polyQ or mutSOD1 was reversed not only by overexpression of HSF1, but also by the aging regulator DAF-16.[38]

The causative connection between proteostasis, aging and cellular dysfunction was further illustrated in *C. elegans* by showing that, unlike in young animals, the ts proteins in older adults kept at the permissive conditions begin to gradually misfold and lose function, coincident with a reduced ability to activate the heat shock response and the unfolded protein response.[143] Increasing the activity of either HSF1, or DAF-16, suppressed the misfolding of metastable proteins and restored cellular proteostasis. Thus, the dysregulation of protein folding homeostasis may represent a set of early molecular events in aging, with an ability to amplify the protein damage cascade in age-related conformational diseases, while the complement of mutations and polymorphisms, together with the life history of an organism (environmental stress exposure, metabolic state, etc.), set the threshold for the onset of dysfunction and direct specific phenotypes.

Consistent with the proposal that a disruption in proteostasis is a key element of mechanism of toxicity in conformational diseases, genetic screens in invertebrate models described above have revealed proteostasis components as modifiers of aggregation/toxicity.

Modifiers of Conformational Disease

Genetic Screens for Modifiers of Disease-Related Phenotypes

The versatility of *C. elegans* as a model system to study molecular processes involved in human disease has been demonstrated via the implementation of genome-wide RNA interference (RNAi) screens to identify genetic modifiers of disease-related phenotypes. Such screens have been facilitated by the availability of RNAi libraries, consisting of *E. coli* clones containing IPTG-inducible double stranded RNAs for the majority of *C. elegans* genes.[144] Feeding of dsRNA-producing *E. coli* to *C. elegans* has proven to be a highly efficient method for targeted gene silencing, making high-throughput RNAi screens relatively straightforward.[145]

One such study identified modifiers of polyQ protein aggregation in body-wall muscle cells in *C. elegans*.[100] The genome-wide screen took advantage of a polyQ length at the threshold for aggregation (Q35), thereby allowing for a sensitized screen aimed at the identification of factors, which, when knocked down in the background of polyQ-YFP, led to the accumulation of visible protein aggregates. This screen identified 186 proteins that normally suppress age-dependent polyQ protein aggregation, including HSF1 and Hsp-1 (Fig. 3C). The authors found that the suppressors fall into five distinct biological classes: RNA metabolism, protein synthesis, protein folding, protein trafficking and protein degradation.

The identification of chaperones and factors involved in protein clearance was expected and consistent with the results of a screen for genetic modifiers of polyQ protein toxicity in *Drosophila*, that uncovered homologs of Hdj1 and Tpr2, both J-domain-containing cochaperones.[99] Additionally, Hsp70 or Hdj1 overexpression significantly ameliorated toxicity of the polyglutamine-containing proteins ataxin-1 and ataxin-3 in *Drosophila* (Fig. 3D).[98, 146] Thus, chaperones have been found consistently as modifiers of aggregation and toxicity across different tissues and aggregation models. The *C. elegans* screen also uncovered six of the eight subunits of cytosolic chaperonin CCT, whose role as a suppressor of polyQ protein aggregation was previously unknown. This finding was later validated by using both *S. cerevisiae* and mammalian tissue culture cells expressing aggregation-prone Htt-polyQ proteins,[147-149] supporting the idea that common mechanisms underlie polyQ protein aggregation/toxicity. These findings underscore the involvement of molecular chaperone activity in modulating the aggregation/toxicity of polyglutamine proteins. However, the identification of factors involved in other biosynthetic processes led to the conclusion that protein homeostasis is more complex than previously understood and likely begins with gene expression, thus explaining the large fraction of modifiers involved in RNA and protein biosynthesis.[100]

An independent RNAi screen was also performed in *C. elegans* for factors that normally suppress tau-induced motility defects.[150] Wild-type and mutant tau protein become hyperphosphorylated, aggregate and form neurofibrilllary tangles that are associated with neurodegeneration in patients suffering from Alzheimer's disease (AD) and a number of related neurodegenerative diseases.[151] Expression of tau in *C. elegans* neurons caused motility defects that were used as the basis to identify factors, via genome-wide RNAi screening, which, when absent, enhanced the motility (unc) phenotype.[150] This analysis led to the identification of 75 suppressors of tau toxicity falling into the following functional categories: kinases, chaperones, proteases and phosphatases, in addition to a number of genes whose function is unknown.[150] Interestingly, the only RNAi hits in common between this and the polyQ screen described above are the Hsp70 molecular chaperone, *hsp-1* and the heat shock transcription factor, *hsf-1*. Consistent with the idea that tau and polyQ proteins interact differentially with the cellular environment, a complementary study in a *Drosophila* tau model identified mostly kinases and phosphatases.[152] Significantly, no HSPs, molecular chaperones, or components of the protein clearance machineries were identified.[152]

A recent screen for suppressors of α-synuclein (α-syn) aggregation in body wall muscle cells revealed factors involved in vesicle trafficking and lipid metabolism.[112] This screen, however, did uncover several regulators of life-span, including *lagr-1* and *sir-2.1*, consistent with aging being a potent modifier of aggregation/toxicity. Interestingly, knock-down of Hsp70 had no effect on α-syn aggregation,[112] suggesting that α-syn has distinctly different mechanisms of aggregation/toxicity than polyQ or tau proteins. In support of α-syn and polyQ having different mechanisms of aggregation, only one modifier was identified in *C. elegans* that was in common for both proteins.[100,112] Alternatively, since this was an aggregation screen that did not directly examine toxicity, it is possible that aggregation and toxicity of α-syn are uncoupled in the presence of Hsp70 RNAi, a finding that would be consistent with one from *Drosophila*.[153]

The striking lack of overlap between modifiers of aggregation/toxicity for unique aggregation-prone proteins could presumably be due to different factors acting on different misfolded species, in addition to different factors acting in different tissue types, for example, body wall muscle cells as compared to neurons. Certainly, the identification of kinases and phosphatases in the screen for suppressors of tau toxicity provides support for the hypothesis that tau hyperphosphorylation is a prerequisite for disease. Consequently, these data also provide evidence that *C. elegans* and *D. melanogaster* are valid genetic models for the identification of factors which are capable of acting on particular human disease-causing proteins to suppress aggregation/toxicity.

In contrast to α-syn, a screen for modifiers of SOD1 aggregation and toxicity in *C. elegans* neurons yielded mostly molecular chaperones and other factors generally involved in protein quality control,[115] more consistent with what was seen previously with respect to modifiers of polyQ protein aggregation/toxicity. Furthermore, the hits obtained by an RNAi screen aimed at the identification of suppressors of osmotic stress-induced gene expression overlapped to a great extent with those identified as suppressors of polyQ protein aggregation.[154] In that study, the authors identified genes, including *gpdh-1*, that are upregulated in response to osmotic stress, which is known to cause generalized, nonspecific protein misfolding. Although *gpdh-1* expression does not respond to stresses other than osmotic stress,[154] almost 30% of all the genes identified in this study overlap with those identified in the screen for suppressors of polyQ protein aggregation. Furthermore, 73% of the overlapping genes are predicted to fall into biological classes usually associated with the regulation of proteostasis, including RNA processing, protein synthesis, protein folding and degradation.[154] Finally, many of the modifiers of ataxin-3 toxicity in *Drosophila* were also able to rescue the toxic phenotypes due to the reduced function of HSP70, which also causes generalized protein misfolding.[142] Ultimately, these data suggest that a core set of factors function generally in response to stress-induced protein damage and others respond specifically to a particular stress, for example the stress of a misfolded, disease-causing protein, to tailor the response to the situation at hand.

However, the extent to which the molecular mechanisms of disease are conserved between aggregation-prone proteins is unknown. To address this, it will be necessary to express unrelated aggregation-prone proteins, such as Aβ, SOD1, tau, α-syn and a polyQ-containing proteins, in the same cell type in a given model organism, for a direct comparison of modifiers of aggregation/toxicity. The expectation is that common modifiers would be those whose molecular function is in the general folding/clearance of misfolded proteins. On the other hand, modifiers that act on one or the other aggregation-prone protein would most likely be more closely associated with the specific function, or mode of disease progression, of a particular protein. For example, a screen performed in yeast expressing mutant Htt or α-synuclein, revealed almost entirely non-overlapping sets of genes, many with human homologs, acting as modifiers either of mutant Htt or α-synuclein toxicity.[155] The authors speculate that their modifiers likely define mechanisms or pathways that are specific for particular disease-causing proteins, such as vesicle transport playing a role in α-synuclein toxicity.[155]

In addition to RNAi and overexpression screens, forward genetic screens have also been performed to identify modulators of polyQ protein aggregation/toxicity in *C. elegans*. One such screen revealed a novel gene, *pqe-1*, which normally functions to suppress the proteotoxicity of an Htt exon 1 fragment with an expanded polyQ tract.[156] Another forward genetic screen was aimed at the identification of genes that normally function to suppress the aggregation of polyQ-YFP in body wall muscle cells.[101] This screen uncovered mutations in *unc-30*, which encodes the transcription factor that regulates the synthesis of the neurotransmitter GABA.[101] The findings described in[101] are of particular interest, because they demonstrate that the ability of an organism to manage proteotoxic stress is not a cell autonomous process as previously thought, but may be affected by cell-cell communication, for example via neuronal cholinergic signaling. Consistent with this, treating Q35-expressing *C. elegans* with small molecules, acting positively or negatively on neuronal signaling, suppressed or enhanced, respectively, polyQ protein aggregation[101] in postsynaptic cells. The identification, via chemical genetic screens, of additional small molecules that alleviate disease-causing protein aggregation/toxicity will be an important step in the development of pharmaceuticals.

Small Molecule Drug Screens

The search for therapeutics to treat neurodegeneration is turning to the identification of small molecule proteostasis regulators. Such small molecules could function by enhancing the expression or activities of molecular chaperones, thereby effectively increasing the rates of folding of client proteins, or, alternatively, they could act by enhancing degradation/clearance[157] or by modulating protein translation rates.

A number of small molecule regulators of the heat shock response have been identified. These compounds include proteasome inhibitors, serine protease inhibitors, Hsp90 inhibitors, non-steroidal anti-inflammatory drugs (NSAIDS), triterpenoids such as celastrol and inhibitors of HSF1, including triptolide.[49,158] Protease inhibitors (DCIC, TPCK and TLCK) and proteasome inhibitors (MG132, lactacystin) induce the heat-shock response by elevating the effective concentrations of misfolfed, damaged, or otherwise no longer needed, proteins that are normally targeted for degradation.[159,160] In contrast, other inducers of the heat shock response act as inhibitors of Hsp90. These include the fungal antibiotic radicicol, the benzoquinone ansamycins geldanamycin and 17AAG. They activate HSF1, in part because Hsp90 is a negative regulator of HSF1.[61,161,162] NSAIDS, including sodium salicylate, have multiple properties. At higher concentrations they partially activate HSF1, while at lower concentrations they synergize with other stress conditions to induce the heat-shock response.[163] Specifically, exposure of human tissue culture cells to sodium salicylate results in activation of HSF1 with respect to in vivo binding to the Hsp70 gene; however, transcriptional induction fails to occur. Salicylate-treated cells, however, are sensitized to stress and readily activate HS genes upon exposure to other mild stress conditions, often not sufficient themselves to activate the heat shock response. In a similar manner, indomethacin induces HSF1 DNA binding with full Hsp70 transcription upon exposure to a secondary stress.[164] Of the inflammatory modulators, arachidonic acid and the cyclopentenone prostaglandins, including PGA1, PGA2 and PGJ2, all induce the full complement of HSF1 activities.[165,166] The triterpenoid celastrol isolated from the Chinese plant *Triptergium wilfordii* represents an herbal medicine class of bioactive molecules that induces two protective stress responses, the heat shock response and the anti-oxidant response.[167,168] The effects of celastrol are rapid like heat shock; however, unlike the heat-shock response that self-attenuates, the celastrol induction of HS genes persists for an extended period.[167] Consistent with small molecule inducers of HSF1 acting to alleviate the toxic effects of misfolded disease-causing proteins, geranylgeranylacetone (GGA)[169] and celastrol[170] treatment of polyQ protein-expressing cells inhibited polyQ-associated cell death in tissue culture and mouse models of polyQ protein misfolding.

The modulation of protein translation rates also plays an important role in inducible stress responses. A screen for small molecule inhibitors of ER stress-induced apoptosis in PC-12 cells yielded salubrinol, a selective inhibitor of eIF2α dephosphorylation.[171] eIF2α is a translation initiation factor that, when in its phosphorylated state, either under conditions of heat stress or salubrinol treatment, is responsible for mediating a general decrease, but a selective increase in the translation of chaperones and other stress-associated proteins. Like small molecule inducers of HSF1, salubrinol, or other translational regulators, might be expected to modulate the toxic effects of misfolded disease-associated proteins.

Most of the chemical genetic analyses that have yielded the small molecules described above have been performed in tissue culture cells. A limited number of studies aimed at the identification of small molecules that act in invertebrate models to alleviate the toxicity of misfolding-prone disease-associated proteins have been performed. More commonly, small molecules that were identified in tissue culture systems as having the properties of inducing the HSR have been tested in invertebrates. For example, it was shown that the treatment of

Htt-expressing *C. elegans* with the antioxidant resveratrol reduced Htt toxicity in a manner dependent on the sirtuin, sir2.1.[172]

A small-scale chemical genetics analysis was performed in a *C. elegans* Htt-polyQ model, using a screening approach which circumvented the inherent problems associated with screening based on motility defects. The authors examined treated worms for lack of neuronal cell death by visualizing the loss, or lack thereof, of GFP fluorescence in ASH neurons of a sensitized line that rapidly undergoes polyQ-dependent neurodegeneration.[173] These conditions were used to validate candidate compounds identified previously in a large-scale, tissue culture-based, screen[174,175] and revealed two compounds, lithium chloride and mithramycin, which suppressed HD neurotoxicity in the *C. elegans* Htt-polyQ model.[173] The use of this and similar assays, should make it possible to rapidly screen large chemical libraries for their effect on toxicity in *C. elegans* models of neurodegenerative diseases.

Because *C. elegans* is a multicellular organism, we would expect that the successful implementation of large-scale chemical genetics screens will be highly effective in identifying novel therapeutic compounds, not previously identified in cell culture models, that act either cell autonomously or cell non-autonomously. Finally, fluorescent labeling of candidate molecules will be instrumental in elucidating the mode of drug action and to determine whether the drug is acting directly or indirectly on the disease-causing proteins. This will be relatively straightforward in *C. elegans* due the easy visualization of fluorescent markers.

Conclusion

The work discussed here highlights the utility of invertebrate models in the study of neurodegenerative diseases of aging. Recent findings from these model systems have strengthened our understanding of protein conformational diseases. Specifically, they have led to the proposal that protein folding homeostasis, while sufficiently robust to manage protein damage/misfolding caused by acute environmental stress such as heat shock, is apparently ineffective when faced with the chronic expression of an aggregation-prone disease-causing protein. This limited capacity of the cell to manage chronic misfolding is especially pronounced under conditions of additional stress on the proteome caused by the expression of metastable proteins, partially compromised in their folding and providing a relatively high demand for folding resources. To combat this inherent limitation for the purpose of treating patients, small molecule drugs are being sought that will enhance the ability of a cell, or organism, to deal with the expression of chronically misfolded proteins.

Acknowledgements

E.A.K was supported by an individual postdoctoral fellowship from the National Institutes of Health (NINDS); research in the laboratory of R.I.M. was supported by grants from the National Institutes of Health (NIGMS and NIA), the HDSA Coalition for the Cure and the ALS Association.

References

1. Rutherford SL, Lindquist S. Hsp90 as a capacitor for morphological evolution. Nature 1998; 396:336-342.
2. Stevens FJ, Argon Y. Pathogenic light chains and the B-cell repertoire. Immunol Today 1999; 20:451-457.
3. Gidalevitz T, Ben-Zvi A, Ho KH et al. Progressive disruption of cellular protein folding in models of polyglutamine diseases. Science 2006; 311:1471-1474.
4. Yeyati PL, Bancewicz RM, Maule J et al. Hsp90 selectively modulates phenotype in vertebrate development. PLoS Genet 2007; 3:e43.
5. Michels AA, Kanon B, Konings AW et al. Hsp70 and Hsp40 chaperone activities in the cytoplasm and the nucleus of mammalian cells. J Biol Chem 1997; 272:33283-33289.
6. Parsell DA, Lindquist S. The function of heat-shock proteins in stress tolerance: degradation and reactivation of damaged proteins. Annu Rev Genet 1993; 27:437-496.
7. Balch WE, Morimoto RI, Dillin A et al. Adapting proteostasis for disease intervention. Science 2008; 319:916-919.

8. Harman D. Aging: a theory based on free radical and radiation chemistry. J Gerontol 1956; 11:298-300.
9. Sohal RS, Weindruch R. Oxidative stress, caloric restriction and aging. Science 1996; 273:59-63.
10. Berlett BS, Stadtman ER. Protein oxidation in aging, disease and oxidative stress. J Biol Chem 1997; 272:20313-20316.
11. Stadtman ER, Oliver CN. Metal-catalyzed oxidation of proteins. Physiological consequences. J Biol Chem 1991; 266:2005-2008.
12. Smith CD, Carney JM, Starke-Reed PE et al. Excess brain protein oxidation and enzyme dysfunction in normal aging and in Alzheimer disease. Proc Natl Acad Sci USA 1991; 88:10540-10543.
13. Stadtman ER. Protein oxidation and aging. Science 1992; 257:1220-1224.
14. Ahmed N, Dobler D, Dean M et al. Peptide mapping identifies hotspot site of modification in human serum albumin by methylglyoxal involved in ligand binding and esterase activity. J Biol Chem 2005; 280:5724-5732.
15. Lo TW, Westwood ME, McLellan AC et al. Binding and modification of proteins by methylglyoxal under physiological conditions. A kinetic and mechanistic study with N alpha-acetylarginine, N alpha-acetylcysteine and N alpha-acetyllysine and bovine serum albumin. J Biol Chem 1994; 269:32299-32305.
16. Kalapos MP. Methylglyoxal in living organisms: chemistry, biochemistry, toxicology and biological implications. Toxicol Lett 1999; 110:145-175.
17. Kuhla B, Boeck K, Schmidt A et al. Age- and stage-dependent glyoxalase I expression and its activity in normal and Alzheimer's disease brains. Neurobiol Aging 2007; 28:29-41.
18. Hipkiss AR. On the mechanisms of ageing suppression by dietary restriction-is persistent glycolysis the problem? Mech Ageing Dev 2006; 127:8-15.
19. Gnerer JP, Kreber RA, Ganetzky B. Wasted away, a Drosophila mutation in triosephosphate isomerase, causes paralysis, neurodegeneration and early death. Proc Natl Acad Sci USA 2006; 103:14987-14993.
20. Moskovitz J. Roles of methionine suldfoxide reductases in antioxidant defense, protein regulation and survival. Curr Pharm Des 2005; 11:1451-1457.
21. Friguet B. Oxidized protein degradation and repair in ageing and oxidative stress. FEBS Lett 2006; 580:2910-2916.
22. Stadtman ER. Oxidation of free amino acids and amino acid residues in proteins by radiolysis and by metal-catalyzed reactions. Annu Rev Biochem 1993; 62:797-821.
23. Hanson SR, Hasan A, Smith DL et al. The major in vivo modifications of the human water-insoluble lens crystallins are disulfide bonds, deamidation, methionine oxidation and backbone cleavage. Exp Eye Res 2000; 71:195-207.
24. Breusing N, Grune T. Regulation of proteasome-mediated protein degradation during oxidative stress and aging. Biol Chem 2008; 389:203-209.
25. Davies KJ. Degradation of oxidized proteins by the 20S proteasome. Biochimie 2001; 83:301-310.
26. Shringarpure R, Grune T, Mehlhase J et al. Ubiquitin conjugation is not required for the degradation of oxidized proteins by proteasome. J Biol Chem 2003; 278:311-318.
27. Shang F, Nowell TR Jr, Taylor A. Removal of oxidatively damaged proteins from lens cells by the ubiquitin-proteasome pathway. Exp Eye Res 2001; 73:229-238.
28. Bota DA, Davies KJ. Lon protease preferentially degrades oxidized mitochondrial aconitase by an ATP-stimulated mechanism. Nat Cell Biol 2002; 4:674-680.
29. Reinheckel T, Grune T, Davies KJ. The measurement of protein degradation in response to oxidative stress. Methods Mol Biol 2000; 99:49-60.
30. Kiffin R, Christian C, Knecht E et al. Activation of chaperone-mediated autophagy during oxidative stress. Mol Biol Cell 2004; 15:4829-4840.
31. Schultz SC, Richards JH. Site-saturation studies of beta-lactamase: production and characterization of mutant beta-lactamases with all possible amino acid substitutions at residue 71. Proc Natl Acad Sci USA 1986; 83:1588-1592.
32. Pakula AA, Sauer RT. Genetic analysis of protein stability and function. Annu Rev Genet 1989; 23:289-310.
33. Ng PC, Henikoff S. Predicting the effects of amino acid substitutions on protein function. Annu Rev Genomics Hum Genet 2006; 7:61-80.
34. DePristo MA, Weinreich DM, Hartl DL. Missense meanderings in sequence space: a biophysical view of protein evolution. Nat Rev Genet 2005; 6:678-687.
35. Suckow J, Markiewicz P, Kleina LG et al. Genetic studies of the Lac repressor. XV: 4000 single amino acid substitutions and analysis of the resulting phenotypes on the basis of the protein structure. J Mol Biol 1996; 261:509-523.

36. Sachidanandam R, Weissman D, Schmidt SC et al. A map of human genome sequence variation containing 1.42 million single nucleotide polymorphisms. Nature 2001; 409:928-933.
37. Drummond DA, Wilke CO. Mistranslation-induced protein misfolding as a dominant constraint on coding-sequence evolution. Cell 2008; 134:341-352.
38. Gidalevitz T, Krupinski T, Garcia S et al. Destabilizing protein polymorphisms in the genetic background direct phenotypic expression of mutant SOD1 toxicity. PLoS Genet 2009; 5:e1000399.
39. Stefani M, Dobson CM. Protein aggregation and aggregate toxicity: new insights into protein folding, misfolding diseases and biological evolution. J Mol Med 2003; 81:678-699.
40. Kopito RR, Ron D. Conformational disease. Nat Cell Biol 2000; 2:E207-209.
41. Lee JW, Beebe K, Nangle LA et al. Editing-defective tRNA synthetase causes protein misfolding and neurodegeneration. Nature 2006; 443:50-55.
42. Jordanova A, Irobi J, Thomas FP et al. Disrupted function and axonal distribution of mutant tyrosyl-tRNA synthetase in dominant intermediate Charcot-Marie-Tooth neuropathy. Nat Genet 2006; 38:197-202.
43. Martin I, Grotewiel MS. Oxidative damage and age-related functional declines. Mech Ageing Dev 2006; 127:411-423.
44. Calderwood SK, Murshid A, Prince T. The Shock of Aging: Molecular Chaperones and the Heat Shock Response in Longevity and Aging—A Mini-Review. Gerontology 2009.
45. Holmberg CI, Staniszewski KE, Mensah KN et al. Inefficient degradation of truncated polyglutamine proteins by the proteasome. EMBO J 2004; 23:4307-4318.
46. Kim S, Nollen EA, Kitagawa K et al. Polyglutamine protein aggregates are dynamic. Nat Cell Biol 2002; 4:826-831.
47. Suhr ST, Senut MC, Whitelegge JP et al. Identities of sequestered proteins in aggregates from cells with induced polyglutamine expression. J Cell Biol 2001; 153:283-294.
48. Kryukov GV, Pennacchio LA, Sunyaev SR. Most rare missense alleles are deleterious in humans: implications for complex disease and association studies. Am J Hum Genet 2007; 80:727-739.
49. Westerheide SD, Morimoto RI. Heat shock response modulators as therapeutic tools for diseases of protein conformation. J Biol Chem 2005; 280:33097-33100.
50. Ron D, Walter P. Signal integration in the endoplasmic reticulum unfolded protein response. Nat Rev Mol Cell Biol 2007; 8:519-529.
51. Banerji SS, Theodorakis NG, Morimoto RI. Heat shock-induced translational control of HSP70 and globin synthesis in chicken reticulocytes. Mol Cell Biol 1984; 4:2437-2448.
52. Xiao X, Zuo X, Davis AA et al. HSF1 is required for extra-embryonic development, postnatal growth and protection during inflammatory responses in mice. EMBO J 1999; 18:5943-5952.
53. Santos SD, Saraiva MJ. Enlarged ventricles, astrogliosis and neurodegeneration in heat shock factor 1 null mouse brain. Neuroscience 2004; 126:657-663.
54. Zhang P, McGrath B, Li S et al. The PERK eukaryotic initiation factor 2 alpha kinase is required for the development of the skeletal system, postnatal growth and the function and viability of the pancreas. Mol Cell Biol 2002; 22:3864-3874.
55. Reimold AM, Iwakoshi NN, Manis J et al. Plasma cell differentiation requires the transcription factor XBP-1. Nature 2001; 412:300-307.
56. Zhang K, Wong HN, Song B et al. The unfolded protein response sensor IRE1alpha is required at 2 distinct steps in B-cell lymphopoiesis. J Clin Invest 2005; 115:268-281.
57. Elefant F, Palter KB. Tissue-specific expression of dominant negative mutant Drosophila HSC70 causes developmental defects and lethality. Mol Biol Cell 1999; 10:2101-2117.
58. Feder JH, Rossi JM, Solomon J et al. The consequences of expressing hsp70 in Drosophila cells at normal temperatures. Genes Dev 1992; 6:1402-1413.
59. Nylandsted J, Rohde M, Brand K et al. Selective depletion of heat shock protein 70 (Hsp70) activates a tumor-specific death program that is independent of caspases and bypasses Bcl-2. Proc Natl Acad Sci USA 2000; 97:7871-7876.
60. Whitesell L, Mimnaugh EG, De Costa B et al. Inhibition of heat shock protein HSP90-pp60v-src heteroprotein complex formation by benzoquinone ansamycins: essential role for stress proteins in oncogenic transformation. Proc Natl Acad Sci USA 1994; 91:8324-8328.
61. Whitesell L, Lindquist SL. HSP90 and the chaperoning of cancer. Nat Rev Cancer 2005; 5:761-772.
62. Duyao M, Ambrose C, Myers R et al. Trinucleotide repeat length instability and age of onset in Huntington's disease. Nat Genet 1993; 4:387-392.
63. Morley JF, Brignull HR, Weyers JJ et al. The threshold for polyglutamine-expansion protein aggregation and cellular toxicity is dynamic and influenced by aging in Caenorhabditis elegans. Proc Natl Acad Sci USA 2002; 99:10417-10422.
64. Morris JZ, Tissenbaum HA, Ruvkun G. A phosphatidylinositol-3-OH kinase family member regulating longevity and diapause in Caenorhabditis elegans. Nature 1996; 382:536-539.

65. Lin K, Dorman JB, Rodan A et al. daf-16: An HNF-3/forkhead family member that can function to double the life-span of Caenorhabditis elegans. Science 1997; 278:1319-1322.
66. Morley JF, Morimoto RI. Regulation of longevity in Caenorhabditis elegans by heat shock factor and molecular chaperones. Mol Biol Cell 2004; 15:657-664.
67. Hsu AL, Murphy CT, Kenyon C. Regulation of aging and age-related disease by DAF-16 and heat-shock factor. Science 2003; 300:1142-1145.
68. Cohen E, Bieschke J, Perciavalle RM et al. Opposing activities protect against age-onset proteotoxicity. Science 2006; 313:1604-1610.
69. Lithgow GJ, White TM, Melov S et al. Thermotolerance and extended life-span conferred by single-gene mutations and induced by thermal stress. Proc Natl Acad Sci USA 1995; 92:7540-7544.
70. Salmon AB, Sadighi Akha AA, Buffenstein R et al. Fibroblasts from naked mole-rats are resistant to multiple forms of cell injury, but sensitive to peroxide, ultraviolet light and endoplasmic reticulum stress. J Gerontol A Biol Sci Med Sci 2008; 63:232-241.
71. Salmon AB, Murakami S, Bartke A et al. Fibroblast cell lines from young adult mice of long-lived mutant strains are resistant to multiple forms of stress. Am J Physiol Endocrinol Metab 2005; 289:E23-29.
72. Broughton SJ, Piper MD, Ikeya T et al. Longer lifespan, altered metabolism and stress resistance in Drosophila from ablation of cells making insulin-like ligands. Proc Natl Acad Sci USA 2005; 102:3105-3110.
73. Tullet JM, Hertweck M, An JH et al. Direct inhibition of the longevity-promoting factor SKN-1 by insulin-like signaling in C. elegans. Cell 2008; 132:1025-1038.
74. Prahlad V, Cornelius T, Morimoto RI. Regulation of the cellular heat shock response in Caenorhabditis elegans by thermosensory neurons. Science 2008; 320:811-814.
75. Bienz M. Developmental control of the heat shock response in Xenopus. Proc Natl Acad Sci USA 1984; 81:3138-3142.
76. Sprang GK, Brown IR. Selective induction of a heat shock gene in fibre tracts and cerebellar neurons of the rabbit brain detected by in situ hybridization. Brain Res 1987; 427:89-93.
77. Shamovsky I, Gershon D. Novel regulatory factors of HSF-1 activation: facts and perspectives regarding their involvement in the age-associated attenuation of the heat shock response. Mech Ageing Dev 2004; 125:767-775.
78. Mathur SK, Sistonen L, Brown IR et al. Deficient induction of human hsp70 heat shock gene transcription in Y79 retinoblastoma cells despite activation of heat shock factor 1. Proc Natl Acad Sci USA 1994; 91:8695-8699.
79. Marcuccilli CJ, Mathur SK, Morimoto RI et al. Regulatory differences in the stress response of hippocampal neurons and glial cells after heat shock. J Neurosci 1996; 16:478-485.
80. Batulan Z, Shinder GA, Minotti S et al. High threshold for induction of the stress response in motor neurons is associated with failure to activate HSF1. J Neurosci 2003; 23:5789-5798.
81. Muchowski PJ, Wacker JL. Modulation of neurodegeneration by molecular chaperones. Nat Rev Neurosci 2005; 6:11-22.
82. Walker GA, Lithgow GJ. Lifespan extension in C. elegans by a molecular chaperone dependent upon insulin-like signals. Aging Cell 2003; 2:131-139.
83. YYokoyama K, Fukumoto K, Murakami T et al. Extended longevity of Caenorhabditis elegans by knocking in extra copies of hsp70F, a homolog of mot-2 (mortalin)/mthsp70/Grp75. FEBS Lett 2002; 516:53-57.
84. Herndon LA, Schmeissner PJ, Dudaronek JM et al. Stochastic and genetic factors influence tissue-specific decline in ageing C. elegans. Nature 2002; 419:808-814.
85. Yang J, Tower J. Expression of hsp22 and hsp70 transgenes is partially predictive of drosophila survival under normal and stress conditions. J Gerontol A Biol Sci Med Sci 2009; 64:828-838.
86. Yun C, Stanhill A, Yang Y et al. Proteasomal adaptation to environmental stress links resistance to proteotoxicity with longevity in Caenorhabditis elegans. Proc Natl Acad Sci USA 2008; 105:7094-7099.
87. Brunet A, Sweeney LB, Sturgill JF et al. Stress-dependent regulation of FOXO transcription factors by the SIRT1 deacetylase. Science 2004; 303:2011-2015.
88. Westerheide SD, Anckar J, Stevens SM et al. Stress-inducible regulation of heat shock factor 1 by the deacetylase SIRT1. Science 2009; 323:1063-1066.
89. Lee SS, Kennedy S, Tolonen AC et al. DAF-16 target genes that control C. elegans life-span and metabolism. Science 2003; 300:644-647.
90. Kakizuka A. Protein precipitation: a common etiology in neurodegenerative disorders? Trends Genet 1998; 14:396-402.
91. Chiti F, Dobson CM. Protein misfolding, functional amyloid and human disease. Annu Rev Biochem 2006; 75:333-366.
92. Warrick JM, Paulson HL, Gray-Board GL et al. Expanded polyglutamine protein forms nuclear inclusions and causes neural degeneration in Drosophila. Cell 1998; 93:939-949.

93. Faber PW, Alter JR, MacDonald ME et al. Polyglutamine-mediated dysfunction and apoptotic death of a Caenorhabditis elegans sensory neuron. Proc Natl Acad Sci USA 1999; 96:179-184.
94. Feany MB, Bender WW. A Drosophila model of Parkinson's disease. Nature 2000; 404:394-398.
95. Krobitsch S, Lindquist S. Aggregation of huntingtin in yeast varies with the length of the polyglutamine expansion and the expression of chaperone proteins. Proc Natl Acad Sci USA 2000; 97:1589-1594.
96. Satyal SH, Schmidt E, Kitagawa K et al. Polyglutamine aggregates alter protein folding homeostasis in Caenorhabditis elegans. Proc Natl Acad Sci USA 2000; 97:5750-5755.
97. Parker JA, Connolly JB, Wellington C et al. Expanded polyglutamines in Caenorhabditis elegans cause axonal abnormalities and severe dysfunction of PLM mechanosensory neurons without cell death. Proc Natl Acad Sci USA 2001; 98:13318-13323.
98. Warrick JM, Chan HY, Gray-Board GL et al. Suppression of polyglutamine-mediated neurodegeneration in Drosophila by the molecular chaperone HSP70. Nat Genet 1999; 23:425-428.
99. Kazemi-Esfarjani P, Benzer S. Genetic suppression of polyglutamine toxicity in Drosophila. Science 2000; 287:1837-1840.
100. Nollen EA, Garcia SM, van Haaften G et al. Genome-wide RNA interference screen identifies previously undescribed regulators of polyglutamine aggregation. Proc Natl Acad Sci USA 2004; 101:6403-6408.
101. Garcia SM, Casanueva MO, Silva MC et al. Neuronal signaling modulates protein homeostasis in Caenorhabditis elegans postsynaptic muscle cells. Genes Dev 2007; 21:3006-3016.
102. Jackson GR, Salecker I, Dong X et al. Polyglutamine-expanded human huntingtin transgenes induce degeneration of Drosophila photoreceptor neurons. Neuron 1998; 21:633-642.
103. Brignull HR, Moore FE, Tang SJ et al. Polyglutamine proteins at the pathogenic threshold display neuron-specific aggregation in a pan-neuronal Caenorhabditis elegans model. J Neurosci 2006; 26:7597-7606.
104. Marsh JL, Walker H, Theisen H et al. Expanded polyglutamine peptides alone are intrinsically cytotoxic and cause neurodegeneration in Drosophila. Hum Mol Genet 2000; 9:13-25.
105. Link CD. Expression of human beta-amyloid peptide in transgenic Caenorhabditis elegans. Proc Natl Acad Sci USA 1995; 92:9368-9372.
106. Finelli A, Kelkar A, Song HJ et al. A model for studying Alzheimer's Abeta42-induced toxicity in Drosophila melanogaster. Mol Cell Neurosci 2004; 26:365-375.
107. Wittmann CW, Wszolek MF, Shulman JM et al. Tauopathy in Drosophila: neurodegeneration without neurofibrillary tangles. Science 2001; 293:711-714.
108. Kraemer BC, Zhang B, Leverenz JB et al. Neurodegeneration and defective neurotransmission in a Caenorhabditis elegans model of tauopathy. Proc Natl Acad Sci USA 2003; 100:9980-9985.
109. Nussbaum RL, Polymeropoulos MH. Genetics of Parkinson's disease. Hum Mol Genet 1997; 6:1687-1691.
110. Chase TN. A gene for Parkinson disease. Arch Neurol 1997; 54:1156-1157.
111. Lakso M, Vartiainen S, Moilanen AM et al. Dopaminergic neuronal loss and motor deficits in Caenorhabditis elegans overexpressing human alpha-synuclein. J Neurochem 2003; 86:165-172.
112. van Ham TJ, Thijssen KL, Breitling R et al. C. elegans model identifies genetic modifiers of alpha-synuclein inclusion formation during aging. PLoS Genet 2008; 4:e1000027.
113. Rosen DR. Mutations in Cu/Zn superoxide dismutase gene are associated with familial amyotrophic lateral sclerosis. Nature 1993; 364:362.
114. Watson MR, Lagow RD, Xu K et al. A drosophila model for amyotrophic lateral sclerosis reveals motor neuron damage by human SOD1. J Biol Chem 2008; 283:24972-24981.
115. Wang J, Farr GW, Hall DH et al. An ALS-linked mutant SOD1 produces a locomotor defect associated with aggregation and synaptic dysfunction when expressed in neurons of Caenorhabditis elegans. PLoS Genet 2009; 5:e1000350.
116. Ahn SG, Thiele DJ. Redox regulation of mammalian heat shock factor 1 is essential for Hsp gene activation and protection from stress. Genes Dev 2003; 17:516-528.
117. Hahn JS, Thiele DJ. Activation of the Saccharomyces cerevisiae heat shock transcription factor under glucose starvation conditions by Snf1 protein kinase. J Biol Chem 2004; 279:5169-5176.
118. Thomson S, Hollis A, Hazzalin CA et al. Distinct stimulus-specific histone modifications at hsp70 chromatin targeted by the transcription factor heat shock factor-1. Mol Cell 2004; 15:585-594.
119. Ananthan J, Goldberg AL, Voellmy R. Abnormal proteins serve as eukaryotic stress signals and trigger the activation of heat shock genes. Science 1986; 232:522-524.
120. Hiromi Y, Okamoto H, Gehring WJ et al. Germline transformation with Drosophila mutant actin genes induces constitutive expression of heat shock genes. Cell 1986; 44:293-301.
121. Parsell DA, Sauer RT. Induction of a heat shock-like response by unfolded protein in Escherichia coli: dependence on protein level not protein degradation. Genes Dev 1989; 3:1226-1232.
122. Lindquist S, Craig EA. The heat-shock proteins. Annu Rev Genet 1988; 22:631-677.

123. Morimoto RI. Regulation of the heat shock transcriptional response: cross talk between a family of heat shock factors, molecular chaperones and negative regulators. Genes Dev 1998; 12:3788-3796.
124. Morimoto RI, Kline MP, Bimston DN et al. The heat-shock response: regulation and function of heat-shock proteins and molecular chaperones. Essays Biochem 1997; 32:17-29.
125. Hartl FU. Molecular chaperones in cellular protein folding. Nature 1996; 381:571-579.
126. Bukau B, Horwich AL. The Hsp70 and Hsp60 chaperone machines. Cell 1998; 92:351-366.
127. Pratt WB, Toft DO. Regulation of signaling protein function and trafficking by the hsp90/hsp70-based chaperone machinery. Exp Biol Med (Maywood) 2003; 228:111-133.
128. Bukau B, Weissman J, Horwich A. Molecular chaperones and protein quality control. Cell 2006; 125:443-451.
129. Kieran D, Kalmar B, Dick JR et al. Treatment with arimoclomol, a coinducer of heat shock proteins, delays disease progression in ALS mice. Nat Med 2004; 10:402-405.
130. Katayama T, Imaizumi K, Honda A et al. Disturbed activation of endoplasmic reticulum stress transducers by familial Alzheimer's disease-linked presenilin-1 mutations. J Biol Chem 2001; 276:43446-43454.
131. Cowan KJ, Diamond MI, Welch WJ. Polyglutamine protein aggregation and toxicity are linked to the cellular stress response. Hum Mol Genet 2003; 12:1377-1391.
132. Hay DG, Sathasivam K, Tobaben S et al. Progressive decrease in chaperone protein levels in a mouse model of Huntington's disease and induction of stress proteins as a therapeutic approach. Hum Mol Genet 2004; 13:1389-1405.
133. Zourlidou A, Gidalevitz T, Kristiansen M et al. Hsp27 overexpression in the R6/2 mouse model of Huntington's disease: chronic neurodegeneration does not induce Hsp27 activation. Hum Mol Genet 2007; 16:1078-1090.
134. Wen FC, Li YH, Tsai HF et al. Down-regulation of heat shock protein 27 in neuronal cells and non-neuronal cells expressing mutant ataxin-3. FEBS Lett 2003; 546:307-314.
135. Chai Y, Koppenhafer SL, Bonini NM et al. Analysis of the role of heat shock protein (Hsp) molecular chaperones in polyglutamine disease. J Neurosci 1999; 19:10338-10347.
136. Stenoien DL, Cummings CJ, Adams HP et al. Polyglutamine-expanded androgen receptors form aggregates that sequester heat shock proteins, proteasome components and SRC-1 and are suppressed by the HDJ-2 chaperone. Hum Mol Genet 1999; 8:731-741.
137. Bence NF, Sampat RM, Kopito RR. Impairment of the ubiquitin-proteasome system by protein aggregation. Science 2001; 292:1552-1555.
138. Nishitoh H, Matsuzawa A, Tobiume K et al. ASK1 is essential for endoplasmic reticulum stress-induced neuronal cell death triggered by expanded polyglutamine repeats. Genes Dev 2002; 16:1345-1355.
139. Tatzelt J, Zuo J, Voellmy R et al. Scrapie prions selectively modify the stress response in neuroblastoma cells. Proc Natl Acad Sci USA 1995; 92:2944-2948.
140. Van Dyk TK, Gatenby AA, LaRossa RA. Demonstration by genetic suppression of interaction of GroE products with many proteins. Nature 1989; 342:451-453.
141. Brown CR, Hong-Brown LQ, Welch WJ. Correcting temperature-sensitive protein folding defects. J Clin Invest 1997; 99:1432-1444.
142. Bilen J, Bonini NM. Genome-wide screen for modifiers of ataxin-3 neurodegeneration in Drosophila. PLoS Genet 2007; 3:1950-1964.
143. Ben-Zvi A, Miller EA, Morimoto RI. Collapse of proteostasis represents an early molecular event in Caenorhabditis elegans aging. Proc Natl Acad Sci USA 2009.
144. Fraser AG, Kamath RS, Zipperlen P et al. Functional genomic analysis of C. elegans chromosome I by systematic RNA interference. Nature 2000; 408:325-330.
145. Kamath RS, Martinez-Campos M, Zipperlen P et al. Effectiveness of specific RNA-mediated interference through ingested double-stranded RNA in Caenorhabditis elegans. Genome Biol 2001; 2:RESEARCH0002.
146. Ghosh S, Feany MB. Comparison of pathways controlling toxicity in the eye and brain in Drosophila models of human neurodegenerative diseases. Hum Mol Genet 2004; 13:2011-2018.
147. Kitamura A, Kubota H, Pack CG et al. Cytosolic chaperonin prevents polyglutamine toxicity with altering the aggregation state. Nat Cell Biol 2006; 8:1163-1170.
148. Behrends C, Langer CA, Boteva R et al. Chaperonin TRiC promotes the assembly of polyQ expansion proteins into nontoxic oligomers. Mol Cell 2006; 23:887-897.
149. Tam S, Geller R, Spiess C et al. The chaperonin TRiC controls polyglutamine aggregation and toxicity through subunit-specific interactions. Nat Cell Biol 2006; 8:1155-1162.
150. Kraemer BC, Burgess JK, Chen JH et al. Molecular pathways that influence human tau-induced pathology in Caenorhabditis elegans. Hum Mol Genet 2006; 15:1483-1496.
151. Lee VM, Goedert M, Trojanowski JQ. Neurodegenerative tauopathies. Annu Rev Neurosci 2001; 24:1121-1159.

152. Shulman JM, Feany MB. Genetic modifiers of tauopathy in Drosophila. Genetics 2003; 165:1233-1242.
153. Auluck PK, Chan HY, Trojanowski JQ et al. Chaperone suppression of alpha-synuclein toxicity in a Drosophila model for Parkinson's disease. Science 2002; 295:865-868.
154. Lamitina T, Huang CG, Strange K. Genome-wide RNAi screening identifies protein damage as a regulator of osmoprotective gene expression. Proc Natl Acad Sci USA 2006; 103:12173-12178.
155. Willingham S, Outeiro TF, DeVit MJ et al. Yeast genes that enhance the toxicity of a mutant huntingtin fragment or alpha-synuclein. Science 2003; 302:1769-1772.
156. Faber PW, Voisine C, King DC et al. Glutamine/proline-rich PQE-1 proteins protect Caenorhabditis elegans neurons from huntingtin polyglutamine neurotoxicity. Proc Natl Acad Sci USA 2002; 99:17131-17136.
157. Powers ET, Morimoto RI, Dillin A et al. Biological and chemical approaches to diseases of proteostasis deficiency. Annu Rev Biochem 2009; 78:959-991.
158. Westerheide SD, Kawahara TL, Orton K et al. Triptolide, an inhibitor of the human heat shock response that enhances stress-induced cell death. J Biol Chem 2006; 281:9616-9622.
159. Mathew A, Mathur SK, Morimoto RI. Heat shock response and protein degradation: regulation of HSF2 by the ubiquitin-proteasome pathway. Mol Cell Biol 1998; 18:5091-5098.
160. Rossi A, Elia G, Santoro MG. Activation of the heat shock factor 1 by serine protease inhibitors. An effect associated with nuclear factor-kappaB inhibition. J Biol Chem 1998; 273:16446-16452.
161. Bagatell R, Paine-Murrieta GD, Taylor CW et al. Induction of a heat shock factor 1-dependent stress response alters the cytotoxic activity of hsp90-binding agents. Clin Cancer Res 2000; 6:3312-3318.
162. Bagatell R, Whitesell L. Altered Hsp90 function in cancer: a unique therapeutic opportunity. Mol Cancer Ther 2004; 3:1021-1030.
163. Jurivich DA, Sistonen L, Kroes RA et al. Effect of sodium salicylate on the human heat shock response. Science 1992; 255:1243-1245.
164. Lee BS, Chen J, Angelidis C et al. Pharmacological modulation of heat shock factor 1 by antiinflammatory drugs results in protection against stress-induced cellular damage. Proc Natl Acad Sci USA 1995; 92:7207-7211.
165. Amici C, Sistonen L, Santoro MG et al. Antiproliferative prostaglandins activate heat shock transcription factor. Proc Natl Acad Sci USA 1992; 89:6227-6231.
166. Jurivich DA, Sistonen L, Sarge KD et al. Arachidonate is a potent modulator of human heat shock gene transcription. Proc Natl Acad Sci USA 1994; 91:2280-2284.
167. Westerheide SD, Bosman JD, Mbadugha BN et al. Celastrols as inducers of the heat shock response and cytoprotection. J Biol Chem 2004; 279:56053-56060.
168. Trott A, West JD, Klaić L et al. Activation of heat shock and antioxidant responses by the natural product celastrol: transcriptional signatures of a thiol-targeted molecule. Mol Biol Cell. 2008;19(3):1104-12. Epub 2008 Jan 16.
169. Katsuno M, Sang C, Adachi H et al. Pharmacological induction of heat-shock proteins alleviates polyglutamine-mediated motor neuron disease. Proc Natl Acad Sci USA 2005; 102:16801-16806.
170. Zhang YQ, Sarge KD. Celastrol inhibits polyglutamine aggregation and toxicity though induction of the heat shock response. J Mol Med 2007; 85:1421-1428.
171. Boyce M, Bryant KF, Jousse C et al. A selective inhibitor of eIF2alpha dephosphorylation protects cells from ER stress. Science 2005; 307:935-939.
172. Parker JA, Arango M, Abderrahmane S et al. Resveratrol rescues mutant polyglutamine cytotoxicity in nematode and mammalian neurons. Nat Genet 2005; 37:349-350.
173. Voisine C, Varma H, Walker N et al. Identification of potential therapeutic drugs for huntington's disease using Caenorhabditis elegans. PLoS ONE 2007; 2:e504.
174. Varma H, Cheng R, Voisine C et al. Inhibitors of metabolism rescue cell death in Huntington's disease models. Proc Natl Acad Sci USA 2007; 104:14525-14530.
175. Varma H, Voisine C, DeMarco CT et al. Selective inhibitors of death in mutant huntingtin cells. Nat Chem Biol 2007; 3:99-100.

Chapter 12

Roles for SUMO Modification during Senescence

Artemisia M. Andreou and Nektarios Tavernarakis*

Abstract

SUMOylation is a reversible post-translational modification, where a small peptide (SUMO) is covalently attached to a target protein and changes its activity, subcellular localization and/or interaction with other macromolecules. SUMOylation substrates are numerous and diverse and modification by SUMO is involved in many biological functions, including the response to stress. The SUMO pathway has recently been implicated in the process of cellular senescence, the irreversible loss of cell replication potential that occurs during aging in vivo and in vitro. SUMO peptides, a SUMO E3 ligase and a SUMO-specific peptidase can induce or hinder the onset of senescence, thus supporting an association of SUMOylation with cell growth arrest and organismal aging. Preliminary results on comparative analysis of proteomics and mRNA levels between young and old human and murine tissues show elevated levels of global protein SUMOylation and a decrease in components of the SUMOylation process with age. Further connections between the SUMO pathway and the aging process remain to be elucidated.

Introduction to the SUMO Modification System

The small ubiquitin-related modifier (SUMO) belongs to a large family of proteins related to ubiquitin (Ub) and the small ubiquitin-like proteins (Ulp). The overall sequence identity with ubiquitin is small (around 18%), but the C-terminus of the protein, which confers most of its activity, is almost super-imposable to the equivalent region of ubiquitin.[1,2] Although SUMO shares similar attachment mechanisms to Ub, its function is completely different and sometimes counteractive to the ubiquitin/proteasome pathway. Conjugation of SUMO can block the sites of attachment for ubiquitin and can thus protect proteins from degradation. Since its discovery just over a decade ago, the list of SUMOylation targets has been growing constantly and is now well into the hundreds. Perhaps a reason why this fairly ubiquitous modification has only recently been identified and studied lies in the reversibility of the process and the fact that, apart from very few exceptions, the amount of any SUMO-modified protein within a cell only makes up a very small percentage of its total amount, thus making detection by various molecular and biochemical methods more challenging.[3]

SUMO Isoforms

SUMO is highly conserved in all eukaryotic cells and in higher organisms it is present in all tissues and developmental stages. It has been shown to be essential for cell viability both at the organism level and in cells in culture. The number of SUMO genes differs between organism families,

*Corresponding Author: Nektarios Tavernarakis, Institute of Molecular Biology and Biotechnology (IMBB), Foundation for Research and Technology Hellas (FORTH), N. Plastira 100, Vassilika Vouton, PO Box 1385, Heraklion GR 70013, Crete, GREECE
Email: tavernarakis@imbb.forth.gr

Protein Metabolism and Homeostasis in Aging, edited by Nektarios Tavernarakis.

with one SUMO species present in yeast, nematodes and fruit flies and four in mammals, termed SUMO-1, SUMO-2, SUMO-3 and SUMO-4. SUMO-4 is the most recently identified isoform, through an 86% similarity to SUMO-2. mRNA transcripts show limited expression compared to the other SUMO genes and are present mainly in kidney, lymph system and spleen.[4] SUMO-1 is 50% identical in sequence to SUMO-2 and -3, while SUMO-2/3 share a 95% identity. This observation translates to functional activity as well; activity of SUMO2/3 is almost indistinct, while SUMO-1 has a dissimilar function. Furthermore, SUMO-1 is rarely found unattached within cells, but there are pools of free SUMO-2/3 available, which can be promptly conjugated to target proteins under specific stress conditions.[5] There is a preference between SUMO-1 and SUMO-2/3 in conjugation to certain proteins, but others can be modified equally well by both SUMO species.

Conjugation of SUMO to Target Proteins

Attachment of SUMO to target proteins occurs via an isopeptide bond between the glycine residue at the C-terminal end of SUMO and the ε-amino group of an internal lysine residue within the substrate. Enzymes analogous to the ubiquitin pathway, but specific for SUMO modification, catalyze the formation of this bond in four steps.[6] The SUMO peptide is initially translated as a precursor, ending with a stretch of sequence at its carboxyl end that follows the active –GG part of the protein. The length of this sequence varies between SUMO species. Proteolytic cleavage of this amino acid sequence converts SUMO to its mature form. SUMO-specific peptidases, for example members of the sentrin-specific protease (SENP) family, catalyse this step. SUMO-activating enzymes, also called E1, activate the mature SUMO in an ATP-dependent reaction. Active SUMO is then transferred onto the E2 SUMO-conjugating enzyme Ubc9 (ubiquitin-like protein SUMO-1 conjugating enzyme 9). Conjugation of SUMO to target proteins occurs through Ubc9 with the aid of an E3 ligase.[7] Substrate specificity for SUMO is conferred by both Ubc9 and various E3 ligases. Ubc9 actually recognizes a consensus motif on the protein substrate and forms the covalent attachment of SUMO to its targets[8] while the SUMO ligases probably interact with other areas of the substrate and provide more specificity (Fig. 1).[6,9]

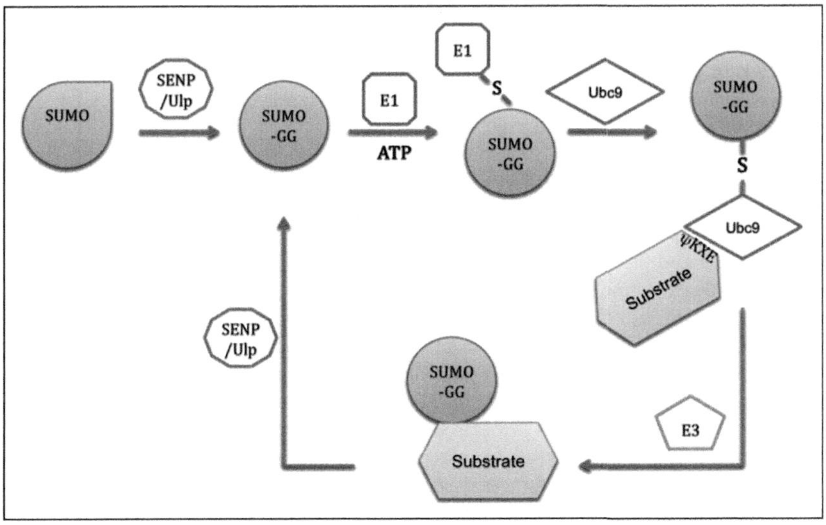

Figure 1. A brief description of the SUMOylation process. Attachment of SUMO to and de-conjugation from target substrate proteins happens in distinct enzymatic steps, similar to those of the Ubiquitin pathway. SUMO-specific enzymes convert SUMO to its mature form, which is then activated and passed on, through subsequent thioester bonds, from the E2-conjugating enzyme Ubc9 to its target protein. SUMO can then be released from the complex via the action of a SUMO isopeptidase of the SENP/Ulp family.

Protein Recognition Sites

A general consensus recognition site for SUMO attachment is ψKXE,[10,11] where ψ is a large hydrophobic amino acid, like leucine, isoleucine or valine; K is the lysine residue at which the attachment takes place; X is any amino acid and E is a glutamic acid. Although this consensus site is conserved, a number of alternatives have been described, where for example a D can be present at the position of the E, or the amino acid before the lysine can be a different one from the three consensus residues. Other sequences may also be involved, since in some cases, proteins have been shown to be SUMOylated that either do not contain the consensus site, use a different motif for SUMO attachment[12] or maintain SUMOylation after the site has been mutated. More extensive SUMO conjugation motifs have also been described; it has been suggested that acidic residues downstream of the core SUMOylation motif have a role in enhancing specificity for substrates.[13]

Links with Other Post-Translational Protein Modifications

Although ubiquitination, acetylation and methylation all take place on lysine residues of their target proteins, it was perhaps more expected (at least initially) to study a possible link between SUMOylation and the Ubiquitin pathway, due to the extensive similarity between the peptides themselves and the enzymes facilitating the two processes. Indeed, IκBα and PCNA can be modified by either SUMO or ubiquitin on the same lysine in their sequence, with each modification resulting in a separate function for the substrate.[14,15]

Besides ubiquitination, other post-translational modifications have been associated with the SUMO modification process. SUMOylation of proteins including GATA1 and heat shock factors HSF1 and HSF4b has been shown to be dependent on their phosphorylation status. A phosphorylation-dependent SUMOylation motif (PDSM) has been described for such proteins.[16] Phosphorylation of this motif results in increased SUMOylation of HSF1 and MEF2, while SUMO attachment to IκBα, p53, c-Fos and c-Jun is negatively regulated by phosphorylation.[17-19] Furthermore, the association of SUMO with histones supports a role in genome organization and stability. Histone deacetylases, including HDAC2 and HDAC6, have been shown to preferentially interact with SUMO-modified substrates.[20,21] It has been suggested that deacetylation of histones by HDAC enzymes may make more lysine (K) residues available for SUMOylation.[7] A SUMOylation switch based on acetylation/deacetylation has been described,[22] and increased SUMOylation of H4 correlates with decreased acetylation of the gene, when Ubc9 is targeted to the promoter region.[23]

De-Conjugation of SUMO from Substrate

SUMO attachment is a reversible and highly transient modification. The same enzymes that facilitate the initial maturation of SUMO molecules also catalyse the cleavage from their substrates.[6] Ulp1 and Ulp2 are SUMO-specific proteases in yeast,[24,25] while six members of the SENP family, SENPs 1-3 and 5-7, have been shown to have this role in human tissues.[9] In contrast to the SUMO-conjugating enzymes, SUMO proteases are not similar to the equivalent enzymes in the ubiquitin pathway, but the homology they share appears closely related to viral proteases.[26] Differences in the sub-cellular localization of the SENP proteins is believed to be dictated by nonconserved N-terminal sequences and provides the specificity for the SUMO-substrate complexes they regulate.[27,28] SENP1 is localized mainly in the nucleus with little, albeit persistent, cytoplasmic presence.[29,30] SENP2 is associated with the nuclear pore,[31] as is Ulp1,[32] SENP3 and SENP5 are nucleolar,[33] while contradictory reports place SENP6 in both the nucleoplasm and the cytosol.[34,35] Since the SUMO targets described to date are present throughout the cell, the differential localization of the SENP proteins is perhaps expected.[36,37] Specificity towards certain conjugates may be achieved through the sub-cellular and sub-nuclear location of each protease.

Physiological Functions of SUMO Modification

SUMO attachment has been implicated in a number of cell processes, such as regulation of transcription, nuclear transport, DNA repair, protein stability, cell cycle and chromatin structure. Although its function is as diverse as its substrates, one generalisation could be that modification of a protein substrate by SUMO alters its interactions with other protein and DNA molecules. SUMOylation can aid or block protein-protein interactions equally well, depending on the substrate proteins involved. For example, during DNA replication in yeast, PCNA needs to be SUMOylated in order to recruit Srs2 helicase to the site of the replication fork and prevent recombination.[38,39] Reversely, interaction of the CtBP corepressor with PDZ domains is hindered when the former is SUMOylated.[40] The most well studied example of SUMO modification altering the DNA-binding ability of a protein is that of the DNA repair enzyme thymine DNA glycosylase (TDG). After TDG has completed the base excision, it needs SUMO association in order to detach itself from the DNA, as SUMO-TDG has much lower affinity for DNA. Once in the nucleoplasm, the SUMO protein is removed from the conjugate by a specific peptidase, thus allowing TDG to participate in another round of DNA base excision repair.[41,42]

Recently, proteins with active roles in the control of cell survival and proliferation have been identified as substrates for SUMOylation and enzymes partaking in the SUMO modification pathway have been associated with the onset of cellular senescence. It is thus very likely that SUMOylation may actively contribute to phenotypes of growth arrest and cellular aging.

Cellular Senescence

Derived from the Latin *senex*, meaning "old man", "old age" or "advanced in age", cellular senescence was first described in the early 1960s as the process that limits indefinite growth of primary fibroblasts in culture.[43] The term replicative senescence was favoured later to more specifically describe the irreversible loss of cell division potential at this final stage in the lifespan of somatic cells in culture.

Senescence as a Model for Aging

Cellular senescence is considered a major tumor suppressor mechanism, as a process for eliminating the proliferation of damaged or dysfunctional cells.[44,45] Besides its importance in tumour suppression, replicative senescence can occur independently of cancer and is regarded as a model for cellular and organismal aging.[46] This correlation has been supported by various observations and experimental data. Senescent cells have been identified in human and other tissues in vivo and become more abundant as the organism ages. In cultured cells, the time period before the cells exit the proliferating state and become growth-arrested is directly related to the maximum lifespan of the species from which they were obtained. In addition, cells derived from patients with premature aging syndromes enter senescence more rapidly in culture. A recent study that characterized the role of p63 showed that deficiency in this molecule in mice resulted in premature aging and a shortened lifespan.[47] Induction of p63 deficiency in a tissue-specific way lead to extensive cellular senescence and accelerated aging phenotypes, thus presenting a causative association between senescence and—at least premature—aging. In vivo, the presence of elevated numbers of senescent cells significantly limits the regeneration potential of animal tissues and changes intra- and inter-tissue communication through the cells' ability to modify their extracellular environment. Senescent cells secrete biochemical signals that may target them for destruction by the immune system, but these signals can be harmful to their environment, especially as numbers of senescent cells start to accumulate. This subsequent alteration in tissue and organ homeostasis may ultimately lead to organismal aging.[48]

Characteristics of Senescent Cells

Besides the significant suppression of DNA synthesis and cell proliferation, senescence is characterized by certain changes in the cell morphology, metabolism and epigenetic state. Cells usually appear enlarged, flattened and more granular, they show higher levels of cellular autofluorescence and senescence-associated beta-galactosidase activity, while a number of proteins involved in cell cycle arrest are up-regulated, as for instance p21 and p16.[49] These morphological characteristics together with SA-β- gal activity (measured at specific pH levels) are used to distinguish senescent cells.[50] After they have entered senescence, cells in culture can remain alive and metabolically active for long periods of time.[51]

Replicative and Stress-Induced Senescence

The cellular mechanism leading to senescence has been mainly associated with the gradual shortening of telomeres.[52,53] Due to the semi-conservative nature of DNA replication, where the polymerase only adds nucleotides in a 5′ to 3′ direction and requires binding of an RNA primer, the ends of chromosomes become shorter with each round of cell division. Telomerase, the enzyme responsible for maintenance of telomeric ends, is normally absent from most human somatic cells. Consequently, when telomere ends reach a critical length, signals are initiated as the cellular response to DNA damage and activate p53-mediated pathways that lead the cell to senescence or apoptosis. This theory is supported by data showing that expression of telomerase increases the proliferating time period of cells.[54] In addition, in mice—which normally possess longer telomeres and show more extensive somatic expression of telomerase than humans[55]—absence of telomerase results in a premature aging phenotype[56] and seriously compromises the replicative lifespan of stem cells.[57]

However, cell proliferation potential can also be hindered in ways that are independent of telomere shortening. Various types of stress, such as DNA damage, chromatin remodelling, activated oncogenes, oxidative stress and chemotherapeutic agents may impede normal cell proliferation and result in a prematurely arrested state that greatly resembles cellular senescence.[58] The time limit of the replicative state and entry into senescence can be clearly manipulated in cultured cells by optimizing physiological conditions of oxygen levels and the presence of serum, in ways unrelated to telomere length.[59] Stress- induced senescence is acute, rapid and homogeneous, in contrast to replicative senescence occurring normally, where the cells may differ in the time they enter senescence and the rate by which they reach growth arrest. Thus, since the pattern of gene expression between the two states is extensively overlapping, induced senescence may be preferably used in experiments for practical reasons.[60,61]

Pathways Mediating Stress Response

When the stress response pathways are activated within a cell, signalling is usually transduced through the p21/p53 and p16/pRB pathways.[62,63] A number of genes associated with regulation of senescence have been shown to be part of either the p53 or pRB pathways.[64] The proteins involved in these signalling cascades, mainly p53 and pRB, will determine how a cell will respond to the stress factor(s). This response is most commonly balanced between cells entering senescence or undergoing apoptosis, as both mechanisms are used to control unsolicited cell growth.[65]

SUMO and Senescence

A good indication that the SUMOylation pathway may be involved in cellular senescence was the identification of a number of senescence-associated proteins as targets for SUMO conjugation. Since this post-translational modification has been shown to greatly influence protein activity, stability and interaction of substrates with other macromolecules, the active participation of the SUMO pathway in the mechanisms of senescence induction and, subsequently, aging was an appealing thought.

SUMOylation has been associated with responses to various types of cellular stress. Altered interactions of a number of proteins with both SUMO-2/3 and/or SUMO-1 upon exposure to such stresses support this involvement. For example, the expression of SUMO-1 has been shown to increase during hypoxia.[66] On the other hand, oxidative stress, induced by treatment of cells with hydrogen peroxide, results in increased levels of SUMO-2/3- modified p53, while levels of p53/SUMO-1 conjugates appear unaffected.[67] Because of the close and often causative, association between environmental and cellular stress and the onset of senescence, involvement of the SUMO pathway in stress response provides an important, though indirect, link between SUMO, senescence and, very likely, the aging process.

Recently, SUMO species themselves as well as a number of enzymes that participate in SUMO attachment and de-conjugation processes have been associated with senescence. SUMO proteins, a SUMO ligase and a SUMO isopeptidase have been shown to either induce or repress the

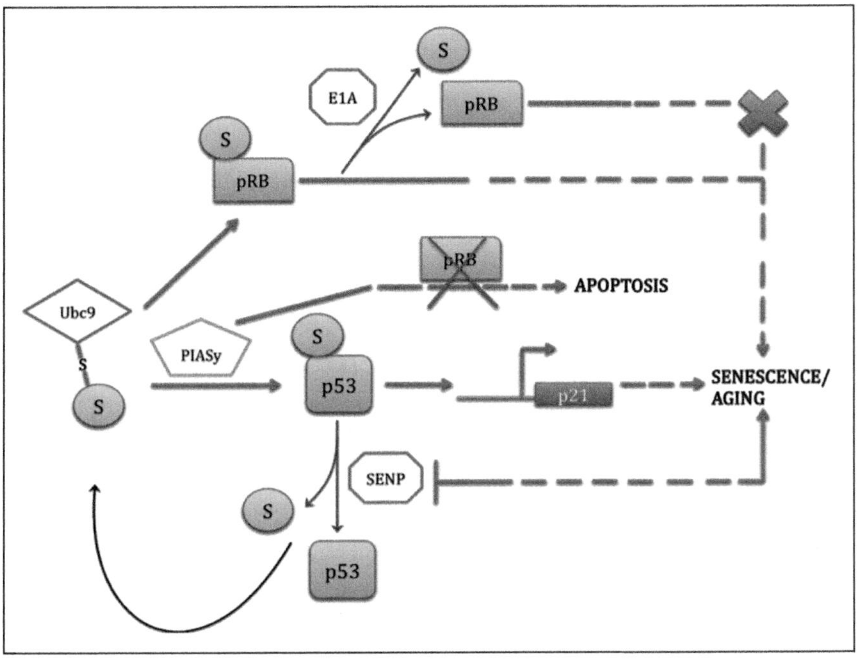

Figure 2. Model of SUMO modification proteins in pathways leading to cellular senescence and aging. The SUMOylation cycle of p53 affects downstream cascades leading the cell to senescence/aging and disturbances at both the SUMO conjugation and de-conjugation steps can inhibit the onset of senescence. SUMOylation of pRB can influence the cell's decision to undergo senescence or apoptosis. The SUMO E3 ligase PIASy and SUMO proteases of the SENP family also appear to have important roles in these pathways.

onset of cellular senescence. Figure 2 presents a model of how these molecules may be involved in senescence-inducing pathways and more details on each of them are described in the text below. In addition, elevated numbers of SUMOylated proteins have been shown to accumulate in senescent cells compared to normal replicating cells,[68] while levels of SUMO isoforms and associated enzymes appear to decrease with age, in a different tissue type.[69]

SUMO Molecules in Senescence

SUMO-1 and SUMO-2/3 species can have different preferences for substrates and thus be involved in different cascades within the cell, even when these belong to the same greater pathway. Recent experiments have shown that over-expression of SUMO-2/3 in cultured cells results in a premature senescence morphology, supported by slow cell growth and early growth arrest, as shown by reporter assays.[67] In contrast, over-expression of SUMO-1 appears not to directly affect the process of cellular senescence.[70] This is perhaps not surprising, considering that SUMO-2/3 is thought to be the isoform(s) mainly and most frequently associated with the response to stress. However, it is interesting that under normal conditions SUMO-1 appears mainly conjugated to target proteins within the cells, while unconjugated pools of SUMO-2/3 species are abundant and available to be used as required.

The major players of the pathways that lead to senescence, p53 and pRB, actively inhibit unsolicited cell growth and this function is regulated by post-translational modifications, including phorpshorylation, acetylation and ubiquitination.[19,71-73] The SUMO modification pathway has also been shown to modulate activity of these proteins, as they are subject to SUMOylation by both SUMO-1 and SUMO-2/3. The transcriptional activity of p53 is positively regulated by SUMO-1 attachment, as shown by up-regulation of *p21* expression, a p53 target gene.[67] In addition, the repressor activity of pRB on E2F-regulated target genes is controlled by components of the SUMO pathway.[74] Under normal cell conditions, pRB is de-SUMOylated by E1A, a viral oncoprotein; [72] it has been suggested that this may be a strategy used by viruses to suppress cell senescence. Over-expression of SUMO-2/3 surpasses this de-SUMOylation, leaving pRB-SUMO-2/3 conjugates free to stimulate senescence.[67]

Down-regulation of p53 and pRB by RNA interference appears to counteract the senescence phenotype seen in SUMO-2/3 over-expressing cells, suggesting that the effect of SUMO-2/3 modification on senescence occurs via p53- and pRB-mediated pathways.[67] p53-induced up-regulation of p21 is a known pathway for activation of senescence.[75] During SUMO-2/3 over-expression, the transcriptional activity of p53 is significantly enhanced. Levels of p53 protein seem to remain unchanged, while the senescence-associated protein p21 is clearly up-regulated. Since p21 is under the transcriptional control of p53, this increase in p21 protein levels may be due to the altered SUMOylation state of p53, modified by SUMO-2/3.[67]

SUMO-Specific E3 Ligase PIASY and Senescence

Protein inhibitors of activated STAT (PIAS) have been shown to function as E3 SUMO ligases and data from various SUMO conjugates make them one of the largest families of SUMO-specific ligases.[76] During replicative senescence the levels of endogenous PIASy, a member of this protein family, are significantly increased compared to presenescent cells, as do levels of hyper-sumoylated proteins.[74]

This effect is considered specific to PIASy, as no other member of the PIAS protein family appears to have similar activity. In addition, over-expression of PIASy can induce cellular senescence or apoptosis.[74] The entry into either process occurs through the p53 and pRB tumour-suppressor pathways and depends on their state within the cell. Induction of senescence is conferred by the E3 ligase activity of PIASy, which facilitates the SUMOylation of p53 and subsequent activation of p53 target promoter regions, as shown by the increase activation of the *p21* promoter upon PIASy overexpression. Mutation of the PIASy E3 ligase active site abolishes the effect. The onset of senescence through PIASy is counteracted by the E6 oncoprotein, through inhibition of the SUMO ligase activity of PIASy.[74] Interestingly, an extended lifespan is observed in cells that over-express mutant forms of PIASy with an inactive E3 ligase site.

In further support of this process, deletion of PIASy in mouse embryo fibroblasts results in a significant delay of the onset of senescence after appropriate signalling. Even after induction of p53 expression by a pro-senescence signal, for example through oncogenic RAS, p53 target genes *p21* and *MDM2* are not up-regulated in the absence of PIASy. In parallel, PIASy can induce p53-dependent apoptosis during pRB deficiency; interestingly inactivation of pRB by hyper-phosphorylation is not enough to give similar results and the effect of PIASy is not there when pRB is present, even in an (as far as we know) inactive state.[74]

All available data to date support an active contribution of PIASy to the senescence process and perhaps also to aging. The difference in the cellular response to PIASy, depending on the pRB status, suggests a possible role for PIASy as one of the factors influencing the cell's decision to undergo senescence or apoptosis. This may occur through an altered binding affinity between pRB and its cofactors, in a SUMOylation-mediated manner. More specifically, it has been suggested that hypophosphorylated pRB enlists its corepressors together with components of the SUMO pathway to the site of genes promoting proliferation. A number of these cofactors could be themselves targets of SUMOylation and these interactions could stabilize the pRB repressor complex. Also, as it has been shown that SUMO attachment increases affinity between protein partners,[77] the presence of SUMO-modified pRB repressor group on DNA could provide a high affinity site for the recruitment of proteins involved in chromatin remodelling and reorganization.[68] This process could ultimately lead to the silencing of genes involved, also facilitated by the activity of histones, which are substrates for SUMOylation and have major roles in transcriptional repression in their modified state.[23,78]

SUMO Proteases (SENP Family) and Senescence

Senescence may be induced either by increased SUMOylation of common target conjugates of the SUMO proteases or by an increase in total levels of SUMOylated proteins.[79] Repression of a number of SUMO proteases, of the SENP family, has been shown to result in a senescence-like phenotype. SUMO proteases may thus be required for the proliferation of normal human cells and have important roles in age-related phenotypes.

Cells that undergo the induced senescence caused by Senp protein repression display all key morphological characteristics and activity (as measured by SA-β-gal activity) associated with a senescent state. In human firbroblasts, SUMO-containing PML bodies that accumulate in the nucleus after many cell passages and during normal replicative senescence are also seen to increase in number upon repression of SUMO proteases Senp1, Senp2 and Senp7. Repression of Senp1 in particular and Senp7 perhaps to a lesser extend, provokes premature senescence through the p53 signalling pathway.[79] Inactivation of the p53 pathway limits the induction of senescence caused by Senp1 repression. This data points to a role of Senp1 in maintaining a balance in the cell after exposure to stress and perhaps preventing premature senescence.

SUMO and Maintenance of Telomere Length

Functional telomeres are required for cells to maintain their replication potential. In addition to the gradual shortening of telomeres after certain numbers of cell division, mutated or depleted proteins that protect telomere ends may also result in telomere dysfunction.[80,81]

In yeast, deficiencies in homologues of SUMO and E3 ligases of the PIAS family have been shown to provoke a significant increase in telomere length.[82,83] These results advocate a role for SUMO in maintaining telomere length and suggest that the SUMO pathway may be a negative modulator of this process. SUMO modification of factors associated with telomeres through DNA damage response pathways further supports an important role for SUMO in telomere maintenance and consequently tumour suppression, cellular senescence and normal aging.[68]

Changes in Global Protein SUMOylation during Aging

Recent experiments have shown that the amount of SUMOylated proteins in rodent spleen tissue increases with age and this increase is not subject to limitation by dietary restriction, as is the case for global protein ubiquitination.[84] Furthermore, mRNA levels of components of the SUMO pathway *sumo-1*, *ubc-9* and *senp1* show significant decrease in aged murine brain tissues, as compared to young ones and the effect is mimicked upon inflammation stimuli in astrocytes.[69] These results suggest a role for protein SUMOylation in aging tissues that is most likely tissue-specific at least in mammals and perhaps leads to altered protein function that is clearly distinct from that of ubiquitin. It would be very interesting to include further tissue and cell types in similar studies, in order to confirm a more general effect of aging on protein SUMOylation.

Conclusion

Damaged, oxidised or mislocalized proteins that would be properly dealt with in healthy proliferating cells, may accumulate in aging cells due to changes in the activities of the SUMOylation and ubiquitination pathways and this may have an important role in the onset of senescence and age-related disease.[79] In regard to the SUMO modification pathway, elevated SUMOylation of certain target proteins can cause premature cellular senescence,[67,74] the E3 SUMO ligase PIASy has recently been associated with the induction of senescence,[74] while the SUMO isopeptidase Senp1 appears to obstruct it.[79]

Targets for SUMOylation have been identified in almost every cell and part of the cell and the SUMO pathway is constantly being linked to more cellular processes. The physiological function of protein modification by SUMO ranges greatly and includes pathways from DNA repair and transcription to protein-protein interactions and subcellular localization. Besides the activity of SUMO substrates, integrity of these molecules has been shown to be controlled by SUMOylation in a number of occasions. For instance, stability and DNA binding of the stem cell transcription factor Oct4 increases upon SUMOylation,[85] as does stability of Pax8,[86] Ku70[87] and Apa-1, the latter being associated with senescence of human fibroblasts.[88] Moreover, HIF1α steady state is regulated by Senp1, in an example of SUMOylation acting synergistically with the Ubiquitin pathway to promote protein degradation.[89] Similarly, the process of Specificity Protein 1 (Sp1) turnover is dependent on its SUMOylation, which increases interaction with proteasome subunits and targets Sp1 for proteolysis.[90]

So, the SUMO pathway affects the levels, stability and half-life of proteins by protecting them from Ubiquitin through occupation of lysine residues in their sequence, or by leading them to proteasome-mediated degradation. Although in the area of cell senescence and tissue aging SUMO has been predominantly associated with the DNA damage response and maintenance of telomere integrity, it is not absurd to imagine, or even expect, that SUMOylation, either as a whole or through its individual components (or both), may also have an active role in regulating protein synthesis and degradation during these processes. Proteins that are important in pathways leading to cellular senescence and organismal aging have already been linked to SUMOylation, as have enzymes facilitating the modification process, but a direct link between SUMO and protein turnover during aging needs to be elucidated. Results from ongoing and future experiments in this area remain to be seen, but since the rates and balance of protein synthesis and degradation are crucial during aging, it seems likely that SUMO will be shown to have an important role in these processes.

Acknowledgements

Dr. Artemisia Andreou is supported by a fellowship from the Bodossaki Foundation.

References

1. Bayer P, Arndt A, Metzger S et al. Structure determination of the small ubiquitin-related modifier SUMO-1. J Mol Biol 1998; 280(2):275-286.
2. Jin C, Shiyanova T, Shen Z et al. Heteronuclear nuclear magnetic resonance assignments, structure and dynamics of SUMO-1, a human ubiquitin-like protein. Int J Biol Macromol 2001; 28(3):227-234.
3. Johnson ES. Protein modification by SUMO. Annu Rev Biochem 2004; 73:355-382.
4. Bohren KM, Nadkarni V, Song JH et al. A M55V polymorphism in a novel SUMO gene (SUMO-4) differentially activates heat shock transcription factors and is associated with susceptibility to type I diabetes mellitus. J Biol Chem 2004; 279(26):27233-27238.
5. Saitoh H, Hinchey J. Functional heterogeneity of small ubiquitin-related protein modifiers SUMO-1 versus SUMO-2/3. J Biol Chem 2000; 275(9):6252-6258.
6. Hay RT. SUMO: a history of modification. Mol Cell 2005; 18(1):1-12.
7. Gill G. Something about SUMO inhibits transcription. Curr Opin Genet Dev 2005; 15(5):536-541.
8. Sampson DA, Wang M, Matunis MJ. The small ubiquitin-like modifier-1 (SUMO-1) consensus sequence mediates Ubc9 binding and is essential for SUMO-1 modification. J Biol Chem 2001; 276(24):21664-21669.
9. Melchior F, Schergaut M, Pichler A. SUMO: ligases, isopeptidases and nuclear pores. Trends Biochem Sci 2003; 28(11):612-618.
10. Minty A, Dumont X, Kaghad M et al. Covalent modification of p73alpha by SUMO-1. Two-hybrid screening with p73 identifies novel SUMO-1-interacting proteins and a SUMO-1 interaction motif. J Biol Chem 2000; 275(46):36316-36323.
11. Rodriguez MS, Dargemont C, Hay RT. SUMO-1 conjugation in vivo requires both a consensus modification motif and nuclear targeting. J Biol Chem 2001; 276(16):12654-12659.
12. Song J, Durrin LK, Wilkinson TA et al. Identification of a SUMO-binding motif that recognizes SUMO-modified proteins. Proc Natl Acad Sci USA 2004; 101(40):14373-14378.
13. Yang SH, Galanis A, Witty J et al. An extended consensus motif enhances the specificity of substrate modification by SUMO. EMBO J 2006; 25(21):5083-5093.
14. Desterro JM, Rodriguez MS, Hay RT. SUMO-1 modification of IkappaBalpha inhibits NF-kappaB activation. Mol Cell 1998; 2(2):233-239.
15. Hoege C, Pfander B, Moldovan GL et al. RAD6-dependent DNA repair is linked to modification of PCNA by ubiquitin and SUMO. Nature 2002; 419(6903):135-141.
16. Hietakangas V, Anckar J, Blomster HA et al. PDSM, a motif for phosphorylation-dependent SUMO modification. Proc Natl Acad Sci USA 2006; 103(1):45-50.
17. Bossis G, Malnou CE, Farras R et al. Down-regulation of c-Fos/c-Jun AP-1 dimer activity by sumoylation. Mol Cell Biol 2005; 25(16):6964-6979.

18. Lin JY, Ohshima T, Shimotohno K. Association of Ubc9, an E2 ligase for SUMO conjugation, with p53 is regulated by phosphorylation of p53. FEBS Lett 2004; 573(1-3):15-18.
19. Muller S, Berger M, Lehembre F et al. c-Jun and p53 activity is modulated by SUMO-1 modification. J Biol Chem 2000; 275(18):13321-13329.
20. Girdwood D, Bumpass D, Vaughan OA et al. P300 transcriptional repression is mediated by SUMO modification. Mol Cell 2003; 11(4):1043-1054.
21. Yang SH, Sharrocks AD. SUMO promotes HDAC-mediated transcriptional repression. Mol Cell 2004; 13(4):611-617.
22. Stankovic-Valentin N, Deltour S, Seeler J et al. An acetylation/deacetylation-SUMOylation switch through a phylogenetically conserved psiKXEP motif in the tumor suppressor HIC1 regulates transcriptional repression activity. Mol Cell Biol 2007; 27(7):2661-2675.
23. Shiio Y, Eisenman RN. Histone sumoylation is associated with transcriptional repression. Proc Natl Acad Sci USA 2003; 100(23):13225-13230.
24. Li SJ, Hochstrasser M. A new protease required for cell-cycle progression in yeast. Nature 1999; 398(6724):246-251.
25. Li SJ, Hochstrasser M. The yeast ULP2 (SMT4) gene encodes a novel protease specific for the ubiquitin-like Smt3 protein. Mol Cell Biol 2000; 20(7):2367-2377.
26. Azuma Y, Tan SH, Cavenagh MM et al. Expression and regulation of the mammalian SUMO-1 E1 enzyme. FASEB J 2001; 15(10):1825-1827.
27. Bailey D, O'Hare P. Characterization of the localization and proteolytic activity of the SUMO-specific protease, SENP1. J Biol Chem 2004; 279(1):692-703.
28. Mukhopadhyay D, Dasso M. Modification in reverse: the SUMO proteases. Trends Biochem Sci 2007; 32(6):286-295.
29. Bailey D, O'Hare P. Herpes simplex virus 1 ICP0 colocalizes with a SUMO-specific protease. J Gen Virol 2002; 83(Pt 12):2951-2964.
30. Gong L, Millas S, Maul GG et al. Differential regulation of sentrinized proteins by a novel sentrin-specific protease. J Biol Chem 2000; 275(5):3355-3359.
31. Hang J, Dasso M. Association of the human SUMO-1 protease SENP2 with the nuclear pore. J Biol Chem 2002; 277(22):19961-19966.
32. Li SJ, Hochstrasser M. The Ulp1 SUMO isopeptidase: distinct domains required for viability, nuclear envelope localization and substrate specificity. J Cell Biol 2003; 160(7):1069-1081.
33. Gong L, Yeh ET. Characterization of a family of nucleolar SUMO-specific proteases with preference for SUMO-2 or SUMO-3. J Biol Chem 2006; 281(23):15869-15877.
34. Kim KI, Baek SH, Jeon YJ et al. A new SUMO-1-specific protease, SUSP1, that is highly expressed in reproductive organs. J Biol Chem 2000; 275(19):14102-14106.
35. Mukhopadhyay D, Ayaydin F, Kolli N et al. SUSP1 antagonizes formation of highly SUMO2/3-conjugated species. J Cell Biol 2006; 174(7):939-949.
36. Ayaydin F, Dasso M. Distinct in vivo dynamics of vertebrate SUMO paralogues. Mol Biol Cell 2004; 15(12):5208-5218.
37. Geiss-Friedlander R, Melchior F. Concepts in sumoylation: a decade on. Nat Rev Mol Cell Biol 2007; 8(12):947-956.
38. Papouli E, Chen S, Davies AA et al. Crosstalk between SUMO and ubiquitin on PCNA is mediated by recruitment of the helicase Srs2p. Mol Cell 2005; 19(1):123-133.
39. Pfander B, Moldovan GL, Sacher M et al. SUMO-modified PCNA recruits Srs2 to prevent recombination during S phase. Nature 2005; 436(7049):428-433.
40. Lin X, Sun B, Liang M et al. Opposed regulation of corepressor CtBP by SUMOylation and PDZ binding. Mol Cell 2003; 11(5):1389-1396.
41. Baba D, Maita N, Jee JG et al. Crystal structure of thymine DNA glycosylase conjugated to SUMO-1. Nature 2005; 435(7044):979-982.
42. Hardeland U, Steinacher R, Jiricny J et al. Modification of the human thymine-DNA glycosylase by ubiquitin-like proteins facilitates enzymatic turnover. EMBO J 2002; 21(6):1456-1464.
43. Hayflick L, Moorhead PS. The serial cultivation of human diploid cell strains. Exp Cell Res 1961; 25:585-621.
44. Campisi J. Cellular senescence as a tumor-suppressor mechanism. Trends Cell Biol 2001; 11(11):S27-31.
45. Green DR, Evan GI. A matter of life and death. Cancer Cell 2002; 1(1):19-30.
46. Jeyapalan JC, Sedivy JM. Cellular senescence and organismal aging. Mech Ageing Dev 2008; 129(7-8):467-474.
47. Keyes WM, Wu Y, Vogel H et al. p63 deficiency activates a program of cellular senescence and leads to accelerated aging. Genes Dev 2005; 19(17):1986-1999.
48. Chen JH, Hales CN, Ozanne SE. DNA damage, cellular senescence and organismal ageing: causal or correlative? Nucleic Acids Res 2007; 35(22):7417-7428.

49. Dimri GP. What has senescence got to do with cancer? Cancer Cell 2005; 7(6):505-512.
50. Dimri GP, Lee X, Basile G et al. A biomarker that identifies senescent human cells in culture and in aging skin in vivo. Proc Natl Acad Sci USA 1995; 92(20):9363-9367.
51. Goldstein S. Replicative senescence: the human fibroblast comes of age. Science 1990; 249(4973):1129-1133.
52. Shay JW, Wright WE. Telomeres and telomerase: implications for cancer and aging. Radiat Res 2001; 155(1 Pt 2):188-193.
53. Stewart SA, Weinberg RA. Telomerase and human tumorigenesis. Semin Cancer Biol 2000; 10(6):399-406.
54. Bodnar AG, Ouellette M, Frolkis M et al. Extension of life-span by introduction of telomerase into normal human cells. Science 1998; 279(5349):349-352.
55. Wright WE, Shay JW. Telomere dynamics in cancer progression and prevention: fundamental differences in human and mouse telomere biology. Nat Med 2000; 6(8):849-851.
56. Chang S, Multani AS, Cabrera NG et al. Essential role of limiting telomeres in the pathogenesis of Werner syndrome. Nat Genet 2004; 36(8):877-882.
57. Choudhury AR, Ju Z, Djojosubroto MW et al. Cdkn1a deletion improves stem cell function and lifespan of mice with dysfunctional telomeres without accelerating cancer formation. Nat Genet 2007; 39(1):99-105.
58. Serrano M, Blasco MA. Putting the stress on senescence. Curr Opin Cell Biol 2001; 13(6):748-753.
59. Ohtani N, Mann DJ, Hara E. Cellular senescence: its role in tumor suppression and aging. Cancer Sci 2009; 100(5):792-797.
60. Hardy K, Mansfield L, Mackay A et al. Transcriptional networks and cellular senescence in human mammary fibroblasts. Mol Biol Cell 2005; 16(2):943-953.
61. Johung K, Goodwin EC, DiMaio D. Human papillomavirus E7 repression in cervical carcinoma cells initiates a transcriptional cascade driven by the retinoblastoma family, resulting in senescence. J Virol 2007; 81(5):2102-2116.
62. Campisi J. Senescent cells, tumor supression and organismal aging: good citizens, bad neighbors. Cell 2005; 120:513-522.
63. Okada H, Mak TW. Pathways of apoptotic and non-apoptotic death in tumour cells. Nat Rev Cancer 2004; 4(8):592-603.
64. Ben-Porath IaW RA. When cells get stressed: an integrative view of cellular senescence. J Clin Invest 2004; 113:8-13.
65. Hickman ES, Moroni MC, Helin K. The role of p53 and pRB in apoptosis and cancer. Curr Opin Genet Dev 2002; 12:60-66.
66. Comerford KM, Leonard MO, Karhausen J et al. Small ubiquitin-related modifier-1 modification mediates resolution of CREB-dependent responses to hypoxia. Proc Natl Acad Sci USA 2003; 100(3):986-991.
67. Li T, Santockyte R, Shen RF et al. Expression of SUMO-2/3 induced senescence through p53- and pRB-mediated pathways. J Biol Chem 2006; 281:36221-36227.
68. Bischof O, DEjean A. SUMO is growing senescent. Cell Cycle 2007; 6:677-681.
69. Akar CA, Feinstein DL. Modulation of inducible nitric oxide synthase expression by sumoylation. J Neuroinflammation 2009; 6:12.
70. Li T, Evdokimov E, Shen RF et al. Sumoylation of heterogeneous nuclear ribonucleoproteins, zinc finger proteins and nuclear pore complex proteins: a proteomic analysis. Proc Natl Acad Sci USA 2004; 101(23):8551-8556.
71. Gostissa M, Hengstermann A, Fogal V et al. Activation of p53 by conjugation to the ubiquitin-like protein SUMO-1. EMBO J 1999; 18:6462-6471.
72. Ledl A, Schmidt D, Muller S. Viral oncoproteins E1A and E7 and cellular LxCxE proteins repress SUMO modification of the retinoblastoma tumor supressor. Oncogene 2005; 24:3810-3818.
73. Rodriguez M, Desterro JM, Lain S et al. SUMO-1 modification activates the transcriptional response of p53. EMBO J 1999; 18:6455-6461.
74. Bischof O, Schwamborn K, Martin N et al. The E3 SUMO ligase PIASy is a regulator of cellular senescence and apoptosis. Mol Cell 2006; 22(6):783-794.
75. Sugrue MM, Shin DY, Lee SW et al. Wild-type p53 triggers a rapid senescence program in human tumor cells lacking functional p53. Proc Natl Acad Sci USA 1997; 97:9648-9653.
76. Schmidt DaM S. PIAS/SUMO: new partners in transcriptional regulation. Cell Mol Life Sci 2003; 60:2561-2574.
77. Uchimura Y, Ichimura T, Uwada J et al. Involvement of SUMO modification in MBD1- and MCAF1-mediated heterochromatin formation. J Biol Chem 2006; 281:23180-23190.

78. Nathan D, Ingvarsdottir K, Sterner DE et al. Histone sumoylation is a negative regulator in Saccharomyces serevisiae and shown dynamic interplay with positive-acting histone modifications. Genes Dev 2006; 20:966-976.
79. Yates KE, Korbel GA, Shtutman M et al. Repression of the SUMO-specific protease Senpl induces p53-dependent premature senescence in normal human fibroblasts. Aging Cell 2008; 7(5):609-621.
80. de Lange T. Shelterin: The protein complex that shapes and safeguards human telomeres. Genes Dev 2005; 19:2100-2110.
81. Stewart SA, Weinberg RA. Telomeres: Cancer to human aging. Annu Rev Cell Dev Biol 2006; 22:531-557.
82. Tanaka K, Nishide J, Okazaki K et al. Characterization of a fission yeast SUMO-1 homologue, pmt3p, required for multiple nuclear events, including the control of telomere length and chromosome segregation. Mol Cell Biol 1999; 19:8660-8672.
83. Xhemalce B, Seeler JS, Thon G et al. Role of the fission yeast SUMO E3 ligase Plilp in centromere and telomere maintenance. EMBO J 2004; 23:3844-3853.
84. Zhang L, Li F, Dimayuga E et al. Effects of aging and dietary restriction on ubiquitination, sumoylation and the proteasome in the spleen. FEBS Lett 2007; 581:5543-5547.
85. Wei F, Scholer HR, Atchison ML. Sumoylation od Oct4 enhances its stability, DNA binding and transactivation. J Biol Chem 2007; 282:21551-21560.
86. de Cristofaro T, Mascia A, Pappalardo A et al. Pax8 protein stability is controlled by sumoylation. J Mol Endocrinol 2009; 42:35-46.
87. Yurchenko V, Xue Z, Gama V et al. Ku70 is stabilized ny increased cellular SUMO. Biochem Biophys Res Commun 2008; 366:263-268.
88. Benanti J, Williams DK, Robinson KL et al. Induction of extracellular matrix-remodelling genes by the senescence-associated protein APA-1. Mol Cell Biol 2002; 22:7385-7397.
89. Cheng J, Kang X, Zhang S et al. SUMO-specific protease 1 is essential for stabilization of hypoxia-inducible factor-1a during hypoxia. Cell 2007; 131:584-595.
90. Wang YT, Chuang JY, Shen MR et al. Sumoylation of specificity protein 1 augments its degradation by changing the localization and increasing the specificity protein 1 proteolytic process. J Mol Biol 2008; 380:869-885.

Chapter 13

Post-Translational Modification of Cellular Proteins by Ubiquitin and Ubiquitin-Like Molecules:

Role in Cellular Senescence and Aging

Johannes Grillari,* Regina Grillari-Voglauer and Pidder Jansen-Dürr

Abstract

Ubiquitination of endogenous proteins is one of the key regulatory steps that guides protein degradation through regulation of proteasome activity. During the last years evidence has accumulated that proteasome activity is decreased during the aging process in various model systems and that these changes might be causally related to aging and age-associated diseases. Since in most instances ubiquitination is the primary event in target selection, the system of ubiquitination and deubiquitination might be of similar importance. Furthermore, ubiquitination and proteasomal degradation are not completely congruent, since ubiquitination confers also functions different from targeting proteins for degradation.

Depending on mono- and polyubiquitination and on how ubiquitin chains are linked together, post-translational modifications of cellular proteins by covalent attachment of ubiquitin and ubiquitin-like proteins are involved in transcriptional regulation, receptor internalization, DNA repair, stabilization of protein complexes and autophagy. Here, we summarize the current knowledge regarding the ubiquitinome and the underlying ubiquitin ligases and deubiquitinating enzymes in replicative senescence, tissue aging as well as in segmental progeroid syndromes and discuss potential causes and consequences for aging.

Introduction

Post-translational modifications of cellular proteins by ubiquitin and ubiquitin-like proteins are important regulatory events involved in several aspects of cellular physiology that have come into focus of aging research recently. One of the most important machineries for degradation of endogenous proteins is the proteasome, whose activity decreases during the aging process. Since several major age-related diseases have been characterized as protein degradation failures, most studies have aimed at elucidating the function of the proteasome and its activity in the course of cell, tissue and organism aging and several excellent reviews are available in this regard[1-4] as well as on age-associated neurodegenerative pathologies.[5,6]

However, little has been reported on age-associated modulation of the primary event, the selection of proteasome targets through polyubiquitination, a major regulatory step in this system resulting from the opposing activities of ubiquitin ligases and deubiquitinylases (DUB).

*Corresponding Author: Johannes Grillari—Institute of Applied Microbiology, Department of Biotechnology, University for Natural Resources and Applied Life Sciences, Vienna, Muthgasse 18, A-1190 Vienna, Austria. Email: johannes.grillari@boku.ac.at

Protein Metabolism and Homeostasis in Aging, edited by Nektarios Tavernarakis.

In addition, during the last years it became quite clear that degradation is not the only possible consequence for proteins after modification by ubiquitin and ubiquitin-like proteins. Depending on the type of ubiquitination, it has been found as to have "noncanonical" functions like regulation of DNA repair (reviewed in ref. 7), transcriptional regulation,[8] membrane trafficking,[9] cell signalling[10] and autophagy.[11]

In this chapter we provide some examples of the current research regarding activity and functions of the ubiquitin system with focus on ubiquitination during cellular aging and aging of tissues. Since, as stated above, many excellent reviews on changes of the proteasome during aging exist, we refer the reader to those for detailed information. A contemporary review of this topic is also found in the chapter by Gonos et al in this book. Similarly, the recent tight links that have been found between aging and autophagy have been reviewed in detail and here in the chapter by G. Mariño et al. Changes in this major cellular pathway are of high interest since they will affect cellular behaviour within tissues and organisms and thus might influence the aging process and might also be causally related to the aging process or pathogenesis of aging related diseases.

Ubiquitin Is Activated and Transferred by a Cascade of E1, E2 and E3 Enzymes

A sequence of reactions guarantees the correct selection of target proteins for ubiquitination. Most targets are ubiquitinated at internal lysine residues, but in some lysine-less proteins also N-terminal ubiquitination has been observed.[12] The first enzyme in this cascade is E1 that activates ubiquitin under consumption of ATP by forming a thiol ester with the C-terminal glycine of ubiquitin. Then the ubiquitin moiety is transferred to an E2 enzyme by transesterification. The E2 enzyme in turn either transfers the ubiquitin to an E3 enzyme that is responsible for the transfer to a specific final target, or it ubiquitinates the target by direct interaction with an E3 enzyme (reviewed in refs. 13-15). In some cases, a novel class of E4 enzymes exists that is responsible for polyubiquitylation, as reviewed by ref. 16.

Specificity of target recognition is guaranteed in a sophisticated and economic way by up to ten E1 enzymes, up to 100 E2 enzymes and an estimated number of over 1000 E3 enzymes in human cells.[17] Current estimations suggest that there are as many E3 enzymes encoded in the human genome as kinases, underlining the universal importance of the ubiquitin proteasome pathway. Therefore we also propose to use the term "ubiquitinome" to refer to the total of ubiquitinated proteins in analogy to the "phosphoproteome" and other well known-omes.

Alternative Linking of Polyubiquitin Chains Decides the Fate of the Target Protein

The "classical" signal for protein degradation is conferred by K48 linkage of several ubiquitin moieties to the target protein. At least 4 ubiquitins are necessary for efficient binding to the proteasome and for degradation.[18] Including the regular K48, there are a total of 7 lysine residues in ubiquitin which can be used for linking poly-ubiquitin chains together. A recent report shows different reactivity of these individual residues: K48, K63 and K6 are approximately equally reactive followed by K33 and K11. The least reactive sites are K27 and K29.[19]

K63 linkage is involved in a variety of different cellular pathways like endocytosis of cell surface receptors,[20] cell signalling,[10] mitochondrial inheritance and morphogenesis,[21] ribosomal function,[22] DNA repair[7] and activation of the NFκB signalling complex,[23,24] whose inactivation is in turn guaranteed by deubiquitination;[25] see also below).

Little is known about the function of K29 ubiquitin chains yet, but it seems to allow proteasomal degradation similar to K48.[26] E2 and E3 enzymes that act together for this type of polyubiquitination have been identified as UbcH5A, a 120-kDa E2,[27] as well as the HECT domain E3 KIAA10.[28] Both proteins can also promote K48 chain assembly and at least for KIAA10 it has been shown that site selection is dependent on the ubiquitin.[29]

One of the few currently known E3 ligases that synthesize K11 linked polyubiquitin is CHIP, a "Really Interesting New Gene" (RING)-domain E3 ligase. Upon formation of a trimer consisting of HSc70, the cochaperone BAG-1 and CHIP, BAG-1 is strongly polyubiquitinated by a K11 ubiquitin chain. This leads to a degradation-independent association of the cochaperone with the proteasome, suggesting that ubiquitination of delivery factors might represent a novel mechanism to regulate protein sorting to the proteasome.[30] Furthermore, CHIP seems by ubiquitination of tau to attenuate the formation of neurofibrillary tangles, which are a major diagnostic marker of Alzheimer's disease.[31] K6 linkage has been observed to counteract proteasomal degradation as competitor.[32] The last noncanonical form of ubiquitin attachment is monoubiquitination that is involved in DNA repair,[7] receptor endocytosis[33] and signalling.[8]

Ubiquitin Is Expressed as Fusion Protein and Is Recycled by Ubiquitin Specific Hydrolases

Ubiquitin is not expressed from a single gene locus but either from multi-copy loci or as linear fusion to ribosomal subunits. Because of the latter, translation and maturation of the ribosome that synthesizes proteins simultaneously provides the signal to destroy them. Furthermore, steady state levels of free ubiquitin are also regulated by the recycling rate from ubiquitinated proteins that are degraded by the proteasome (reviewed by ref. 14). The proteins involved in providing free ubiquitin are categorized as deubiquitinating enzymes (DUBs). All of them are cysteine proteases that hydrolyze the amide bond immediately after the COOH-terminal Gly76. Two different classes of enzymes are discriminated: ubiquitin COOH-terminal hydrolyses (UCHs) and ubiquitin-specific proteases (UBPs). The discrimination is based on molecular weight and sequence similarity. While the smaller UCHs cut peptides and flexible small proteins, the larger UBPs are processing ubiquitinated proteins. Genome sequencing projects have identified many candidate deubiquitinating enzymes, making them the largest family of enzymes in the ubiquitin system.[34]

Ubiquitin-Like Proteins

In addition to ubiquitin, several ubiquitin-like molecules have been described in the past of which SUMO is the molecule that is most studied. Although SUMOylation affects a similar large number of substrates than ubiquitination, the pathway is much simpler and involves one E1 ligase, a single E2-ligase (UBC9) and only a few E3 ligases. UBC9 usually binds the substrates directly, but the E3 ligases specific for SUMOylation also seem to contribute to substrate specificity. SUMOylation often targets a lysine residue within a consensus sequence, but other lysine residues can be modified as well. In contrast to yeast, the vertebrate genomes encode four SUMO variants, referred to as SUMO-1, -2 ,-3 and -4. Both ubiquitin and SUMO can form polymodifier chains and then can be conjugated to lysine residues of an already substrate-conjugated molecule. In contrast to polyubiquitin chains, little is known about the general function of polySUMO chains. The situation is even more complex, since recently it was discovered that branched polyubiquitin modifications occur in vivo and that SUMO itself may be modified by ubiquitin, resulting in mixed ubiquitin/SUMO chains, although the relevance of these modifications still have to be demonstrated. Importantly, both ubiquitination and SUMOylation are reversible by specific hydrolases, which can remove the modifications (for review, see ref. 35).

Further ubiquitin-like modifiers the enzymology of which is highly similar to that of ubiqutin transfer are atg8 and atg12, necessary for autophagy (reviewed in ref. 36), Nedd8[37] and Isg15 (interferon stimulated gene 15), involved in anti-viral responses.[38]

Role of Ubiquitination, SUMOylation and ISG15 in Cellular Senescence

In 1961, Hayflick and Moorhead described that normal human cells, after a defined number of cell divisions, enter a phase of irreversible growth arrest termed replicative senescence.[39] Since then replicative senescence has been widely studied and discussed for its value as model system for aging research. Recently replicative senescence has been associated with cellular aging and tumor suppression,[40] since cellular senescence was shown to inhibit unwanted cell proliferation; moreover senescent cells in vivo might provide a microenvironment in tissues that favours proliferation of tumor cells.[41] In addition, replicative senescent cells were observed in a large variety of tissues of aged mice[42] and different human tissues such as skin,[43,44] liver,[45] kidney[46,47] and vasculature[48,49] and preliminary evidence suggests that they are implicated in aging and aging associated disease.

Several lines of evidence indicate that a prominent stimulus for entering the phase of replicative senescence is progressive telomere erosion due to the "end-replication problem".[50] Critically short telomeres induce DNA damage response pathways and subsequent the induction of a permanent cell cycle arrest,[51] which is executed by the p53/p21 or the $p16^{INK4a}$ pathway in a cell type specific manner (reviewed by refs. 52, 53). The hypothesis of telomere dependent induction of senescence is supported by the fact that overexpression of the catalytic subunit of human telomerase (hTERT) and subsequent telomere stabilization immortalizes a large variety of normal human cells.[54,55]

In addition to telomere erosion, other stimuli can induce (premature) cellular senescence, which in most cases is not linked to telomere shortening. Other events that lead to senescent-like phenotypes are oxidative stress,[56] DNA damage[57] and aberrant signalling (e.g., ras).[58]

A direct link of ubiquitination/SUMOylation and cellular senescence has been established by experiments regarding phenotypic consequences of the depletion of proteins involved in post-translational modification by ubiquitin and/or SUMO.

Furthermore, this link seems to be conserved during evolution, since direct interaction between replicative and chronological aging with the ubiquitin-dependent proteasome system have been identified by a systems biology approach.[59]

In human cells, it was shown that knockdown of ROC1, a key component of the SCF E3 ubiquitin ligases, inhibited the growth of multiple human cell lines by the induction of senescence and/or apoptosis.[60] Similarly, inactivation of the VHL tumor suppressor gene, coding for a subunit of E3 ubiquitin ligase, induces a senescent-like phenotype in human cancer cell lines. This phenotype is independent of p53 and HIF-1, but dependent on the retinoblastoma protein.[61] In mice, knock-out of Cdh1, an adaptor protein of the anaphase promoting complex (APC), which is an E3 ligase, induces premature replicative senescence in MEFs.[62]

Furthermore, upregulation of the ubiquitin ligase SMURF2 induced senescence in a variety of human cell types.[63] However, the ability of SMURF2 to induce senescence did not require its ubiquitin ligase activity.[64] With respect to SUMOylation, it was shown that overexpression of the E3 SUMO ligase PIASy in normal human fibroblasts recruits the p53 and Rb tumor suppressor pathways to provoke a senescence arrest. By contrast, in Rb-deficient fibroblasts, expression of PIASy leads to p53-dependent apoptosis and PIASy was shown to stimulate SUMOylation and transcriptional activity of p53.[65] Interestingly PIASy has been shown to interact with pyruvate kinase M2 (M2-PK), a key regulatory enzyme controlling proliferation and tumorigenesis by a modulation of cellular metabolism. Previous evidence implicated M2PK as a regulator of senescence-associated growth arrest and senescent cells have been shown to accumulate preferentially the active tetrameric form of the enzyme.[66]

In addition, stable cell lines overexpressing processed forms of SUMO-2/3 (SUMO-2/3GG) showed a premature senescence phenotype. Both p53 and pRB were found to be modified by SUMO-2/3, suggesting that such modification of p53 and pRB may play roles in premature senescence and stress response,[67] reviewed by ref. 68.

Other experiments have revealed an important role for the SUMO specific proteases SENP1, SENP2 and SENP7 in cellular senescence. Knockdown of these genes in human fibroblasts induced senescence and a global increase in SUMOylated proteins. The data suggest that SENP1 repression induces p53 mediated premature senescence and that SUMO proteases are required for the proliferation of normal human cells.[69]

How could ubiquitination or SUMOylation be involved in onset or bypass of senescence? The first approach to answer this question was to study global changes in the ubiquitinome during cellular aging and was published by Pan and coworkers in 1993, reporting a decrease of free ubiquitin, while the pool of ubiquitinated proteins was increased in senescent fibroblasts. The most

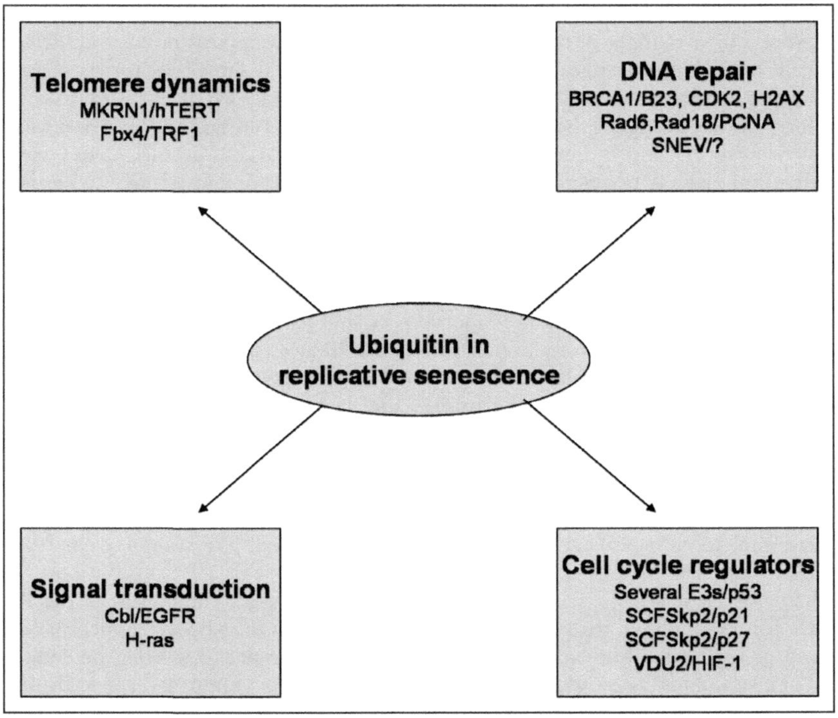

Figure 1. Implications of the ubiquitin system on pathways that are involved in induction of replicative senescence. See text for description. Reproduced with permission from Grillari J et al. Exp Gerontol 2006; 41(11):1067-79.

prominent ubiquitin conjugate was a discrete band of 55 kD, most probably the E2-55 ligase.[70] The ubiquitin-proteasome system also seems to play a role in vascular senescence and atherosclerotic progression of elderly patients. Thus, atherosclerotic plaques from elderly patients had increased ubiquitin levels and reduced proteasome activity compared to controls, suggesting that reduction in the activity of the ubiquitin-proteasome system promotes vascular cell senescence, thereby contributing to the pathogenesis of human atherosclerosis.[71] Experiments summarized below have identified several individual proteins, that might link the SUMO/ubiquitin system to cellular aging (Fig. 1). Specifically it was shown, that key processes regulated by ubiquitination are telomere maintenance, DNA repair and cell cycle control, as is detailed in the following sections.

Ubiquitin, Telomeres and Telomerase

Telomerase activity itself is regulated by ubiquitination. The catalyitic subunit hTERT directly interacts with the chaperone Hsp90.[72] If this interaction is disrupted, e.g., by geldamycin, hTERT is targeted to ubiquitylation and subsequent degradation by the RING finger E3 ligase MKRN1. This leads to telomere shortening in cells with active telomerase. This finding was confirmed by overexpression of MKRN1 which as well results in shortened telomeres.[73] Furthermore, the telomeric repeat binding factor 1 (TRF1), a negative regulator of telomere length,[74] is targeted to ubiquitin dependent degradation. The responsible protein is Fbx4, an adaptor to Cul1 type E3 ligases, that upon overexpression is sufficient to stabilize telomeres, while its knock-down by siRNA leads to increased telomere shortening and to accelerated entry into senescence-like growth arrest.[75] It was shown that RLIM, a Ring H2 zinc finger protein with intrinsic ubiquitin ligase activity interacts with TRF1 and increases its turnover by targeting it for degradation by the proteasome. Interestingly, this activity is independent of FBX4. Depletion of endogenous RLIM by shRNA mediated knockdown increased the level of TRF1 and led to telomere shortening and growth arrest.[76] In fission yeast, PLI1B has been described as a SUMO E3 ligase, which upon mutation leads to deregulated homologous recombination and marked defects in chromosome segregation and centromere silencing grown, with a consistent increase in telomere length. It was shown that telomere lengthening induced by lack of SUMOylation was not due to unscheduled telomere recombination, instead SUMOylation was shown to increase telomerase activity.[77] A role for SUMOylation of telomere binding proteins in alternative lengthening of telomeres (ALT) has also been demonstrated. The hallmark of ALT cells is the recruitment of telomeres to PML bodies. It was shown that the SUMO-ligase MNS21, part of the SMC5/6 complex, SUMOylates multiple telomere binding proteins, including TRF1 and TRF2, which prevents recruitment of telomeres to PML bodies. Depletion of FMC5/6 subunits by shRNA was shown to induce telomere shortening and senescence in ALT cells, suggesting that SUMOylation of telomere binding proteins facilitates telomere homologous recombination and elongation in ALT cells.[78]

Concerning ubiquitination, it was shown that the human orthologue of the yeast protein EST1B is a target for ubiquitin mediated degradation. EST1B was shown to bind histone deacetylase 8 (HDAC8) and phosphorylated HDAC8 inhibits degradation of HEST1B by the E3 ligase CHIP. Regulation of HEST1B protein stability by HDAC8 modulates the enzymatic activity of telomerase.[79] Furthermore, poly-ubiquitin conjugates of yet unknown nature are found in PML bodies of senescent cells.[80]

A surprising link between another ubiquitin like modifier, ISG15 and telomere length has been observed only recently. The ISG15 gene is located in the subtelomeric region of chromosome 1 and its mRNA and protein expression levels are tightly correlated with telomere length. While telomere position effects have been found in yeast a couple of years ago,[81] ISG15 is the first known human gene whose expression is regulated by telomere length.[82]

Regulation of DNA Repair and Growth Arrest in Cell Culture

The regulatory function of ubiquitination in DNA repair might be of high importance for aging, not only in regard to replicative senescence, but also in regard to segmental progeroid syndromes whose symptoms are largely caused by defects in the DNA repair system.[83] Recent data suggest that virtually all the major DNA repair pathways and DNA damage response mechanisms and checkpoint responses are regulated in some way by ubiquitination, SUMOylation or both. Strikingly, genes of the RAD6 damage tolerance pathway encode mostly enzymes involved in the ubiquitin pathway and the interstrand crosslink repair pathway linked to Fanconi anemia seems to consist largely of enzymes and substrates of the ubiquitin system (for recent review see ref. 35). DNA damage usually induces the activation of DNA repair or damage avoidance pathway and often a checkpoint response that triggers cell cycle arrest to allow time for the repair. This reaction, referred to as a DNA damage response (DDR), is typically initiated by proteins that recognize DNA lesions and is followed by the recruitment and activation of proteins that trigger checkpoint signaling or directly perform the necessary repair steps. This involves large protein assemblies, which are microscopically identifiable as repair foci. Once the DNA is repaired, the machinery needs to be disassembled and the DNA damage response turned off. It has been shown that both ubiquitination and SUMOylation are key steps controlling consecutive events required for the DNA damage response (for a more detailed description the reader is referred to 35). In the following section, individual components of this pathway,[84] which are of particular importance for senescence and aging, are addressed in detail.

Breast Cancer Associated Protein 1 (BRCA1)

Most combined familial breast and ovarian cancers and ~40% of familial breast cancer cases have been linked to mutations in BRCA1. A mouse model carrying a homozyogous mutant of $Brca1^{\Delta 11}$ and haploidity for p53, shows p53-dependent senescence in mouse embryonic fibroblasts as well as a premature aging syndrome at the organismal level.[85] The heterodimer of BRCA1-BARD1 is one of the E3 enzymes that catalyzes autoubiquitination by K6 linkage in vitro and in vivo, that is not degraded by the 26S proteasome in vitro.[86] Rather, it leads to 20-fold activation of its E3 ligase activity,[87] as well as to relocalization to nuclear foci upon DNA damage.[88]

One target of BRCA1-BARD1 ubiquitination is nucleophosmin/B23,[89] a nucleolar-cytoplasm shuttling protein that induces premature senescence upon overexpression.[90] Another protein in this regulatory pathway is the central cell cycle regulator CDK2-cyclin E1 as a negative regulator of BRCA1-BARD1 activity.[91] Since CDK2 protein levels and activity are markedly downregulated during replicative senescence of endothelial cells,[92] it is intriguing to think that this might cause high BRCA1-BARD1 activity. As a consequence, this might lead to K6 polyubiquitination of nucleophosmin/B23, activating and most probably also stabilizing it, since K6 chains are also known to counteract ubiquitin dependent hydrolysis.[32] At the end of this signalling high nucleophosmin levels might help to induce or at least to maintain the senescent phenotype by directly interacting with p53 and inducing p21.[90]

Another target of BRCA1-BARD1 E3 ligase activity is the histone H2AX, which is monoubiquitinated upon DNA damage and colocalizes with BRCA1 at foci of DNA damage.[87] Formation of the phosphorylated H2AX (γH2AX) foci is also one of the hallmarks of replicative senescence in cell culture as reviewed by ref. 93 and also the phosphorylated form of H2AX colocalizes with BRCA1.[94] Therefore, it seems plausible that these foci also contain the monoubiquitinated γH2AX, since phosphorylation is a often a signal for consecutive ubiquitination.[95]

Proliferating Cell Nuclear Antigen (PCNA)

Other proteins that are monoubiquitinated upon DNA damage are histone 2A (H2A)[96] and proliferating cell nuclear antigen (PCNA). PCNA is involved in a large variety of DNA transactions (reviewed by ref. 97) like DNA replication, repair of interstrand cross links (ICL) and translesion synthesis.[98] The E2 and E3 ligases transferring the monoubiquitin moietiy to PCNA are Rad6 and Rad18.[99] Only the monoubiquitinated form of PCNA can physically interact with polymerase η (pol η)[100] and this interaction is necessary to bypass pyrimidine dimers after UV damage in DNA replication.[101] A clinical phenotype of mutations in this DNA repair function is known: mutations in pol η result in patients suffering from hypersensitivity to UV-light phenotypically related to the segmental progeroid syndrome xeroderma pigmentosa. Indeed this mutation has been classified to the XP variant (XPV) complementation group.[102] Regulation of PCNA function by post-translational modification has been worked out recently and provides a striking example for the regulatory interplay between ubiquitin and SUMO modifications. PCNA homotrimeres act as processing factors for DNA polymerases and as a moving platform for factors that mediate replication linked functions, such as chromatin assembly or sister chromatid cohesion.[103] Remarkably, PCNA can be modified on the same conserved lysine residue (Lys164) either by monoubiquitylation, Lys63 linked polyubiquitylation chains or by SUMOylation. Both ubiquitination and SUMOylation of PCNA occur in S-phase, but ubiquitination specifically occurs, when DNA is damaged. SUMOylation of PCNA predominantly on Lys164 attracts the antirecombinogenic helicase SRS2 to inhibit unwanted recombination during DNA synthesis. SUMOylation on Lys127 inhibits the interaction with certain PCNA binding proteins, such as Eco1. SUMOylation and ubiquitination are carried out by a complex system of conjugating enzymes, such as RAD15, RAD18, RAD6 and these modifications regulate the protein-protein interactions within DNA synthesis and DNA repair complexes. Because Lys164 of PCNA is a target for ubiquitin and SUMO, the two modifiers compete for the substrate. This has led to the development of a model for assembly and disassembly of PCNA multiprotein complexes, which is regulated by ubiquitination and SUMOylation, referred to as the ubiquitin/SUMO "switchboard".[35]

An additional link to aging might be provided by PCNA's function in DNA replication that is inhibited by p21 binding in the early phase of replicative senescence[104] reviewed by ref. 105. Since besides monoubiquitination, PCNA can be K63-polyubiquitinated as well as SUMOylated at the same lysine residue (K164), it is well possible that these modifications might as well be involved in inducing or maintaining the terminal growth arrest at the Hayflick limit.

Senescence Evasion Factor (SNEV$^{Prp19/Pso4}$)

In a previous screening for differentially expressed genes after induction of replicative senescence we have identified the highly conserved protein SNEV$^{Prp19/Pso4}$.[106] Overexpression of SNEV$^{Prp19/Pso4}$ increases the life span of human endothelial cells, which is correlated with higher resistance to stress and lower levels of DNA damage.[107] Furthermore, SNEV$^{Prp19/Pso4}$ is a U-box E3 ligase[108] and interacts with the proteasome, by directly binding to the beta 7 subunit,[109] while in mouse an interaction to the regulatory subunit SUG1 was reported.[110] It is of note that the beta subunits β1, β2 and β5 of the proteasome are markedly downregulated in senescent fibroblasts[111] and that overexpression of β5 is sufficient to confer higher stress resistance and survival to fibroblasts as well as a delay in onset of replicative senescence for 4-5 population doublings.[112] In mouse, SNEV is essential for embryonic development, since its knock-out is lethal at the blastocyst stage.[113] MEFs isolated from SNEV+/- heterozygous mice undergo senescence earlier than MEFs derived from litter mate controls.[113] Furthermore, SNEV is necessary for hematopoietic progenitor cell self renewal.[114] Although it is possible that SNEV$^{Prp19/Pso4}$ and the proteasome act in the same pathway resulting in the observed increase of the life span, also different explanations for the underlying mechanisms in the case of SNEV$^{Prp19/Pso4}$ are possible, since SNEV$^{Prp19/Pso4}$ is also involved in DNA repair,[115,116] specifically in DNA interstrand cross link repair, where SNEV acts in a complex containing WRN, the gene mutated in Werner syndrome.[117] Although the DNA repair function seems to be as plausible a reason for life span extension in endothelial cells as the proteasome connection, it cannot be excluded that other functions like SNEV$^{Prp19/Pso4}$'s role in pre-mRNA splicing[118] are involved, before the substrate of SNEV$^{Prp19/Pso4}$ or the domains that are involved in DNA repair have been identified.

p53 Senescence and Tumor Suppressor Pathway

p53

The regulation of p53 tumor suppressor protein plays a major role in signalling of DNA damage, protecting the integrity of the genome and is largely involved in aging of cells and organisms.[119,120] Excellent reviews are available on its highly complex and sophisticated regulation by the ubiquitin and sumoylation system and its multiple regulators Mdm2, vHL, ARF, ARF-BP1, HAUSP and MdmX to which readers are referred.[121,122] Still, also ubiquitin independent degradation of p53 is reported.[123]

p19(ARF)

Besides the CDK inhibitor p16INK4A (see below), the INK4 locus codes for a second protein generated by alternative splicing, which leads to changes in the open reading frame, therefore this protein has been referred to as ARF (for alternative reading frame). Early reports suggested that ARF binds to p53 and stabilizes this tumor suppressor protein, however more recent data have identified that ARF can also promote SUMOylation of a variety of cellular target proteins, including MDM2.[124,125] Similar to p16INK4A, ARF is N-terminally ubiquitinated and degraded in proteasomes. Recent studies have revealed an antagonism between SUMO deconjugating protease SENP3 and ARF, both of which interact with the abundant nucleolar protein nucleophosmin (NPM). It was shown that SENP3 and ARF antagonize each other's function in regulating the SUMOylation of target proteins, including NPM itself. The data suggest that the p53 independent tumor suppressor functions of ARF may be mediated by its ability to antagonize SENP3 thereby elevating cellular levels of SUMOylated proteins and inducing subsequent cell cycle arrest.[126] ARF was also found to promote the polyubiquitination, through Lys63 of ubiquitin, of COMMD1, triggering this protein for proteasome dependent proteolysis. However, the enzymology of COMMD1 polyubiquitination needs to be clarified.[127] One question of interest in regard to p53, p21 and p16 is if and which of the pathways to degrade or stabilize these key molecules are changed during the aging process.

Cell Cycle Inhibitors are Mainly Regulated by Ubiquitination Followed by Degradation

p21(WAF1/Cip1)

One downstream target of p53, the cdk inhibitor, p21 (WAF1), which is necessary for induction of replicative senescence,[128] can as well either be degraded in a ubiquitin independent way or can be regulated by SCF/Skp2 ubiquitin ligase activity,[129] which attaches the ubiquitin its N-terminus. Recent data have revealed molecular mechanisms underlying ubiquitin-dependent degradation of p21. It was shown that p21 is degraded in mitosis subsequently to APC/C mediated ubiquitination, suggesting that degradation of p21 contributes to the full activation of CDK1 necessary for mitosis and prevents mitotic slippage during spindle checkpoint activation.[130] p21 is also degraded during S-phase and in response to low doses of ultraviolet light and this involves PCNA, which promotes the ubiquitination and degradation of p21. This involves the ubiquitin ligase Cul4A-DDB1(Cdt2). It was suggested that the ubiquitin ligases CRL4 (Cdt2) and SCF are redundant with each other to promote the degradation of p21 during S-phase.[131,132] p21 can also be degraded in a ubiquitin and ATP independent way in response to high doses of UV,[133] a process that may involve the proteasome activator REGgamma.[134] Ubiquitin dependent degradation of p21 can also be triggered by reactive oxygen species.[135]

p16(INK4)

p16(INK4a) encodes another cell cycle inhibitor, involved in induction of both replicative and oncogenic ras-induced senescence.[57,136] It is ubiquitinated by N-terminal linkage, since p16 does not contain any lysine residues.[12] The CDK inhibitor p19INK4D, which bears structural and functional similarity with p16, was shown to be a target for ubiquitin proteasome dependent degradation, which explains the oscillation of the protein, but not the mRNA, during a normal cell cycle. Ubiquitination of p19INK4D was dependent on the integrity of K62.[137]

p27(KIP1)

Another important regulator of the cell cycle involved in Rb-mediated senescence is p27 that regulates cdk2 kinase activity.[138] In senescent fibroblasts, p27 is stabilized by decrease in the F-box protein Skp2, which forms part of the SCF E3 ligase complex responsible for p27 ubiquitination and consequent degradation.[139] Similarly, in cells of the Ewing sarcoma tumor family (EFT), depletion of the tumour-causing, rearranged EWS-Fli1 protein by siRNA elicits a senescent-like phenotype which is dependent on stability of p27, since Skp2 knockdown reversed the senescent phenotype.[140]

The stability of p27 is regulated at several levels, predominantly including two distinct pathways for ubiquitin proteasome mediated proteolysis. On the one hand, Serin10 phosphorylation increases in the early G1-phase of the cell cycle, allowing nuclear export of p27 and proteolysis of p27 in a KPC-mediated reaction. Secondly, phosphorylation of p27 at threonin187 occurs in S-phase, which triggers proteolysis through SCF ubiquitin ligase. Major pathways for p27 ubiquitination depend on KPC and SCF (SKP2) ubiquitin ligase complexes (reviewed in ref. 22).[141]

Deubiquitinating Enzymes and Their Influence on the Cell Cycle

UchL1/PGP9.5

An alternative possibility to stabilize p27 is its deubiquitination and indeed, the DUB UchL1/PGP9.5 was found to form a heterotrimeric complex with Jab1 and p27. This complex forms after serum restimulation of cells. In contrast, UchL1/PGP9.5 localizes to the perinuclear region and the cytoplasm in contact inhibited cells, but p27 remains nuclear,[142] suggesting that it does not get deubiquitinated and therefore is not stabilized to induce cell cycle arrest under these conditions.

Furthermore, UchL1/PGP9.5 transcription is largely increased in pituitary glands of aged mice[143] and has been implicated in neurodegeneration in Parkinson's, Alzheimer's and Huntington's disease patients. In all these diseases, reduced UCHL1 function may jeopardize the survival of central nervous system neurons.[144]

Potential Role of Other DUBs in Cellular Proliferation and Growth Arrest

No direct role of other DUBs in cellular aging has yet been reported, however, several of them are involved in regulating cell cycle progression. DUB-1 for example inhibits the cell cycle in G1 phase, if expressed from an inducible promoter in the pre B cell line Ba/F3.[145] Similarly, DUB-3, a cytokine-inducible deubiquitinating enzyme blocks proliferation in Ba/F3, but also in mouse embryonic fibroblasts upon overexpression.[146] CYLD, a DUB that negatively affects NFκB signalling by removing K63-linked ubiquitin chains from tumour necrosis factor familiy members,[147,148] inhibits proliferation of keratinocytes by deubiquinating Bcl-3.[149] Furthermore, von Hippel Lindau (VHL) interacting deubiquitinating enzyme (VDU2) deubiquitinates and thus stabilizes hypoxia inducible factor 1 (HIF-1),[150] which is causally involved in several types of cancer. In contrast, HIF-1 downregulation induces cellular senescence in endometrial cancer cells.[151] In keeping with this, hypoxia has been reported to extend the replicative life span of rat aortic smooth muscle cells.[152] Thus, deregulation of HIF-1 might contribute to cellular senescence or tumor pathogenesis. Since HIF-1 is also involved in the reduced response of aged organisms to hypoxic stress,[153,154] its ubiquitination status might be of importance also during organismal aging.

Signal Transduction, Receptor Endocytosis—EGFR and Ras

Epidermal growth factor receptor (EGFR) is one prominent example for the role of monoubiquitination in receptor endocytosis (reviewed by ref. 20). EGFR signalling is necessary for epithelial cell proliferation and differentiation[155] and resistance to EGFR ligand signalling in senescent cells is well documented[156] and is considered to be involved in slower wound healing in the elderly.

Upon monoubiquitination, EGFR is internalized and sorted to the lysosomes, where it is degraded.[33] The ubiquitin ligases Cbl[157] as well Sts2[158] are involved in this pathway. Deubiquitination by UBPY/USP8 can antagonize this and slow down the EGFR internalization.[159]

The mechanisms leading to decreased EGFR signalling in senescent cells described so far include diminished EGFR mRNA transcription, enhanced phosphatase activitiy that dephosphorylates EGFR[160] as well as altered endocytosis. Dependent on the dose of ligand EGFR is either internalized by ubiquitin-independent mechanism via caveolae at high EGF doses or by ubiquitin dependent clathrin-coated pits at lower doses.[8] Both receptor internalization pathways are affected in senescent cells: Enhanced EGFR internalization by caveolin has been shown as a major mechanism to downregulate EGF signalling in senescent cells, since knockdown of caveolins led to restoration of EGF signalling in senescent fibroblasts and overexpression of caveolins in young cells led to a senescent-like phenotype.[161,162] Similarly, the clathrin-dependent pathway is downregulated in senescent fibroblasts since one of the essential proteins, amphiphysin-1, is not internalized.[163]

One of the branches of EGFR signalling leads via ras to the MAP kinase pathway. Since constant signalling of ras, e.g., by overexpression of oncogenic ras (H-ras), leads to induction of premature senescence,[58] also ubiquitin might be involved in its regulation. Indeed, mono or deubiquitination of H-ras is important for docking to endosomes and to activate the downstream MAPK/raf signal cascade.[164]

Ubiquitin-Dependent and Independent Mitochondrial Protein Quality Control

Ubiquitin-independent protein quality control of mitochondrial matrix proteins is well characterized and until recently the mitochondria were considered ubiquitination-free organelles. However, several recent studies indicate multiple roles of the ubiquitin proteasome pathway in the regulation and maintenance of mitochondrial integrity. Of particular interest is the finding of a mitochondrial ubiquitin dependent protein quality control, which shares similarity to the endoplasmic reticulum associated degradation (ERAD) pathway that acts to eliminate misfolded proteins from the lumen of the endoplasmic reticulum. Given the key role of mitochondrial functionality for healthy aging and the role of mitochondrial dysfunction in various aging processes, it can be expected that further insight into this field will elucidate new roles of ubiquitination in the aging process.[165]

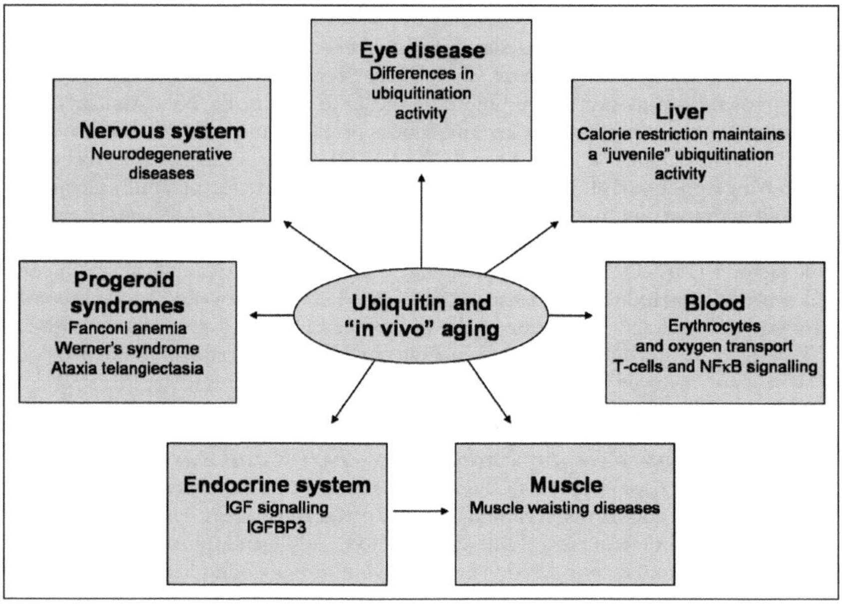

Figure 2. Age related changes in the ubiquitin system affect several different tissues. Reproduced with permission from Grillari J et al. Exp Gerontol 2006; 41(11):1067-79.

Ubiquitination in Tissues during Aging

Several tissues are subject to intensive investigations in regard to the activity and to single components of the ubiquitin system. In the following section, a summary on the most prominent examples is presented (Fig. 2).

Ubiquitination in the Nervous System

Since ubiquitination and protein degradation failure are involved in the pathogenesis of several neurodegenerative diseases and has been extensively reviewed, the reader is referred to some of the excellent current reviews.[2,6,166] Concerning Parkinson's disease some recent data shed new light on the role of ubiquitination in neuronal degeneration. Specifically Parkin, which is mutated in familial forms of PD, encodes a ubiquitin E3 ligase,[167] the inactivation of which leads to a dysfunction of the ubiquitin proteasome system and to the accumulation of aggregated alpha synuclein in the cytosol of dopaminergic cells. Recent data suggest that parkin can also translocate to the nucleus upon DNA damage[168] where it participates in the regulation of DNA repair, probably through an interaction with PCNA.[169] Another gene that is relevant for Parkinson's disease codes for ubiquitin c-terminal hydrolase L1, which degrades the polyubiquitin chain and provides monoubiquitin to the cell. Similar to Parkin, dysfunction of UCHL1 leads to a dysfunction of the ubiquitin proteasome system and the accumulation of protein aggregates.[170] Expression of a human Parkin missense mutant in Drosophila led to the degeneration of specific dopaminergic neuronal clusters and concommittant locomotor deficits that accelerated with age. These results provide in vivo evidence that Parkin mutants may directly exert neurotoxicity in vivo.[171]

Ubiquitination, Eye Diseases and Cataracts

Similar to neurodegenerative diseases, accumulation of damaged protein is also involved in the formation of cataracts, in which high molecular weight protein aggregates form as an aging associated phenomenon.[172] Since this might be related to the ubiquitinome of epithelial lens cells,[173,174] several studies have explored the amount of ubiquitin conjugates, or the ubiquitination activity in either in vitro cultivated lens cells or in the epithelial lens tissue. Aging and cellular maturation cause decreases in ubiquitination activity as well as in the amount of free and conjugated ubiquitin in bovine lens epithelial tissue.[175] In response to oxidative stress the increase in ubiquitination activity was as well reduced.[176] Furthermore, gene expression profiling in human age-related nuclear cataracts found, besides others, two ubiquitin-conjugating enzymes[177] as well as an E3 ligase[178] as downregulated in the aged tissue. Furthermore, an N-terminal protease-cleaved α-crystallin fragment which is increases during cataract formation is resistant against ubiquitin-proteasome mediated degradation.[179]

Taken together, this would suggest that the selection of degradation targets is decreased and that a slower turn-over of proteins might result in accumulation of damaged proteins and protein complexes. The role of the ubiquitin proteasome pathway in the length retinal cornea was recently reviewed.[180] Recently it was shown, that inactivation of the ubiquitin c-terminal hydrolase 3 (UCH-L3) leads to retinal degeneration along with muscular degeneration. In a normal retina UCH-L3 was enriched in the photoreceptor inner segment and loss of UCH-L3 led to mitochondrial oxidative stress related photoreceptor cell apoptosis in a caspase independent manner.[181] Proteins modified by a highly reactive lipid peroxide, 4-hydroxynonenal (HNE) are also ubiquitinated in cultured epithelial lens cells. Surprisingly, these damaged proteins are degraded in a novel ubiquitin dependent pathway that degrades them within the lysosomes.[182] To what extent this pathway might be deregulated and contribute to accumulation of nonfunctional protein during aging has not been investigated yet.

Ubiquitination in the Liver: Aging and Calorie Restriction

Calorie restriction increases the life span of rodents[183] and agents mimicking calorie restriction might be a target for anti-aging drugs.[184] Effect of aging and calorie restriction on the ubiquitination potential in liver was first analysed by Scrofano and coworkers. They found increased levels of ubiquitin conjugates, correspondingly higher E1 and E2 activities in the livers of 23-months versus 6-months old mice.[185] Calorie restriction, however, maintained a "young" phenotype also in this regard and prevented the increase. Surprisingly, the activity of ATP dependent and independent proteolysis did not change with age when using β-lactoglobulin and oxidized RNAse as model substrates.[186] Similarly, in a study aiming at the validation of housekeeping genes for normalization of gene expression levels in the aging liver of Fischer 344 rats, the amount of ubiquitin transcripts was found to change dramatically.[187]

In a consecutive study, the response to oxidative stress in livers of 23-months old Emory mice after injection of paraquat, a generator of superoxide radicals, was analysed.[186] Again, calorie restricted animals showed significantly lower levels of E1 and E2 activity as well as of ubiquitin conjugates. After exposure to paraquat, however, E1, E2 activity was induced to the same high amounts observed in ad libitum fed mice and ubiquitin-conjugates were even significantly higher in the calorie restricted mice, suggesting an increased turnover of protein and thus improved cellular repair. However, it is not clear if and how these ubiquitinated proteins are degraded, since Davies and coworkers have shown that oxidatively damaged proteins are degraded by the 20S proteasome in a ubiquitin independent manner.[188,189] Using lipopolysaccharides (LPS) as a stressor, calorie restriction also attenuated liver injury, did not increase a pro-inflammatory response by NFκB pathway and slightly influenced the mRNA levels of proteasome subunits like beta 2 and 3, as well as several ATPases of the proteasome lid in the liver of Fisher 344 rats.[190]

Additional experiments performed in mouse models suggest that dietary restriction significantly reduced age-related impairments in proteasome mediated protein degradation and reduced age-related increases in ubiquitinated, oxidized and SUMOylated protein in the heart. These results indicate that dietary restriction has many beneficial effects towards the ubiquitin proteasome system and suggest that a preservation of the UPP may be a potential mechanism, by which dietary restriction mediates beneficial effects on the cardiovascular system.[191]

Insulin-Like Growth Factor 1 (IGF-1) Signalling as Example of the Endocrine System

The highly conserved members of the IGF-1 signalling pathway are implicated in regulating the life span of species as diverse as *C. elegans*, *D. melanogaster* and mouse.[192] This pathway includes the hormone IGF-1 whose bio-availability is regulated by IGF1-binding proteins (IGFBPs) and signal transduction through the cell membrane by the IGF-1 receptor (IGFR), which forms dimers and shows tyrosine kinase activity upon ligand binding. One branch of the signalling then leads to the MAP kinase pathway via ras, another one via insulin receptor substrate 1 and 2 to akt, both branches specifically activating different genes in the nucleus.[193]

In regard to its regulation by the ubiquitin system, one of the IGFBPs, IGFBP3 has been found not only to be secreted, but also to be present in the nucleus,[194] where it is subject to polyubiquitination and degradation by the proteasome.[195] IGFBP3 internalization and translocation, however, might be independent from IGF-1.[196] Thus, it is not clear to which extent IGF-1 signalling might be affected by aging related changes of the ubiquitin system.

Ubiquitin System and the Aging Muscle

Certainly related to the IGF signalling pathway, however, is the ubiquitination and down-regulation of IGFR by β-arrestin which acts as adaptor for Mdm2 as the specific E3 ligase.[197] Furthermore, IGF-1 triggers ubiquitination and proteasomal degradation of its downstream signalling molecules insulin receptor substrate 1 (IRS-1),[198] IRS-2[199] and Akt.[200] IGF-1 stimulates muscle growth by suppressing the atrophy related ubiquitin ligases atrogin-1 and muscle ring finger-1 (MuRF1).[201] Similarly, it reduces the levels of free ubiquitin as well as that of various ubiquitin ligases.[202] Decrease in the growth hormone/IGF-1 axis thus plays a major role in muscle wasting associated with diseases like diabetes,[203] chronic heart failure[204] and cancer.[205]

Sarcopenia, an aging related form of muscle wasting,[206] is also associated with decrease in IGF-1 signalling (reviewed by ref. 207), followed by significantly higher free ubiquitin amounts in human and rat skeletal muscle of aged as compared to young individuals. These high levels of free ubiquitin seem to be directly linked to sarcopenia, since direct injection of ubiquitin into young healthy rat muscle induces muscle degeneration.[208,209] Although a different study does not report changes in total ubiquitin conjugate contents in aging rat soleus muscle, a decline in mRNAs coding for E2(14K) and MuRF-1[210] are observed. In any case, a direct link between loss of muscle protein mass due to changes is ubiquitination and aging seems obvious.

Recent studies have highlighted a key role of two E3 ubiquitin ligases, namely atrogin-1/MAFbx and MuRF1 as major factors responsible for skeletal muscle atrophy. Both enzymes are upregulated in many conditions of atrophy and utilized for protein degradation during muscle atrophy. It was reported that this is not the case in age-related loss of muscle mass (sarcopenia). On the contrary, Atrogin-1/MAFbx and MuRF1 were found downregulated in skeletal muscle of 30-month-old rats, probably due to AKT (protein kinase B)-mediated inactivation as well as by MDM2 mediated degradation[211] of the forkhead box O 4 (FOXO4) transcription factor. Dietary restriction was found to impede both sarcopenia as well as the effects of aging on AKT phosphorylation, FOXO4 phosphorylation and Atrogin-1/MAFbx and MuRF1 transcript regulation. Hence, sarcopenia appears mechanistically different from acute atrophies induced by disuse, disease and denervation.[212] It appears that a major age-dependent alteration in muscle proteolysis is a lack of responsiveness of the ubiquitin proteasome dependent proteolytic pathway to anabolic and catabolic stimuli (reviewed in ref. 213).

In mice overexpressing a mitochondrial T3 receptor, which acts as a mitochondrial transcription factor, a progressive decrease of mitochondrial DNA content was observed, which led to muscle atrophy probably through a stimulation of atrogin-1 and MURF1.[214] In a separate study overexpression of MURF1 did not lead to elevated ubiquitination of myosins instead the mice displayed lower levels of several metabolic enzymes required for glycolysis and glycogen metabolism. Whereas these data suggest that MURF1 expression in skeletal muscle redirects glycogen synthesis to the liver, molecular mechanisms are not clear at the moment.[215]

Age Related Changes in the Blood and Ubiquitination

In erythrocytes, spectrins are membrane proteins responsible for shape and mechanical properties. α-spectrin is a chimeric E2/E3 enzyme,[216] which is also ubiquitinated. A marked decrease in α-spectrin ubiquitination due to age-dependent changes in the erythrocyte membrane have been observed[217] and might influence the stability and deformability as well as the oxygen transport properties, all of which are reported to change with age.[218-220]

Aging associated changes in the behaviour of T-cells is also of major importance for organisms. One of the reasons of reduced defense by the aging immune system is involution of the thymus and consequently, reduction of naïve T-cells.[221,222] But also functional changes of T-lymphocytes in the elderly are observed, especially the NFκB pathway that leads to activation by cytokines and to inflammatory response of T-cells undergoes major changes in T-cells of the elderly reviewed by ref. 223. One of the underlying mechanisms is a change in proteasomal as well as in deubiquitination activity.[224] Since various types of ubiquitination play a major regulatory part in at least 3 different steps of NFκB pathway,[23] all of these might be of importance in regard to the aging process.

Phosphorylation of inhibitors of NFκB (IκB) by a kinase complex termed IκB kinase (IKK) is the event that consequently allows K48 polyubiquitination and proteasomal degradation of IκB. This ultimately leads to activation of NFκB.[225] At a different level, K63 polyubiquitin linkage is involved as well, since the regulatory subunit of IKK, NEMO, binds to K63 polyubiquitin chains. If the interaction site in NEMO is disrupted by point mutations, the activation of NFκB is inhibited.[24] Finally, also monoubiquitination plays a role in the NFκB pathway. Site-specific monoubiquitination of IKK down regulates continuous signalling by pro-inflammatory cytokines as well as by the oncogenic viral protein Tax.[226,227]

The discovery that two other inhibitors of NFκB signaling, A20 and CYLD1, are deubiquitinating enzymes, has emphasized the importance of these modifications. From these studies a general model of activation of NFκB has emerged. Whereas the details of this regulation vary depending on the particular signal triggering the signaling events, the ubiquitin dependent regulation of the NFκB pathway consists of the following steps: receptor engagement activates members of the TRAF family of E3 ligases to assemble a K63 linked polyubiquitin chain, which recruits the TAK1 protein kinase and its substrate IKB kinase (IKK). The complex catalyzes the K63 polyubiquitination of NEMO, the regulatory γ-subunit of IKK. The phosphorylated and ubiquitinated IKK is now active to phosphorylate IκB resulting in its K48 polyubiquitination to trigger its degradation by the proteasome. Finally the removal of the K63 linked polyubiquitination by the DUB A20 downregulates the signaling response.[228] In this way this pathway presents an excellent example, how subsequent ubiquitination and deubiquitination steps cooperate in the execution of a biological response such as NFκB activation.[229]

Besides its role in the decline of immune functions, NFκB deregulation is involved in loss of hearing and audionerve degeneration during aging in a mouse model,[229] its activity is dramatically increased in livers of old rats in comparison to those of young,[230] while DNA binding activity of NFκB is significantly reduced in skeletal muscle of old rats.[231] However, a direct role of ubiquitin dependent regulation of these processes has yet to be established.

Segmental Progeroid Syndromes and the Ubiquitin System

The ubiquitin system might also play a role in several segmental progeroid syndromes that are regarded as model systems for aging research.[232,233] Several of these syndromes are caused by mutations in DNA damage repair genes, which might provide one link to aging as outlined above. A connection to the ubiquitin system is presented by interaction of these proteins with and their regulation by the ubiquitin system. Some examples are listed in the following.

Although not initially included in Martin's list of segmental progeroid syndromes, Fanconi Anemia (FA) is postulated to induce several progeroid features in patients (reviewed in ref. 234). Monoubiquitination plays a specially important role in FA,[235] since the so far known catalytic activity of the FA complex is that of an E3 ligase. Mutations that interfere with monoubiquitination of one of the FA pathway proteins, FANCD2, lead to defects in homologous recombination and translesion synthesis resulting in increased chromosome instability and leukaemia[236] and is classified as segmental progeroid syndrome.[232] Furthermore, FA derived hematopoietic bone marrow cells undergo premature senescence upon hypoxia/reoxigenization stress treatment[237] and bone marrow failure is one of the major causes of death in these patients at the age of around 16. The E3 ligase complex responsible for this ubiquitination contains 8 FA proteins so far.[7] In addition, BRCA1-BARD1 can act as FANCD2 E3 ligase in vitro, however, its significance in vivo is not clear yet.[238]

Other examples of proteins mutated in progeroid syndromes and linked to the ubiquitin system comprise WRN, the DNA helicases that is mutated in Werner syndrome patients. WRN associates with VCP/p97, an AAA ATPase implicated in the ubiquitin/proteasome pathway.[239,240] RECQL4, the dysfunctional protein in Rothmund-Thomson syndrome, interacts with UBR1 and UBR2, the E3 ligases of the N-end rule pathway but is not ubiquitinated nor degraded.[241] The responsible mutation of Hutchinson-Gilford progeria has recently been identified to lie within lamin A/C[242,243] and lamin A/C interacts with the sumoylation specific E2 ligase UBC9.[244] Mutations in the sumoylation site of lamin A lead to cardiomyopathy.[245]

Ataxia telangiectasia (A-T) is due to lack of functional ATM kinase that responds to DNA damage and oxidative stress.[246] ATM is furthermore essential in inducing growth arrest upon reaching critically short telomeres.[247] Nonetheless, the absence of functional ATM in fibroblasts derived from A-T patients leads to a reduced life span in vitro.[248] Lymphoblastoid A-T-cells have higher endogenous ubiquitin conjugate levels and enhanced ubiquitination activity, but a muted response to H_2O_2.[249]

Role of Ubiquitinylation and SUMOylation in Aging of Short-Lived Model Organisms

Accumulating evidence, in particular in the nematode C. elegans, suggests that there is a role of ubiquitinylation for aging also in short-lived model organisms. It has been shown that the transcription factor DAF16, which mediates signaling through the insulin IGF-like signaling pathway in C.elegans, is the target for ubiquitin mediated proteasomal degradation (reviewed by ref. 250). On the one hand, DAF16 is polyubiquitinated by a recently identified E3 ligase called RLE-1, which is the C. elegans orthologue of a murine E3 ubiquitin ligase called Roquin. DAF16 is driving the expression of genes favoring longevity and it was reported that RLE1 mutants are long-lived and that lifespan extension requires DAF16.[251] On the other hand, DAF16 activity, but not expression level, is also affected by depletion of the C. elegans homologue of Cullin1 and Cullin1 depletion results in a specific reduction of the lifespan in IIS long-lived mutants, whereas it has little effect on the lifespan of normal worms. Since Cul1 depletion did not affect the level of DAF16, it was concluded that the Cullin based E3 ligase complex targets a so far unidentified repressor of DAF16, thereby explaining the negative effects on lifespan.[252] In addition, it was shown that loss of the C. elegans homologue of the von Hippel-Lindau tumor suppressor gene VHL1, a Cullin based E3 ubiquitin ligase that negatively regulates the hypoxic response, increases the lifespan of C. elegans. It was also observed that hypoxia-inducible factor (HIF1) acts downstream of VHL1 to modulate aging and proteotoxicity in C. elegans and that both genes control longevity by a mechanism distinct from both dietary restriction and insulin-like signaling.[253]

In mice, CHIP deficiency reduces the life span with accelerated age-related pathophysiology and premature senescence of in-vitro cultivated cells of these mice.[254]

Conclusion

The universal importance of the ubiquitin-proteasome system for molecular and cellular biology, as well as for medical sciences is still increasing and there are only few cellular pathways left that at one step or another are not regulated by ubiquitin. In regard to aging of cells and tissues, the examples presented in this chapter (summarized in Fig. 3) suggest that deregulation and changes of the ubiquitinome might have vast implications for aging of organisms as well as for aging associated

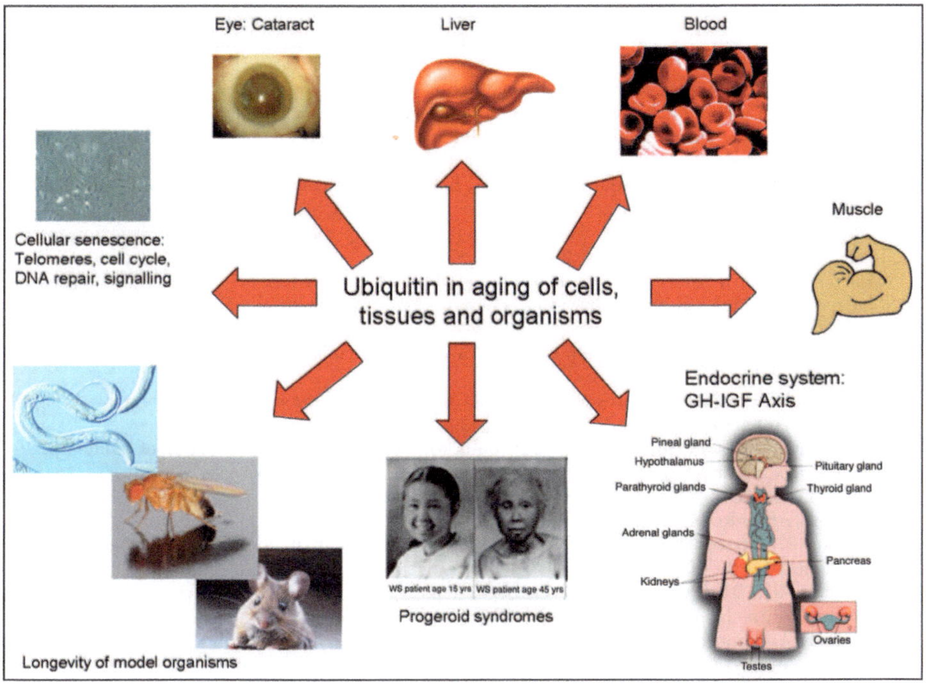

Figure 3. Overview on the so far known implications of ubiquitin in aging.

diseases, although in some cases direct evidence is still missing. In any case, further understanding of the influence of the ubiquitin system and its related molecules like SUMO, Nedd8, ISG15, atg8 or atg12 on the aging process might help to identify targets for prevention of deleterious loss of cell and tissue function and pathogenesis of aging related diseases.

Acknowledgements

This work was supported by NRN grant S093-B06 of the Austrian Science fund.

References

1. Carrard G, Bulteau AL, Petropoulos I et al. Impairment of proteasome structure and function in aging. Int J Biochem Cell Biol 2002; 34(11):1461-1474.
2. Martinez-Vicente M, Sovak G, Cuervo AM. Protein degradation and aging. Exp Gerontol 2005; 40(8-9):622-633.
3. Farout L, Lamare M, Clavel S et al. Differential expression of ubiquitin and proteasome-dependent pathway components in rat tissues. Comp Biochem Physiol B Biochem Mol Biol 2003; 134(2):297-305.
4. Bregegere F, Milner Y, Friguet B. The ubiquitin-proteasome system at the crossroads of stress-response and ageing pathways: a handle for skin care? Ageing Res Rev 2006; 5(1):60-90.
5. Ciechanover A, Schwartz AL. Ubiquitin-mediated degradation of cellular proteins in health and disease. Hepatology 2002; 35(1):3-6.
6. Layfield R, Lowe J, Bedford L. The ubiquitin-proteasome system and neurodegenerative disorders. Essays Biochem 2005; 41:157-171.
7. Huang TT, D'Andrea AD. Regulation of DNA repair by ubiquitylation. Nat Rev Mol Cell Biol 2006; 7(5):323-334.
8. Sigismund S, Polo S, Di Fiore PP. Signaling through monoubiquitination. Curr Top Microbiol Immunol 2004; 286:149-185.
9. Chen ZJ, Sun LJ. Nonproteolytic functions of ubiquitin in cell signaling. Mol Cell 2009; 33(3):275-286.

10. Haglund K, Dikic I. Ubiquitylation and cell signaling. EMBO 2005; 24(19):3353-3359.
11. Kirkin V, McEwan DG, Novak I et al. A role for ubiquitin in selective autophagy. Mol Cell 2009; 34(3):259-269.
12. Ben-Saadon R, Fajerman I, Ziv T et al. The tumor suppressor protein p16(INK4a) and the human papillomavirus oncoprotein-58 E7 are naturally occurring lysine-less proteins that are degraded by the ubiquitin system. Direct evidence for ubiquitination at the N-terminal residue. J Biol Chem 2004; 279(40):41414-41421.
13. Pickart CM, Eddins MJ. Ubiquitin: structures, functions, mechanisms. Biochim Biophys Acta 2004; 1695(1-3):55-72.
14. Glickman MH, Ciechanover A. The ubiquitin-proteasome proteolytic pathway: destruction for the sake of construction. Physiol Rev 2002; 82(2):373-428.
15. Fang S, Weissman AM. A field guide to ubiquitylation. Cell Mol Life Sci 2004; 61(13):1546-1561.
16. Hoppe T. Multiubiquitylation by E4 enzymes: 'one size' doesn't fit all. Trends Biochem Sci 2005; 30(4):183-187.
17. Staub O, Rotin D. Role of ubiquitylation in cellular membrane transport. Physiol Rev 2006; 86(2):669-707.
18. Thrower JS, Hoffman L, Rechsteiner M et al. Recognition of the polyubiquitin proteolytic signal. EMBO J 2000; 19(1):94-102.
19. Novak P, Kruppa GH, Young MM et al. A top-down method for the determination of residue-specific solvent accessibility in proteins. J Mass Spectrom 2004; 39(3):322-328.
20. Hicke L, Dunn R. Regulation of membrane protein transport by ubiquitin and ubiquitin-binding proteins. Annu Rev Cell Dev Biol 2003; 19:141-172.
21. Altmann K, Westermann B. Role of essential genes in mitochondrial morphogenesis in Saccharomyces cerevisiae. Mol Biol Cell 2005; 16(11):5410-5417.
22. Spence J, Gali RR, Dittmar G et al. Cell cycle-regulated modification of the ribosome by a variant multiubiquitin chain. Cell 2000; 102(1):67-76.
23. Chen ZJ. Ubiquitin signalling in the NF-kappaB pathway. Nat Cell Biol 2005; 7(8):758-765.
24. Wu CJ, Conze DB, Li T et al. Sensing of Lys 63-linked polyubiquitination by NEMO is a key event in NF-kappaB activation (corrected). Nat Cell Biol 2006; 8(4):398-406.
25. Krappmann D, Scheidereit C. A pervasive role of ubiquitin conjugation in activation and termination of IkappaB kinase pathways. EMBO Rep 2005; 6(4):321-326.
26. Johnson ES, Ma PC, Ota IM et al. A proteolytic pathway that recognizes ubiquitin as a degradation signal. J Biol Chem 1995; 270(29):17442-17456.
27. Mastrandrea LD, You J, Niles EG et al. E2/E3-mediated assembly of lysine 29-linked polyubiquitin chains. J Biol Chem 1999; 274(38):27299-27306.
28. Wang M, Pickart CM. Different HECT domain ubiquitin ligases employ distinct mechanisms of polyubiquitin chain synthesis. EMBO J 2005; 24(24):4324-4333.
29. Wang M, Cheng D, Peng J et al. Molecular determinants of polyubiquitin linkage selection by an HECT ubiquitin ligase. EMBO J 2006; 25(8):1710-1719.
30. Alberti S, Demand J, Esser C et al. Ubiquitylation of BAG-1 suggests a novel regulatory mechanism during the sorting of chaperone substrates to the proteasome. J Biol Chem 2002; 277(48):45920-45927.
31. Sahara N, Murayama M, Mizoroki T et al. In vivo evidence of CHIP up-regulation attenuating tau aggregation. J Neurochem 2005; 94(5):1254-1263.
32. Shang F, Deng G, Liu Q et al. Lys6-modified ubiquitin inhibits ubiquitin-dependent protein degradation. J Biol Chem 2005; 280(21):20365-20374.
33. Haglund K, Sigismund S, Polo S et al. Multiple monoubiquitination of RTKs is sufficient for their endocytosis and degradation. Nat Cell Biol 2003; 5(5):461-466.
34. D'Andrea A, Pellman D. Deubiquitinating enzymes: a new class of biological regulators. Crit Rev Biochem Mol Biol 1998; 33(5):337-352.
35. Bergink S, Jentsch S. Principles of ubiquitin and SUMO modifications in DNA repair. Nature 2009; 458(7237):461-467.
36. Geng J, Klionsky DJ. The Atg8 and Atg12 ubiquitin-like conjugation systems in macroautophagy. 'Protein modifications: beyond the usual suspects' review series. EMBO Rep 2008; 9(9):859-864.
37. Xirodimas DP. Novel substrates and functions for the ubiquitin-like molecule NEDD8. Biochem Soc Trans 2008; 36(Pt 5):802-806.
38. Sadler AJ, Williams BR. Interferon-inducible antiviral effectors. Nat Rev Immunol 2008; 8(7):559-568.
39. Hayflick L, Moorhead PS. The serial cultivation of human diploid cell strains. Exp Cell Res 1961; 25:585-621.

40. Campisi J. Aging, tumor suppression and cancer: high wire-act! Mech Ageing Dev 2005; 126(1):51-58.
41. Krtolica A, Campisi J. Cancer and aging: a model for the cancer promoting effects of the aging stroma. Int J Biochem Cell Biol 2002; 34(11):1401-1414.
42. Wang C, Jurk D, Maddick M et al. DNA damage response and cellular senescence in tissues of aging mice. Aging Cell 2009; 8(3):311-323.
43. Dimri GP, Lee X, Basile G et al. A biomarker that identifies senescent human cells in culture and in aging skin in vivo. Proc Natl Acad Sci USA 1995; 92(20):9363-9367.
44. Herbig U, Ferreira M, Condel L et al. Cellular senescence in aging primates. Science 2006; 311(5765):1257.
45. Krizhanovsky V, Yon M, Dickins RA et al. Senescence of activated stellate cells limits liver fibrosis. Cell 2008; 134(4):657-667.
46. Koppelstaetter C, Schratzberger G, Perco P et al. Markers of cellular senescence in zero hour biopsies predict outcome in renal transplantation. Aging Cell 2008; 7(4):491-497.
47. Melk A. Senescence of renal cells: molecular basis and clinical implications. Nephrol Dial Transplant 2003; 18(12):2474-2478.
48. Erusalimsky JD, Kurz DJ. Cellular senescence in vivo: Its relevance in ageing and cardiovascular disease. Exp Gerontol 2005; 40(8-9):634-642.
49. Minamino T, Komuro I. Role of telomeres in vascular senescence. Front Biosci 2008; 13:2971-2979.
50. Olovnikov AM. (Principle of marginotomy in template synthesis of polynucleotides). Dokl Akad Nauk SSSR 1971; 201(6):1496-1499.
51. d'Adda di Fagagna F, Reaper PM, Clay-Farrace L et al. A DNA damage checkpoint response in telomere-initiated senescence. Nature 2003; 426(6963):194-198. Epub 2003, 2005.
52. Jansen-Dürr P. The making and the breaking of senescence: changes of gene expression during cellular aging and immortalization. Exp Gerontol 1998; 33(4):291-301.
53. Herbig U, Sedivy JM. Regulation of growth arrest in senescence: telomere damage is not the end of the story. Mech Ageing Dev 2006; 127(1):16-24.
54. Bodnar A, Ooellette M, Frolkis M et al. Extension of life-span by introduction of telomerase into normal human cells. Science 1998; 279:349-352.
55. Chang MW, Grillari J, Mayrhofer C et al. Comparison of early passage, senescent and hTERT immortalized endothelial cells. Exp Cell Res 2005; 309(1):121-136.
56. Toussaint O, Medrano EE, von Zglinicki T. Cellular and molecular mechanisms of stress-induced premature senescence (SIPS) of human diploid fibroblasts and melanocytes. Exp Gerontol 2000; 35(8):927-945.
57. Robles SJ, Adami GR. Agents that cause DNA double strand breaks lead to p16INK4a enrichment and the premature senescence of normal fibroblasts. Oncogene 1998; 16(9):1113-1123.
58. Serrano M, Lin AW, McCurrach ME et al. Oncogenic ras provokes premature cell senescence associated with accumulation of p53 and p16INK4a. Cell 1997; 88(5):593-602.
59. Barea F, Bonatto D. Aging defined by a chronologic-replicative protein network in Saccharomyces cerevisiae: an interactome analysis. Mech Ageing Dev 2009; 130(7):444-460.
60. Jia L, Soengas MS, Sun Y. ROC1/RBX1 E3 ubiquitin ligase silencing suppresses tumor cell growth via sequential induction of G2-M arrest, apoptosis and senescence. Cancer Res 2009; 69(12):4974-4982.
61. Young AP, Schlisio S, Minamishima YA et al. VHL loss actuates a HIF-independent senescence programme mediated by Rb and p400. Nat Cell Biol 2008; 10(3):361-369.
62. Li M, Shin YH, Hou L et al. The adaptor protein of the anaphase promoting complex Cdh1 is essential in maintaining replicative lifespan and in learning and memory. Nat Cell Biol 2008; 10(9):1083-1089.
63. Zhang H, Cohen SN. Smurf2 up-regulation activates telomere-dependent senescence. Genes Dev 2004; 18(24):3028-3040.
64. Zhang H, Teng Y, Kong Y et al. Suppression of human tumor cell proliferation by Smurf2-induced senescence. J Cell Physiol 2008; 215(3):613-620.
65. Bischof O, Schwamborn K, Martin N et al. The E3 SUMO ligase PIASy is a regulator of cellular senescence and apoptosis. Mol Cell 2006; 22(6):783-794.
66. Mazurek S, Zwerschke W, Jansen-Durr P et al. Effects of the human papilloma virus HPV-16 E7 oncoprotein on glycolysis and glutaminolysis: role of pyruvate kinase type M2 and the glycolytic-enzyme complex. Biochem J 2001; 356(Pt 1):247-256.
67. Li T, Santockyte R, Shen RF et al. Expression of SUMO-2/3 induced senescence through p53- and pRB-mediated pathways. J Biol Chem 2006; 281(47):36221-36227.
68. Bischof O, Dejean A. SUMO is growing senescent. Cell Cycle 2007; 6(6):677-681.

69. Yates KE, Korbel GA, Shtutman M et al. Repression of the SUMO-specific protease Senp1 induces p53-dependent premature senescence in normal human fibroblasts. Aging Cell 2008; 7(5):609-621.
70. Pan JX, Short SR, Goff SA et al. Ubiquitin pools, ubiquitin mRNA levels and ubiquitin-mediated proteolysis in aging human fibroblasts. Exp Gerontol 1993; 28(1):39-49.
71. Marfella R, Di Filippo C, Laieta MT et al. Effects of ubiquitin-proteasome system deregulation on the vascular senescence and atherosclerosis process in elderly patients. J Gerontol A Biol Sci Med Sci 2008; 63(2):200-203.
72. Forsythe HL, Jarvis JL, Turner JW et al. Stable association of hsp90 and p23, but Not hsp70, with active human telomerase. J Biol Chem 2001; 276(19):15571-15574.
73. Kim JH, Park SM, Kang MR et al. Ubiquitin ligase MKRN1 modulates telomere length homeostasis through a proteolysis of hTERT. Genes Dev 2005; 19(7):776-781.
74. Smogorzewska A, van Steensel B, Bianchi A et al. Control of human telomere length by TRF1 and TRF2. Mol Cell Biol 2000; 20(5):1659-1668.
75. Lee TH, Perrem K, Harper JW et al. The F-box protein FBX4 targets PIN2/TRF1 for ubiquitin-mediated degradation and regulates telomere maintenance. J Biol Chem 2006; 281(2):759-768.
76. Her YR, Chung IK. Ubiquitin ligase RLIM modulates telomere length homeostasis through a proteolysis of TRF1. J Biol Chem 2009; 284(13):8557-8566.
77. Xhemalce B, Riising EM, Baumann P et al. Role of SUMO in the dynamics of telomere maintenance in fission yeast. Proc Natl Acad Sci USA 2007; 104(3):893-898.
78. Potts PR, Yu H. The SMC5/6 complex maintains telomere length in ALT cancer cells through SUMOylation of telomere-binding proteins. Nat Struct Mol Biol 2007; 14(7):581-590.
79. Lee H, Sengupta N, Villagra A et al. Histone deacetylase 8 safeguards the human ever-shorter telomeres 1B (hEST1B) protein from ubiquitin-mediated degradation. Mol Cell Biol 2006; 26(14):5259-5269.
80. Condemine W, Takahashi Y, Le Bras M et al. A nucleolar targeting signal in PML-I addresses PML to nucleolar caps in stressed or senescent cells. J Cell Sci 2007; 120(Pt 18):3219-3227.
81. Gottschling DE, Aparicio OM, Billington BL et al. Position effect at S. cerevisiae telomeres: reversible repression of Pol II transcription. Cell 1990; 63(4):751-762.
82. Lou Z, Wei J, Riethman H et al. Telomere length regulates ISG15 expression in human cells. Aging 2009; 1(7):1-14.
83. Kyng KJ, May A, Stevnsner T et al. Gene expression responses to DNA damage are altered in human aging and in Werner syndrome. Oncogene 2005; 24(32):5026-5042.
84. Hoeijmakers JH. Genome maintenance mechanisms for preventing cancer. Nature 2001; 411(6835):366-374.
85. Cao L, Li W, Kim S et al. Senescence, aging and malignant transformation mediated by p53 in mice lacking the Brca1 full-length isoform. Genes Dev 2003; 17(2):201-213.
86. Nishikawa H, Ooka S, Sato K et al. Mass spectrometric and mutational analyses reveal Lys-6-linked polyubiquitin chains catalyzed by BRCA1-BARD1 ubiquitin ligase. J Biol Chem 2004; 279(6):3916-3924.
87. Mallery DL, Vandenberg CJ, Hiom K. Activation of the E3 ligase function of the BRCA1/BARD1 complex by polyubiquitin chains. EMBO J 2002; 21(24):6755-6762.
88. Morris JR, Solomon E. BRCA1: BARD1 induces the formation of conjugated ubiquitin structures, dependent on K6 of ubiquitin, in cells during DNA replication and repair. Hum Mol Genet 2004; 13(8):807-817.
89. Sato K, Hayami R, Wu W et al. Nucleophosmin/B23 is a candidate substrate for the BRCA1-BARD1 ubiquitin ligase. J Biol Chem 2004; 279(30):30919-30922.
90. Colombo E, Marine JC, Danovi D et al. Nucleophosmin regulates the stability and transcriptional activity of p53. Nat Cell Biol 2002; 4(7):529-533.
91. Hayami R, Sato K, Wu W et al. Down-regulation of BRCA1-BARD1 ubiquitin ligase by CDK2. Cancer Res 2005; 65(1):6-10.
92. Freedman DA, Folkman J. CDK2 translational down-regulation during endothelial senescence. Exp Cell Res 2005; 307(1):118-130.
93. von Zglinicki T, Saretzki G, Ladhoff J et al. Human cell senescence as a DNA damage response. Mech Ageing Dev 2005; 126(1):111-117.
94. Celeste A, Fernandez-Capetillo O, Kruhlak MJ et al. Histone H2AX phosphorylation is dispensable for the initial recognition of DNA breaks. Nat Cell Biol 2003; 5(7):675-679.
95. Brooks CL, Gu W. Ubiquitination, phosphorylation and acetylation: the molecular basis for p53 regulation. Curr Opin Cell Biol 2003; 15(2):164-171.
96. Bergink S, Salomons FA, Hoogstraten D et al. DNA damage triggers nucleotide excision repair-dependent monoubiquitylation of histone H2A. Genes Dev 2006; 20(10):1343-1352.

97. Prosperi E. The fellowship of the rings: distinct pools of proliferating cell nuclear antigen trimer at work. FASEB J 2006; 20(7):833-837.
98. Zhang N, Lu X, Legerski RJ. Partial reconstitution of human interstrand cross-link repair in vitro: characterization of the roles of RPA and PCNA. Biochem Biophys Res Commun 2003; 309(1):71-78.
99. Watanabe K, Tateishi S, Kawasuji M et al. Rad18 guides poleta to replication stalling sites through physical interaction and PCNA monoubiquitination. EMBO J 2004; 23(19):3886-3896.
100. Kannouche PL, Wing J, Lehmann AR. Interaction of human DNA polymerase eta with monoubiquitinated PCNA: a possible mechanism for the polymerase switch in response to DNA damage. Mol Cell 2004; 14(4):491-500.
101. Bienko M, Green CM, Crosetto N et al. Ubiquitin-binding domains in Y-family polymerases regulate translesion synthesis. Science 2005; 310(5755):1821-1824.
102. Masutani C, Kusumoto R, Yamada A et al. The XPV (xeroderma pigmentosum variant) gene encodes human DNA polymerase eta. Nature 1999; 399(6737):700-704.
103. Moldovan GL, Pfander B, Jentsch S. PCNA, the maestro of the replication fork. Cell 2007; 129(4):665-679.
104. Stein GH, Drullinger LF, Soulard ADV. Differential roles for cyclin-dependent kinase inhibitors p21 and p16 in the mechanisms of senescence and differentiation in human fibroblasts. Mol Cell Biol 1999; 19(3):2109-2117.
105. Fotedar R, Bendjennat M, Fotedar A. Role of p21WAF1 in the cellular response to UV. Cell Cycle 2004; 3(2):134-137.
106. Grillari J, Hohenwarter O, Grabherr RM et al. Subtractive hybridization of mRNA from early passage and senescent endothelial cells. Exp Gerontol 2000; 35(2):187-197.
107. Voglauer R, Chang MW, Dampier B et al. SNEV overexpression extends the life span of human endothelial cells. Exp Cell Res 2006; 312(6):746-759.
108. Hatakeyama S, Yada M, Matsumoto M et al. U-box proteins as a new family of ubiquitin-protein ligases. J Biol Chem 2001; 276(35):33111-33120.
109. Loscher M, Fortschegger K, Ritter G et al. Interaction of U-box E3 ligase SNEV with PSMB4, the beta7 subunit of the 20S proteasome. Biochem J 2005; 388(Pt 2):593-603.
110. Sihn CR, Cho SY, Lee JH et al. Mouse homologue of yeast Prp19 interacts with mouse SUG1, the regulatory subunit of 26S proteasome. Biochem Biophys Res Commun 2007; 356(1):175-180.
111. Chondrogianni N, Stratford FL, Trougakos IP et al. Central role of the proteasome in senescence and survival of human fibroblasts: induction of a senescence-like phenotype upon its inhibition and resistance to stress upon its activation. J Biol Chem 2003; 278(30):28026-28037.
112. Chondrogianni N, Tzavelas C, Pemberton AJ et al. Overexpression of proteasome beta5 assembled subunit increases the amount of proteasome and confers ameliorated response to oxidative stress and higher survival rates. J Biol Chem 2005; 280(12):11840-11850.
113. Fortschegger K, Wagner B, Voglauer R et al. Early embryonic lethality of mice lacking the essential protein SNEV. Mol Cell Biol 2007; 27(8):3123-3130.
114. Schraml E, Voglauer R, Fortschegger K et al. Haploinsufficiency of senescence evasion factor causes defects of hematopoietic stem cells functions. Stem Cells Dev 2008; 17(2):355-366.
115. Mahajan KN, Mitchell BS. Role of human Pso4 in mammalian DNA repair and association with terminal deoxynucleotidyl transferase. Proc Natl Acad Sci USA 2003; 100(19):10746-10751.
116. Beck BD, Park SJ, Lee YJ et al. Human PSO4 is a Metnase (SETMAR) binding partner that regulates Metnase' function in DNA repair. J Biol Chem 2008.
117. Zhang N, Kaur R, Lu X et al. The PSO4 MRNA splicing and DNA repair complex interacts with WRN for processing of DNA interstrand cross-links. J Biol Chem 2005; 280(49):40559-40567.
118. Grillari J, Ajuh P, Stadler G et al. SNEV is an evolutionarily conserved splicing factor whose oligomerization is necessary for spliceosome assembly. Nucleic Acids Res 2005; 33(21):6868-6883.
119. Bringold F, Serrano M. Tumor suppressors and oncogenes in cellular senescence. Exp Gerontol 2000; 35(3):317-329.
120. Tyner SD, Venkatachalam S, Choi J et al. p53 mutant mice that display early ageing-associated phenotypes. Nature 2002; 415(6867):45-53.
121. Bode AM, Dong Z. Post-translational modification of p53 in tumorigenesis. Nat Rev Cancer 2004; 4(10):793-805.
122. Brooks CL, Gu W. p53 ubiquitination: Mdm2 and beyond. Mol Cell 2006; 21(3):307-315.
123. Asher G, Tsvetkov P, Kahana C et al. A mechanism of ubiquitin-independent proteasomal degradation of the tumor suppressors p53 and p73. Genes Dev 2005; 19(3):316-321.
124. Ghosh M, Weghorst K, Berberich SJ. MdmX inhibits ARF mediated Mdm2 sumoylation. Cell Cycle 2005; 4(4):604-608.
125. den Besten W, Kuo ML, Tago K et al. Ubiquitination of and sumoylation by, the Arf tumor suppressor. Isr Med Assoc J 2006; 8(4):249-251.

126. Kuo ML, den Besten W, Thomas MC et al. Arf-induced turnover of the nucleolar nucleophosmin-associated SUMO-2/3 protease Senp3. Cell Cycle 2008; 7(21):3378-3387.
127. Huang Y, Wu M, Li HY. Tumor suppressor ARF promotes nonclassic proteasome-independent polyubiquitination of COMMD1. J Biol Chem 2008; 283(17):11453-11460.
128. Brown JP, Wei W, Sedivy JM. Bypass of senescence after disruption of p21CIP1/WAF1 gene in normal diploid human fibroblasts. Science 1997; 277:831-834.
129. Bornstein G, Bloom J, Sitry-Shevah D et al. Role of the SCFSkp2 ubiquitin ligase in the degradation of p21Cip1 in S phase. J Biol Chem 2003; 278(28):25752-25757.
130. Amador V, Ge S, Santamaria PG et al. APC/C(Cdc20) controls the ubiquitin-mediated degradation of p21 in prometaphase. Mol Cell 2007; 27(3):462-473.
131. Abbas T, Sivaprasad U, Terai K et al. PCNA-dependent regulation of p21 ubiquitylation and degradation via the CRL4Cdt2 ubiquitin ligase complex. Genes Dev 2008; 22(18):2496-2506.
132. Nishitani H, Shiomi Y, Iida H et al. CDK inhibitor p21 is degraded by a proliferating cell nuclear antigen-coupled Cul4-DDB1Cdt2 pathway during S phase and after UV irradiation. J Biol Chem 2008; 283(43):29045-29052.
133. Lee H, Zeng SX, Lu H. UV Induces p21 rapid turnover independently of ubiquitin and Skp2. J Biol Chem 2006; 281(37):26876-26883.
134. Li X, Amazit L, Long W et al. Ubiquitin- and ATP-independent proteolytic turnover of p21 by the REGgamma-proteasome pathway. Mol Cell 2007; 26(6):831-842.
135. Hwang CY, Kim IY, Kwon KS. Cytoplasmic localization and ubiquitination of p21(Cip1) by reactive oxygen species. Biochem Biophys Res Commun 2007; 358(1):219-225.
136. Alcorta DA, Xiong Y, Phelps D et al. Involvement of the cyclin-dependent kinase inhibitor p16 (INK4a) in replicative senescence of normal human fibroblasts. Proc Nat Acad Sci USA 1996; 93:13742-13747.
137. Thullberg M, Bartkova J, Khan S et al. Distinct versus redundant properties among members of the INK4 family of cyclin-dependent kinase inhibitors. FEBS Lett 2000; 470(2):161-166.
138. Alexander K, Hinds PW. Requirement for p27(KIP1) in retinoblastoma protein-mediated senescence. Mol Cell Biol 2001; 21(11):3616-3631.
139. Wagner M, Hampel B, Hutter E et al. Metabolic stabilization of p27 in senescent fibroblasts correlates with reduced expression of the F-box protein Skp2. Exp Gerontol 2001; 37:41-55.
140. Matsunobu T, Tanaka K, Nakamura T et al. The possible role of EWS-Fli1 in evasion of senescence in Ewing family tumors. Cancer Res 2006; 66(2):803-811.
141. Chu IM, Hengst L, Slingerland JM. The Cdk inhibitor p27 in human cancer: prognostic potential and relevance to anticancer therapy. Nat Rev Cancer 2008; 8(4):253-267.
142. Caballero OL, Resto V, Patturajan M et al. Interaction and colocalization of PGP9.5 with JAB1 and p27(Kip1). Oncogene 2002; 21(19):3003-3010.
143. Marzban G, Grillari J, Reisinger E et al. Age-related alterations in the protein expression profile of C57BL/6J mouse pituitaries. Exp Gerontol 2002; 37(12):1451-1460.
144. Lombardino AJ, Li XC, Hertel M et al. Replaceable neurons and neurodegenerative disease share depressed UCHL1 levels. Proc Natl Acad Sci USA 2005; 102(22):8036-8041.
145. Zhu Y, Carroll M, Papa FR et al. DUB-1, a deubiquitinating enzyme with growth-suppressing activity. Proc Natl Acad Sci USA 1996; 93(8):3275-3279.
146. Burrows JF, McGrattan MJ, Rascle A et al. DUB-3, a cytokine-inducible deubiquitinating enzyme that blocks proliferation. J Biol Chem 2004; 279(14):13993-14000.
147. Kovalenko A, Chable-Bessia C, Cantarella G et al. The tumour suppressor CYLD negatively regulates NF-kappaB signalling by deubiquitination. Nature 2003; 424(6950):801-805.
148. Trompouki E, Hatzivassiliou E, Tsichritzis T et al. CYLD is a deubiquitinating enzyme that negatively regulates NF-kappaB activation by TNFR family members. Nature 2003; 424(6950):793-796.
149. Massoumi R, Chmielarska K, Hennecke K et al. Cyld inhibits tumor cell proliferation by blocking Bcl-3-dependent NF-kappaB signaling. Cell 2006; 125(4):665-677.
150. Li Z, Wang D, Messing EM et al. VHL protein-interacting deubiquitinating enzyme 2 deubiquitinates and stabilizes HIF-1alpha. EMBO Rep 2005; 6(4):373-378.
151. Kato H, Inoue T, Asanoma K et al. Induction of human endometrial cancer cell senescence through modulation of HIF-1alpha activity by EGLN1. Int J Cancer 2006; 118(5):1144-1153.
152. Minamino T, Mitsialis SA, Kourembanas S. Hypoxia extends the life span of vascular smooth muscle cells through telomerase activation. Mol Cell Biol 2001; 21(10):3336-3342.
153. Frenkel-Denkberg G, Gershon D, Levy AP. The function of hypoxia-inducible factor 1 (HIF-1) is impaired in senescent mice. FEBS Lett 1999; 462(3):341-344.
154. Rivard A, Berthou-Soulie L, Principe N et al. Age-dependent defect in vascular endothelial growth factor expression is associated with reduced hypoxia-inducible factor 1 activity. J Biol Chem 2000; 275(38):29643-29647.

155. Sibilia M, Wagner EF. Strain-dependent epithelial defects in mice lacking the EGF receptor. Science 1995; 269(5221):234-238.
156. Shiraha H, Gupta K, Drabik K et al. Aging fibroblasts present reduced epidermal growth factor (EGF) responsiveness due to preferential loss of EGF receptors. J Biol Chem 2000; 275(25):19343-19351.
157. Haglund K, Shimokawa N, Szymkiewicz I et al. Cbl-directed monoubiquitination of CIN85 is involved in regulation of ligand-induced degradation of EGF receptors. Proc Natl Acad Sci USA 2002; 99(19):12191-12196.
158. Hoeller D, Crosetto N, Blagoev B et al. Regulation of ubiquitin-binding proteins by monoubiquitination. Nat Cell Biol 2006; 8(2):163-169.
159. Mizuno E, Iura T, Mukai A et al. Regulation of epidermal growth factor receptor down-regulation by UBPY-mediated deubiquitination at endosomes. Mol Biol Cell 2005; 16(11):5163-5174.
160. Tran KT, Rusu SD, Satish L et al. Aging-related attenuation of EGF receptor signaling is mediated in part by increased protein tyrosine phosphatase activity. Exp Cell Res 2003; 289(2):359-367.
161. Park WY, Cho KA, Park JS et al. Attenuation of EGF signaling in senescent cells by caveolin. Ann N Y Acad Sci 2001; 928:79-84.
162. Park SC, Park JS, Park WY et al. Down-regulation of receptor-mediated endocytosis is responsible for senescence-associated hyporesponsiveness. Ann N Y Acad Sci 2002; 959:45-49.
163. Park JS, Park WY, Cho KA et al. Down-regulation of amphiphysin-1 is responsible for reduced receptor-mediated endocytosis in the senescent cells. FASEB J 2001; 15(9):1625-1627.
164. Jura N, Scotto-Lavino E, Sobczyk A et al. Differential modification of Ras proteins by ubiquitination. Mol Cell 2006; 21(5):679-687.
165. Germain D. Ubiquitin-dependent and -independent mitochondrial protein quality controls: implications in ageing and neurodegenerative diseases. Mol Microbiol 2008; 70(6):1334-1341.
166. Schulz JB. Update on the pathogenesis of Parkinson's disease. J Neurol 2008; 255(Suppl 5):3-7.
167. Shimura H, Hattori N, Kubo S et al. Familial Parkinson disease gene product, parkin, is a ubiquitin-protein ligase. Nat Genet 2000; 25(3):302-305.
168. Kao SY. DNA damage induces nuclear translocation of parkin. J Biomed Sci 2009; 16:67.
169. Kao SY. Regulation of DNA repair by parkin. Biochem Biophys Res Commun 2009; 382(2):321-325.
170. Gong B, Leznik E. The role of ubiquitin C-terminal hydrolase L1 in neurodegenerative disorders. Drug News Perspect 2007; 20(6):365-370.
171. Wang C, Lu R, Ouyang X et al. Drosophila overexpressing parkin R275W mutant exhibits dopaminergic neuron degeneration and mitochondrial abnormalities. J Neurosci 2007; 27(32):8563-8570.
172. Harding AJ, Stimson E, Henderson JM et al. Clinical correlates of selective pathology in the amygdala of patients with Parkinson's disease. Brain 2002; 125(Pt 11):2431-2445.
173. Jahngen JH, Haas AL, Ciechanover A et al. The eye lens has an active ubiquitin-protein conjugation system. J Biol Chem 1986; 261(29):13760-13767.
174. Jahngen JJ, Eisenhauer D, Taylor A. Lens proteins are substrates for the reticulocyte ubiquitin conjugation system. Curr Eye Res 1986; 5(10):725-733.
175. Jahngen JH, Lipman RD, Eisenhauer DA et al. Aging and cellular maturation cause changes in ubiquitin-eye lens protein conjugates. Arch Biochem Biophys 1990; 276(1):32-37.
176. Shang F, Gong X, Palmer HJ et al. Age-related decline in ubiquitin conjugation in response to oxidative stress in the lens. Exp Eye Res 1997; 64(1):21-30.
177. Ruotolo R, Grassi F, Percudani R et al. Gene expression profiling in human age-related nuclear cataract. Mol Vis 2003; 9:538-548.
178. Hawse JR, Hejtmancik JF, Horwitz J et al. Identification and functional clustering of global gene expression differences between age-related cataract and clear human lenses and aged human lenses. Exp Eye Res 2004; 79(6):935-940.
179. Zhang L, Li F, Dimayuga E et al. Effects of aging and dietary restriction on ubiquitination, sumoylation and the proteasome in the spleen. FEBS Lett 2007; 581(28):5543-5547.
180. Shang F, Taylor A. Function of the ubiquitin proteolytic pathway in the eye. Exp Eye Res 2004; 78(1):1-14.
181. Sano Y, Furuta A, Setsuie R et al. Photoreceptor cell apoptosis in the retinal degeneration of Uchl3-deficient mice. Am J Pathol 2006; 169(1):132-141.
182. Marques C, Pereira P, Taylor A et al. Ubiquitin-dependent lysosomal degradation of the HNE-modified proteins in lens epithelial cells. FASEB J 2004; 18(12):1424-1426.
183. Anisimov VN. Life span extension and cancer risk: myths and reality. Exp Gerontol 2001; 36(7):1101-1136.
184. Ingram DK, Zhu M, Mamczarz J et al. Calorie restriction mimetics: an emerging research field. Aging Cell 2006; 5(2):97-108.

185. Scrofano MM, Shang F, Nowell TR Jr et al. Aging, calorie restriction and ubiquitin-dependent proteolysis in the livers of Emory mice. Mech Ageing Dev 1998; 101(3):277-296.
186. Scrofano MM, Shang F, Nowell TR Jr et al. Calorie restriction, stress and the ubiquitin-dependent pathway in mouse livers. Mech Ageing Dev 1998; 105(3):273-290.
187. Chen J, Rider DA, Ruan R. Identification of valid housekeeping genes and antioxidant enzyme gene expression change in the aging rat liver. J Gerontol A Biol Sci Med Sci 2006; 61(1):20-27.
188. Davies KJ. Degradation of oxidized proteins by the 20S proteasome. Biochimie 2001; 83(3-4):301-310.
189. Davies KJ, Shringarpure R. Preferential degradation of oxidized proteins by the 20S proteasome may be inhibited in aging and in inflammatory neuromuscular diseases. Neurology 2006; 66(2 Suppl 1):S93-96.
190. Tsuchiya T, Higami Y, Komatsu T et al. Acute stress response in calorie-restricted rats to lipopolysaccharide-induced inflammation. Mech Ageing Dev 2005; 126(5):568-579.
191. Li F, Zhang L, Craddock J et al. Aging and dietary restriction effects on ubiquitination, sumoylation and the proteasome in the heart. Mech Ageing Dev 2008; 129(9):515-521.
192. Holzenberger M, Kappeler L, De Magalhaes Filho C. IGF-1 signaling and aging. Exp Gerontol 2004; 39(11-12):1761-1764.
193. LeRoith D, Werner H, Beitner-Johnson D et al. Molecular and cellular aspects of the insulin-like growth factor I receptor. Endocr Rev 1995; 16(2):143-163.
194. Lee KW, Cohen P. Nuclear effects: unexpected intracellular actions of insulin-like growth factor binding protein-3. J Endocrinol 2002; 175(1):33-40.
195. Santer FR, Bacher N, Moser B et al. Nuclear insulin-like growth factor binding protein-3 induces apoptosis and is targeted to ubiquitin/proteasome-dependent proteolysis. Cancer Res 2006; 66(6):3024-3033.
196. Lee KW, Liu B, Ma L et al. Cellular internalization of insulin-like growth factor binding protein-3: distinct endocytic pathways facilitate re-uptake and nuclear localization. J Biol Chem 2004; 279(1):469-476.
197. Girnita L, Shenoy SK, Sehat B et al. {beta}-Arrestin is crucial for ubiquitination and down-regulation of the insulin-like growth factor-1 receptor by acting as adaptor for the MDM2 E3 ligase. J Biol Chem 2005; 280(26):24412-24419.
198. Lee AV, Gooch JL, Oesterreich S et al. Insulin-like growth factor I-induced degradation of insulin receptor substrate 1 is mediated by the 26S proteasome and blocked by phosphatidylinositol 3'-kinase inhibition. Mol Cell Biol 2000; 20(5):1489-1496.
199. Rui L, Fisher TL, Thomas J et al. Regulation of insulin/insulin-like growth factor-1 signaling by proteasome-mediated degradation of insulin receptor substrate-2. J Biol Chem 2001; 276(43):40362-40367.
200. Adachi M, Katsumura KR, Fujii K et al. Proteasome-dependent decrease in Akt by growth factors in vascular smooth muscle cells. FEBS Lett 2003; 554(1-2):77-80.
201. Sacheck JM, Ohtsuka A, McLary SC et al. IGF-I stimulates muscle growth by suppressing protein breakdown and expression of atrophy-related ubiquitin ligases, atrogin-1 and MuRF1. Am J Physiol Endocrinol Metab 2004; 287(4):E591-601.
202. Fang CH, Li BG, Sun X et al. Insulin-like growth factor I reduces ubiquitin and ubiquitin-conjugating enzyme gene expression but does not inhibit muscle proteolysis in septic rats. Endocrinology 2000; 141(8):2743-2751.
203. Dehoux M, Van Beneden R, Pasko N et al. Role of the insulin-like growth factor I decline in the induction of atrogin-1/MAFbx during fasting and diabetes. Endocrinology 2004; 145(11):4806-4812.
204. Schulze PC, Spate U. Insulin-like growth factor-1 and muscle wasting in chronic heart failure. Int J Biochem Cell Biol 2005; 37(10):2023-2035.
205. Argiles JM, Busquets S, Felipe A et al. Molecular mechanisms involved in muscle wasting in cancer and ageing: cachexia versus sarcopenia. Int J Biochem Cell Biol 2005; 37(5):1084-1104.
206. Marzetti E, Anne Lees H, Eva Wohlgemuth S et al. Sarcopenia of aging: underlying cellular mechanisms and protection by calorie restriction. Biofactors 2009; 35(1):28-35.
207. Attaix D, Ventadour S, Codran A et al. The ubiquitin-proteasome system and skeletal muscle wasting. Essays Biochem 2005; 41:173-186.
208. Cai D, Frantz JD, Tawa NE Jr et al. IKKbeta/NF-kappaB activation causes severe muscle wasting in mice. Cell 2004; 119(2):285-298.
209. Cai D, Li M, Lee K et al. Age-related changes of aqueous protein profiles in rat fast and slow twitch skeletal muscles. Electrophoresis 2000; 21(2):465-472.
210. Deruisseau KC, Kavazis AN, Powers SK. Selective downregulation of ubiquitin conjugation cascade mRNA occurs in the senescent rat soleus muscle. Exp Gerontol 2005; 40(6):526-531.

211. Brenkman AB, de Keizer PL, van den Broek NJ et al. Mdm2 induces mono-ubiquitination of FOXO4. PLoS One 2008; 3(7):e2819.
212. Edstrom E, Altun M, Hagglund M et al. Atrogin-1/MAFbx and MuRF1 are downregulated in aging-related loss of skeletal muscle. J Gerontol A Biol Sci Med Sci 2006; 61(7):663-674.
213. Combaret L, Dardevet D, Bechet D et al. Skeletal muscle proteolysis in aging. Curr Opin Clin Nutr Metab Care 2009; 12(1):37-41.
214. Casas F, Pessemesse L, Grandemange S et al. Overexpression of the mitochondrial T3 receptor induces skeletal muscle atrophy during aging. PLoS One 2009; 4(5):e5631.
215. Hirner S, Krohne C, Schuster A et al. MuRF1-dependent regulation of systemic carbohydrate metabolism as revealed from transgenic mouse studies. J Mol Biol 2008; 379(4):666-677.
216. Hsu YJ, Zimmer WE, Goodman SR. Erythrocyte spectrin's chimeric E2/E3 ubiquitin conjugating/ligating activity. Cell Mol Biol (Noisy-le-grand) 2005; 51(2):187-193.
217. Corsi D, Paiardini M, Crinelli R et al. Alteration of alpha-spectrin ubiquitination due to age-dependent changes in the erythrocyte membrane. Eur J Biochem 1999; 261(3):775-783.
218. Waugh RE, Narla M, Jackson CW et al. Rheologic properties of senescent erythrocytes: loss of surface area and volume with red blood cell age. Blood 1992; 79(5):1351-1358.
219. Simpson LO, O'Neill DJ. Red cell shape changes in the blood of people 60 years of age and older imply a role for blood rheology in the aging process. Gerontology 2003; 49(5):310-315.
220. Samaja M, Rovida E, Motterlini R et al. The relationship between the blood oxygen transport and the human red cell aging process. Adv Exp Med Biol 1991; 307:115-123.
221. Wick G, Romen M, Amberger A et al. Atherosclerosis, autoimmunity and vascular-associated lymphoid tissue. FASEB J 1997; 11(13):1199-1207.
222. Grubeck-Loebenstein B, Wick G. The aging of the immune system. Adv Immunol 2002; 80:243-284.
223. Ponnappan U. Regulation of transcription factor NF kappa B in immune senescence. Front Biosci 1998; 3:d152-168.
224. Ponnappan U. Ubiquitin-proteasome pathway is compromised in CD45RO+ and CD45RA+ T-lymphocyte subsets during aging. Exp Gerontol 2002; 37(2-3):359-367.
225. Li X, Stark GR. NFkappaB-dependent signaling pathways. Exp Hematol 2002; 30(4):285-296.
226. Carter RS, Pennington KN, Arrate P et al. Site-specific monoubiquitination of IkappaB kinase IKKbeta regulates its phosphorylation and persistent activation. J Biol Chem 2005; 280(52):43272-43279.
227. Carter RS, Pennington KN, Ungurait BJ et al. In vivo identification of inducible phosphoacceptors in the IKKgamma/NEMO subunit of human IkappaB kinase. J Biol Chem 2003; 278(22):19642-19648.
228. Wertz IE, O'Rourke KM, Zhou H et al. De-ubiquitination and ubiquitin ligase domains of A20 downregulate NF-kappaB signalling. Nature 2004; 430(7000):694-699.
229. Reyes-Turcu FE, Ventii KH, Wilkinson KD. Regulation and cellular roles of ubiquitin-specific deubiquitinating enzymes. Annu Rev Biochem 2009; 78:363-397.
230. Zaidi G, Panda H, Supakar PC. Increased phosphorylation and decreased level of IkappaBalpha during aging in rat liver. Biogerontology 2005; 6(2):141-145.
231. Hollander J, Bejma J, Ookawara T et al. Superoxide dismutase gene expression in skeletal muscle: fiber-specific effect of age. Mech Ageing Dev 2000; 116(1):33-45.
232. Martin GM. Genetic syndromes in man with potential relevance to the pathobiology of aging. Birth Defects Orig Artic Ser 1978; 14(1):5-39.
233. Martin GM. Genetic modulation of senescent phenotypes in Homo sapiens. Cell 2005; 120(4):523-532.
234. Grillari J, Katinger H, Voglauer R. SNEV$^{Prp19/Pso4}$ is a conserved, multifaceted E3 ligase involved in replicative senescence, DNA repair and pre-mRNA splicing. PRP19 Targeted Proteins Database; 2007.
235. Alpi AF, Patel KJ. Monoubiquitylation in the Fanconi anemia DNA damage response pathway. DNA Repair (Amst) 2009; 8(4):430-435.
236. Taniguchi T, Garcia-Higuera I, Andreassen PR et al. S-phase-specific interaction of the Fanconi anemia protein, FANCD2, with BRCA1 and RAD51. Blood 2002; 100(7):2414-2420.
237. Zhang X, Li J, Sejas DP et al. Hypoxia-reoxygenation induces premature senescence in FA bone marrow hematopoietic cells. Blood 2005; 106(1):75-85.
238. Garcia-Higuera I, Taniguchi T, Ganesan S et al. Interaction of the Fanconi anemia proteins and BRCA1 in a common pathway. Mol Cell 2001; 7(2):249-262.
239. Hartmann-Petersen R, Wallace M, Hofmann K et al. The Ubx2 and Ubx3 cofactors direct Cdc48 activity to proteolytic and nonproteolytic ubiquitin-dependent processes. Curr Biol 2004; 14(9):824-828.

240. Wojcik C, Yano M, DeMartino GN. RNA interference of valosin-containing protein (VCP/p97) reveals multiple cellular roles linked to ubiquitin/proteasome-dependent proteolysis. J Cell Sci 2004; 117(Pt 2):281-292.
241. Yin J, Kwon YT, Varshavsky A et al. RECQL4, mutated in the Rothmund-Thomson and RAPADILINO syndromes, interacts with ubiquitin ligases UBR1 and UBR2 of the N-end rule pathway. Hum Mol Genet 2004; 13(20):2421-2430.
242. Eriksson M, Brown WT, Gordon LB et al. Recurrent de novo point mutations in lamin A cause Hutchinson-Gilford progeria syndrome. Nature 2003; 423(6937):293-298.
243. De Sandre-Giovannoli A, Bernard R, Cau P et al. Lamin a truncation in Hutchinson-Gilford progeria. Science 2003; 300(5628):2055.
244. Zhong N, Radu G, Ju W et al. Novel progerin-interactive partner proteins hnRNP E1, EGF, Mel 18 and UBC9 interact with lamin A/C. Biochem Biophys Res Commun 2005; 338(2):855-861.
245. Zhang YQ, Sarge KD. Sumoylation regulates lamin A function and is lost in lamin A mutants associated with familial cardiomyopathies. J Cell Biol 2008; 182(1):35-39.
246. Taylor A, Shang F, Nowell T et al. Ubiquitination capabilities in response to neocarzinostatin and H(2)O(2) stress in cell lines from patients with ataxia-telangiectasia. Oncogene 2002; 21(28):4363-4373.
247. Herbig U, Jobling WA, Chen BP et al. Telomere shortening triggers senescence of human cells through a pathway involving ATM, p53 and p21(CIP1), but not p16(INK4a). Mol Cell 2004; 14(4):501-513.
248. Weirich-Schwaiger H, Weirich HG, Gruber B et al. Correlation between senescence and DNA repair in cells from young and old individuals and in premature aging syndromes. Mutat Res 1994; 316(1):37-48.
249. Shang F, Nowell T, Gong X et al. Sex-linked differences in cataract progression in Emory mice. Exp Eye Res 2002; 75(1):109-111.
250. Bowerman B. C. elegans aging: proteolysis cuts both ways. Curr Biol 2007; 17(13):R514-516.
251. Li W, Gao B, Lee SM et al. RLE-1, an E3 ubiquitin ligase, regulates C. elegans aging by catalyzing DAF-16 polyubiquitination. Dev Cell 2007; 12(2):235-246.
252. Ghazi A, Henis-Korenblit S, Kenyon C. Regulation of Caenorhabditis elegans lifespan by a proteasomal E3 ligase complex. Proc Natl Acad Sci USA 2007; 104(14):5947-5952.
253. Mehta R, Steinkraus KA, Sutphin GL et al. Proteasomal regulation of the hypoxic response modulates aging in C. elegans. Science 2009; 324(5931):1196-1198.
254. Min JN, Whaley RA, Sharpless NE et al. CHIP deficiency decreases longevity with accelerated aging phenotypes accompanied by altered protein quality control. Mol Cell Biol 2008.

Chapter 14

Sensory Influence on Homeostasis and Lifespan:

Molecules and Circuits

Joy Alcedo,* Wolfgang Maier and Queelim Ch'ng*

Abstract

The animal's ability to maintain homeostasis in response to different environments can influence its survival. This chapter will discuss the mechanisms by which environmental cues act through sensory pathways to influence hormone secretion and homeostasis. Interestingly, recent studies also show that there is a sensory influence on lifespan that requires the modulation of hormonal signaling activities. Thus, this raises the possibility that the sensory influence on homeostasis underlies the sensory influence on lifespan.

Introduction

To optimize survival, animals must maintain relative constancy within their internal environment, also known as homeostasis, by adjusting their physiology in response to their changing external environment. Thus, animals employ many mechanisms to: (1) define set points at which different physiological processes function most efficiently under given conditions; and (2) prevent large deviations from these defined set points.

Internal vs. External Sensors

Specification of these set points, such as blood glucose levels, is subject to internal and external stimuli that are detected by different types of sensors. The sensors that detect internal stimuli typically monitor quantitative differences between set points and the existing internal environment. Since many of these internal sensors are involved in transmitting signals through negative feedback systems that correct deviations back to predetermined set points, these homeostatic sensors are required to function with a high level of precision.[1,2] An example of a homeostatic internal sensor is the enzyme glucokinase in the pancreatic β cell, which is a sensor for internal glucose levels to control ATP concentrations and thus insulin secretion from pancreatic β cells.[2,3] The mechanisms by which internal sensors regulate animal homeostasis will not be discussed further here and are reviewed elsewhere.[2,4-6] Rather, this chapter will focus on the role of external cues and their sensors in influencing homeostasis by modulating defined set points.

The sensors that detect external stimuli generally play a more important role in assessing qualitative differences between various external environments.[7,8] Accordingly, external sensors can function with a relatively lower degree of precision to perceive a wider range of concentrations of specific

*Corresponding Authors: Joy Alcedo—Friedrich Miescher Institute for Biomedical Research, Maulbeerstrasse 66, CH-4058 Basel, Switzerland. Email: joy.alcedo@fmi.ch.
Queelim Ch'ng—MRC Centre for Developmental Neurobiology, King's College, London SE1 1UL, UK. Email: queelim.ch'ng@kcl.ac.uk

Protein Metabolism and Homeostasis in Aging, edited by Nektarios Tavernarakis.

stimuli.[7,8] Indeed, such external sensors are less likely involved in rectifying fluctuations from set points but are more likely involved in modulating or resetting the set points.

This chapter will discuss (1) how external sensory inputs are recognized and transduced by their respective sensory cells; and (2) how such sensory information is further processed to modulate the secretion of peptide hormones that maintain homeostasis by regulating different physiological processes. Since external sensory cues and the sensory system have recently been shown to influence lifespan,[9-11] this chapter will also address possible mechanisms involved in the sensory influence on aging, which includes progressive impairment of animal homeostasis.

How External Sensory Cues Influence Homeostasis

Although the different roles of external sensory cues in affecting behavior are well-documented, the influence of these cues on animal homeostasis is less appreciated. Since some of these cues have been shown to affect different physiological processes, it is not surprising that there will also be a sensory influence on homeostasis, as demonstrated by a few examples discussed in subsequent sections.

Sensory cues influence homeostasis by modulating hormone secretion. This can be summarized in a simple regulatory motif consisting of three steps (Fig. 1). In step 1, sensory cues change the activity of sensory neurons. In step 2, the sensory information is processed and transmitted, which leads to step 3, where neuro- or nonneuronal endocrine cells secrete hormones required in maintaining homeostasis. All three steps can occur via intracellular signaling within the sensory neuron itself (e.g., release of insulin-like-peptides from *C. elegans* sensory neurons[12]). Alternatively, these steps could occur over many cells: the sensory neuron could activate sensory neural circuits that signal target cells to secrete the hormone. The target cells may be other neurons where hormone secretion is activated by neurotransmitters that are released from presynaptic neurons (e.g., hormone secretion from hypothalamic areas downstream of visual sensation[13,14]); or nonneuronal endocrine organs where the neurotransmitters activate secretion via intercellular signaling cascades (e.g., cephalic-phase insulin secretion from β-cells in response to cholinergic innervations[15,16]).

The subsequent sections will first describe the molecular components of signaling pathways involved in the reception and transmission of sensory signals and then illustrate examples of sensory and neuroendocrine circuits across which these molecular signaling mechanisms operate.

Molecular Mechanisms Linking Sensory Transduction and Hormonal Outputs

In sensory transduction and hormone secretion, information is converted from one form to another, e.g., from light or chemicals to neuronal activity, or from neurotransmitters to hormone secretion. We summarize the signal transduction pathways involved in these forms of information conversion: from reception of sensory cues, processing and transmission of sensory information, to hormone secretion (Fig. 1).

Receptors That Link External Inputs to Regulated Secretion

The detection of extracellular signals in sensory transduction and hormone secretion commonly relies on two types of receptors: (1) ionotropic receptors that are ion channels that open to allow ion flow when stimulated by an external cue, thereby changing the membrane potential[17]; and (2) G-protein coupled receptors (GPCRs) that are seven transmembrane proteins that activate heterotrimeric G-proteins subsequent to its own activation[18] (Fig. 1). Heterotrimeric G-protein complexes consists of Gα, Gβ and Gγ subunits and are associated with GPCRs in the inactive state, where GDP is bound to the Gα subunit.[18,19] Upon GPCR activation, GDP is exchanged with GTP and the G-protein complex dissociates into active Gα and Gβγ subunits that signal to downstream effectors; mechanisms of G-protein activation and its associated proteins are detailed elsewhere.[18-21]

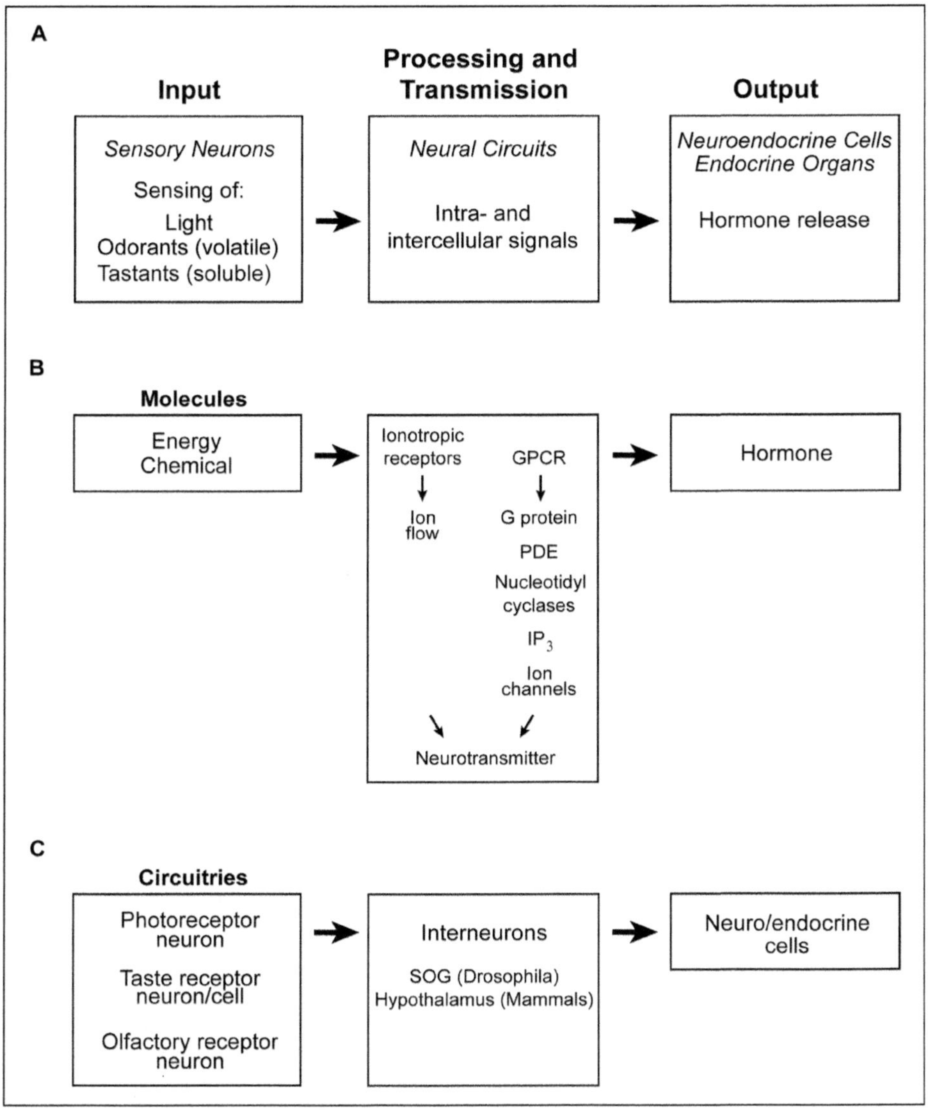

Figure 1. A regulatory motif for the sensory influence on homeostasis. A) Sensory neurons convert various cues to neural activity. Sensory neuron activity is propagated through neural circuits, where information processing and transmission occurs via inter- and intracellular signals. Ultimately, the neural circuits signal to neuroendocrine cells or endocrine organs, which lead to the release of hormones that alter the physiological state of the animal. Multiple regulatory motifs may be combined into a signaling network, allowing cross-talk between physiological and sensory processes[117]. B) A schematic diagram showing the molecular nature of sensory transduction pathways that lead to neurotransmitter or hormone release. C) A scheme demonstrating the flow of sensory information that leads to the secretion of hormones that regulate homeostasis. In some cases, the activated sensory neuron releases the hormone itself through an intracellular signaling cascade, like the one depicted in panel B.

Secretory Pathways for Chemical Signals

G-protein signaling and changes in membrane potential regulate the secretion of neurotransmitters that signal across synapses, as well as that of hormones that regulate physiology and homeostasis. Secretion occurs via the fusion of either one of two general types of vesicles with the plasma membrane: synaptic vesicles (SVs) and dense core vesicles (DCVs).[22,23] SVs and DCVs differ in their cargo: SVs typically deliver small molecule neurotransmitters, such as γ-aminobutyric acid (GABA), acetylcholine or glutamate, that signal to their cognate receptors across a synapse, whereas DCVs typically release neuropeptides, hormones, such as insulin, and other neurotransmitters.[22,23] SVs and DCVs also differ in their biogenesis, trafficking, cell biology and regulation.[22,23]

The secretion from SVs and DCVs are both tightly controlled; these vesicles progress through a sequence of docking (where vesicles become tightly apposed to the plasma membrane), priming (where vesicles become fusion-competent) and a SNARE-dependent vesicle fusion triggered by intracellular calcium (Ca^{2+}).[24-26] The secretory machinery of SVs and DCVs can be regulated at multiple points: the structure and function of release sites (active zones); the activity states of the exocytotic machinery that influence progression through docking, priming and fusion; and the membrane potential that affects Ca^{2+} entry for triggering vesicle fusion.[24-26] G-protein pathways may influence any of these processes, whereas ionotropic receptors primarily affect membrane potential and hence Ca^{2+} entry.[24-26]

Regulation of Secretion by Different G-Proteins and Their Effectors

Different Gα subunits regulate different sets of downstream effectors that target different steps in secretion.[19] $G\alpha_q$, $G\alpha_s$ and $G\alpha_{i/o}$ are the major types of Gα subunits involved in this process.[27-29] Generally, secretion is promoted by $G\alpha_q$ and $G\alpha_s$ and inhibited by $G\alpha_{i/o}$.[27-29] Below, we describe some of the more prominent G-protein targets in secretion, not as a comprehensive list but to provide a flavor for the molecules involved.

$G\alpha_q$ activates phospholipase C-β (PLCβ), which cleaves phosphatidylinositol bisphosphate (PIP_2) to generate the second messengers diacylglycerol (DAG) and inositol 1,4,5-trisphosphate (IP_3).[29] To stimulate secretion, DAG activates protein kinase C (PKC) and UNC-13 family members,[23,29-32] while IP_3 activates the IP_3 receptor.[29] UNC-13 family members have been proposed to affect both priming and additional unknown step(s) in vesicle fusion,[31-33] whereas PKC is thought to influence secretion by phosphorylating targets involved in priming and fusion pore formation.[34] On the other hand, IP_3 receptor activation leads to Ca^{2+} release from the endoplasmic reticulum and the opening of a cation-selective (Na^+/Ca^{2+} exchanger) channel on the cell membrane.[35,36]

$G\alpha_s$ activates adenyl cyclase to produce cyclic AMP (cAMP), which activates protein kinase A (PKA) and the exchange protein activated by cAMP (Epac) to stimulate secretion. Downstream targets of PKA and Epac include proteins involved in secretory SNARE complex formation during priming (e.g., tomosyn), proteins in SV-release sites (e.g., RIM1) and ion channel subunits that regulate membrane potential and Ca^{2+} entry for exocytosis.[37,38] $G\alpha_{olf}$ are mammalian olfactory G-proteins related to $G\alpha_s$ and act on similar targets.[39,40]

$G\alpha_{i/o}$ inhibits the activity of adenyl cyclase and also targets calcium and potassium channels. These pathways also influence secretion by regulating active zone proteins[41,42] and the Rho family of small GTPases that in turn affect actin rearrangements.[28] On the other hand, transducin, which is related to the $G\alpha_{i/o}$, activates cGMP phosphodiesterase (PDE), which leads to the closing of cGMP-gated cation channels, a block in sodium cation (Na^+) influx, hyperpolarization of the cell membrane and neurotransmitter release at synapses.[35,36]

Regulatory interactions between these distinct G-proteins and their effectors allow cells to integrate inputs from multiple signals. For example, in *C. elegans*, double mutant analyses have revealed opposing endogenous interactions between $G\alpha_q$ and $G\alpha_o$.[43] These pathways target different synaptic proteins within *C. elegans* motorneurons, suggesting that their effects on different rate-limiting steps of the secretory process are integrated by the neuron.[30,41,42]

Specificity in Signaling and Secretion

Specific GPCRs are coupled to specific G-proteins with specific targets, leading to different secretory outcomes. For example, pancreatic β cells express a whole plethora of GPCRs involved in regulating insulin secretion,[44] including the M3 muscarinic receptors and the Y1 receptor. Activation of M3 muscarinic receptors by acetylcholine released from parasympathetic nerves that innervate the pancreatic cells, promotes insulin secretion, because M3 muscarinic receptors are coupled to $G\alpha_q$.[44] In contrast, activation of the Y1 GPCRs by its ligand neuropeptide Y leads to inhibition of insulin secretion, because Y1 receptors are coupled to inhibitory G proteins, like $G\alpha_i$.[44] Thus, the outcome of GPCR activation depends on the G-protein to which the GPCR is coupled. A review by Oldham and Hamm[20] further discusses the specificity determinants of GPCR/G-protein coupling and the mechanisms of G protein activation.

Similarly, the ion-selectivity of an ionotropic receptor allows it to alter the membrane potential and thus promote or inhibit Ca^{2+} entry via voltage-gated Ca^{2+} channels, which subsequently modulate neurotransmitter/hormone secretion.[17] For example, at synapses, $GABA_A$ receptors on the postsynaptic neuron are inhibitory because they are permeable to chloride anions (Cl^-), whereas the AMPA receptor for glutamate are excitatory because they are permeable to cations.[17]

Examples of Sensory Signal Transduction

Visual cues (light): Light is detected in photoreceptor cells by rhodopsins, which are composed of the G protein-coupled receptor opsin and a chromophore.[35,36] Light energy isomerizes the chromophore, which causes receptor conformational changes and activation of a G protein: transducin in vertebrates and a $G\alpha_q$ in invertebrates.[35,36] In vertebrates, transducin/PDE activation results in membrane hyperpolarization via inhibition of cGMP-gated channels, whereas in invertebrates $G\alpha_q$/PLC/IP_3 activation leads to depolarization via Ca^{2+} signals and Na^+/Ca^{++} exchangers.[35,36] Despite these differences, these signals lead to changes in neurotransmitter secretion that are propagated through the visual neural circuitry.[35,36]

Olfactory cues: In many species, odors are detected by a large repertoire of olfactory GPCRs that signal to downstream targets with species specificity.[45] In mammals, stimulating olfactory GPCRs lead to the sequential activation of $G\alpha_{olf}$, membrane adenyl cyclases and cAMP to activate cyclic nucleotide-gated channels.[39] In *C. elegans*, different olfactory neurons utilize different signaling pathways. Certain olfactory neurons signal via ODR-3 (related to $G\alpha_{i/o}$), cGMP and the TAX-2/TAX-4 cGMP-gated ion channel; whereas other olfactory neurons signal via phospholipids, polyunsaturated fatty acids and the OSM-9 TRPV cation channel.[46] In contrast, *Drosophila* utilizes a different family of olfactory seven-pass membrane receptors,[47,48] as well as a class of ionotropic glutamate receptor-related proteins,[49] which act as ion channels.

Gustatory cues: In mammals, salty and sour tastants are thought to be sensed by ionotropic receptors, whereas sweet, bitter and umami are sensed by GPCRs that couple to gustducin or $G\alpha_i$.[50] In *Drosophila*, the gustatory receptors (GRs) are related to *Drosophila* olfactory receptors.[51,52] GR5a and Gr64a act in sugar sensing, but neither the downstream signal tranduction pathways nor the receptors for other tastants are fully known.[51,52] In *C. elegans*, gustatory receptors have not been identified, but Na^+ and Cl^- ions are sensed separately by bilaterally asymmetric ASE sensory neurons via a guanyl cyclase signaling pathway[46] that is modulated by G-protein signaling in other sensory neurons.[53]

The Circuitry Underlying the Processing of Sensory Information

The architecture and function of the nervous system of one animal species can vary significantly to those of others and their description lies beyond the scope of this chapter. Rather, this section will focus on examples of neural circuitry involved in sensory processing to illustrate how visual and chemosensory cues can elicit systemic changes by regulating hormonal secretions from target tissues (Fig. 1), which affect animal homeostasis.

Processing of Visual Information to Synchronize Circadian Rhythms

Circadian behavioral and physiological rhythms have evolved in eukaryotes as a means to anticipate environmental changes that occur regularly with the earth's rotation around its axis. In mammals, the time-keeping mechanism that generates these rhythms involves circadian clocks that function within a hierarchy.[8] A master clock resides in neurons of the suprachiasmatic nucleus (SCN) of the hypothalamus, which synchronizes the clocks of all peripheral cells within the animal.[54-56] The master clock in the SCN is capable of self-sustained function,[54] whereas the peripheral clocks oscillate only in the presence of the master clock.[55,56] On the molecular level, each clock consists of an oscillatory gene-regulatory network that has been reviewed extensively by others[57] and will not be addressed here further.

A major stimulus in resetting the circadian clock is the cycle between the presence and absence of light within the animal's daily environment.[8] This visual cue is communicated directly to the master clock in the SCN from a subset of photoreceptor cells that express melanopsin, a rhodopsin-related molecule.[58,59] The axons of the melanopsin-expressing cells relay photic information by releasing the neurotransmitter glutamate that acts on the SCN to promote molecular changes that resynchronize the clock with the animal's light/dark environmental cycle.[60] The SCN secretes signals, like the peptide hormones prokineticin 2[13] and transforming growth factor α (TGF-α),[14] that can then act on other areas of the hypothalamus.[61] These other hypothalamic areas in turn control the secretion of hormones from target endocrine tissues, such as the pituitary, pineal and adrenal glands, which allows for the circadian modulation of many physiological processes required for homeostasis.[61]

Glucose homeostasis is one example that undergoes circadian rhythmicity: blood glucose levels are higher during the light phase of the cycle and lower during the dark phase of the cycle.[62,63] Although these diurnal changes in blood glucose levels could be explained simply by a circadian schedule in the animal's food intake, the perception of light has been shown to affect the levels of melatonin,[64,65] which can stimulate insulin pathway activity and glucose transport into cells.[62,63] The synthesis of the hormone melatonin in the pineal gland is inhibited by the presence of light:[64] there is less circulating melatonin during the light phase of the cycle and more melatonin during the dark phase of the cycle. In addition, this diurnal cycling in melatonin synthesis can be abolished by the removal of retinal photoreceptors or by the disruption of the circuit between the photoreceptor cells and the pineal gland.[65] Thus, these observations together suggest a role for light perception in glucose homeostasis.

Processing of Chemosensory Information to Alter Behavior and Metabolism

Gustatory and olfactory cues are not only perceived by anatomically distinct receptor cells but are also processed by separate circuits in different species.

Worms: In *C. elegans*, a subset of gustatory neurons[66,67] senses a pheromone mixture of glycosides[68,69] that promotes under harsh environments dauer formation, which is an alternative developmental program.[70-72] Harsh conditions, like overcrowding and hence low food availability, can be signified by increasing environmental quantities of the dauer pheromone mixture, which is secreted by each animal throughout its life.[70,71] Accordingly, high concentrations of dauer pheromone have been shown to inhibit the secretion of peptide hormones, such as an insulin-like peptide[12] and TGF-β,[67] from the pheromone-sensing neurons, which in turn can act directly on neuronal (e.g., interneurons) and/or nonneuronal cells. The downregulation of the insulin[70,73] or the TGF-β[67,70] pathway promotes formation of the developmentally arrested dauers. Hence, inhibiting the release of these hormones from the chemosensory neurons is believed to promote a shift not only in the animal's metabolic homeostasis but also in its stress-responsiveness that prepares the dauer for long-term survival under harsh conditions.[74,75] At present, the cells on which these peptides act to regulate dauer formation are unknown.

On the other hand, the role of *C. elegans* olfactory neurons in regulating homeostasis remains unclear, although these neurons have been shown to affect lifespan[10] (discussed below), which could involve a change in the animal's homeostasis. The neuronal circuitries that process olfactory cues in contrast to gustatory cues are also only beginning to be elucidated. For example, the map of postsynaptic partners of gustatory and olfactory neurons shows considerable overlap.[76] However, olfactory neurons synapse more extensively onto one set of interneurons, whereas gustatory neurons synapse more onto another set of neurons.[76]

Fruitflies: In *Drosophila*, the neurons that detect gustatory and olfactory stimuli are found in anatomically different structures.[77] Taste inputs are communicated either directly or indirectly to the subesophageal ganglion (SOG), which further relays taste information not only to higher brain centers, the ventral nerve cord and nonneuronal tissues but also to neuroendocrine cells.[77,78] This raises the possibility that gustatory cues can also modulate the release of hormones that regulate fly homeostasis. Interestingly, a small cluster of SOG neurons that express the neuropeptide *hugin*, which controls food intake and fly growth, not only receive direct inputs[78] from gustatory receptor neurons (that innervate structures within the fly's mouthparts)[79] but also send projections to or near insulin-producing cells.[78] In addition, consistent with the hypothesis that SOG peptides are regulated by gustatory cues that affect homeostasis, microarray analysis of *hugin* mutants demonstrates altered expression of insulin-like peptides.[78]

Recently, fly olfactory mutants have also been shown to have altered lipid content and respiration,[11] which suggests a role for olfactory cues in maintaining this animal's metabolic homeostasis. The cholinergic olfactory receptor neurons convey odor inputs directly to individual glomeruli in the fly antennal lobe, where olfactory information is further relayed to second-order cholinergic projection neurons (PNs).[77] PNs link the antennal lobe with higher brain centers, like the mushroom body, which is involved not only in associative olfactory learning[77] but also in other processes, including sleep homeostasis.[80,81]

Mammals: The homeostatic mechanisms that are activated upon food intake are influenced, not surprisingly, by chemosensory cues.[82] Food intake elicits a wide range of responses that serve to ensure effective digestion, maintenance of metabolic homeostasis upon the availability of new nutrients and termination of feeding upon the release of a satiety signal.[83] The physiological responses during mammalian food intake can be divided into three phases, depending on the tissues that are stimulated during the process: the neural/cephalic phase, the gastric phase and the intestinal phase.[82] The cephalic phase of food intake is studied by subjecting the animal to food odors alone or to mock feedings that involve tasting but not swallowing or digesting the food. This initial phase of food intake implicates odorants and tastants in triggering secretions from the salivary glands, stomach and pancreas.[36,82] For example, there is an increased secretion of saliva,[84] the gastrin hormone and gastric acid[85] upon smelling or tasting palatable food. Moreover, after a mock feeding, there is increased secretion of the hormones leptin[86] and insulin, which occurs in the absence of increased blood glucose levels,[15,16] in rodents, dogs or humans. These pre-absorptive, anticipatory responses that are triggered by gustatory and olfactory stimuli have been shown to be required for normal feeding behavior and optimal digestion[82] and suggest how taste and olfaction influence an animal's metabolic homeostasis.

Gustatory information from taste receptors on the tongue is carried through three different cranial nerves—the facial, the glossopharyngeal and the vagus nerves—to the nucleus of the solitary tract (NST) in the mammalian brain.[36,87] From the NST, which also receives internal viscerosensory inputs, gustatory information is relayed to the parabrachial nucleus (PbN), which further projects to the thalamus and gustatory cortex.[36,87] In primates, the thalamocortical part of the circuit that comes from the NST bypasses the PbN neurons.[36,87] Interestingly, gustatory information can also be relayed in a parallel circuit not only to the amygdala but also to the hypothalamus,[36,87] which links the mammalian nervous system to the endocrine system.

On the other hand, olfactory information is transmitted by olfactory receptor neurons (ORNs) to the first relay point within the brain, the olfactory bulb.[36,88] From there, olfactory information is relayed to cortical regions, as well as amygdalar and hypothalamic regions.[36] Indeed, retrograde labeling of neurons has demonstrated that a discrete set of ORNs relays information to hypothalamic neurons that express the luteinizing hormone release hormone (LHRH), which is a key hormone that regulates mammalian reproductive homeostasis.[89] Consistent with the existence of this neural circuitry, olfactory cues have also been shown to stimulate the secretion of LHRH, which not only controls the release of other hormones required for gonadal development and function but may also directly promote mating behavior.[90,91]

Thus, together these gustatory and olfactory circuits can allow the hypothalamus to integrate both external and internal sensory information to control animal homeostasis in response to the changing quality of the environment.

Sensory Influence on Lifespan

Recent studies have shown that the sensory systems of *C. elegans* and *Drosophila* not only influence the behavior and physiology of these animals but also their lifespan.[9-11] A subset of gustatory and olfactory neurons in *C. elegans*[10] and olfaction in *Drosophila*[11] have been found to exert different effects on lifespan, which suggests that food-derived cues affect longevity. Since dietary restriction has previously been shown to increase lifespan,[92] it is possible that the sensory system influences lifespan by regulating the animal's general food intake. However, the observation that not every gustatory neuron affects lifespan[10] suggests that the animal adjusts its rate of aging not only in response to food levels but also to more specific cues derived from different food sources.

Gustatory Influence on the Insulin/IGF-1 Pathway That Affects Lifespan

Laser ablation of certain gustatory neurons (ASI and ASG) extends *C. elegans* lifespan, which depends on the activities of other gustatory neurons (ASJ and ASK).[10] This suggests that there are two classes of gustatory neurons: one that shortens lifespan and another that lengthens lifespan. These neurons appear to influence lifespan by modulating the activity of the *C. elegans* insulin/IGF-1 pathway,[10] which has been shown previously to modulate longevity[73,93-96] (Fig. 2).

The reduction in function of the *C. elegans* insulin/IGF-1 receptor DAF-2 can increase lifespan by as much as two-fold.[73,93,94] This lifespan extension requires the activity of DAF-16,[93,94] a FOXO transcription factor[95,96] that is activated and translocated into the nucleus upon low DAF-2 activity.[97-99] Mutations that impair *C. elegans* sensory function decrease the activity of the DAF-2 pathway,[9] leading to the nuclear localization of DAF-16.[98] In addition, ablation of the ASI neurons causes lifespan extension that is *daf-16*-dependent, whereas ablation of the ASJ and ASK neurons can partly suppress the lifespan extension observed in *daf-2* reduction-of-function mutants (Fig. 2).[10] Thus, these findings, together with the expression of insulin-like peptides in sensory neurons,[12,100] are in keeping with the idea that the activity of this pathway is subject to modulation by sensory cues.

Olfactory Influence on Signal(s) from the Reproductive System That Affect Lifespan

Ablation of the germline precursor cells extends *C. elegans* lifespan, which can be suppressed by either ablation of the somatic gonad precursor cells or a null mutation in *daf-16*.[101] In contrast, *daf-16* null mutants in which the somatic gonad has been ablated live shorter than *daf-16* mutants with an intact gonad,[101] suggesting that the somatic gonad can act in parallel to *daf-16* to affect lifespan. These observations have led to the model that the germ line of *C. elegans* generates a longevity-inhibiting, *daf-16*-dependent signal, which is counterbalanced by a longevity-promoting, *daf-16*-independent signal from the somatic gonad (Fig. 2).[101]

Germline ablation can further extend the lifespan of *C. elegans* that have sensory defects,[9] which can either be gustatory or olfactory in nature.[10] Surprisingly, however, impairment of olfactory, but not gustatory, function prevents somatic gonad ablation from suppressing the lifespan extension caused by germline ablation.[10] This suggests that, in contrast to gustatory neurons, olfactory neurons modulate the activity of the somatic gonad signal (Fig. 2).[10] Consistent with this observation, the olfactory influence on lifespan is also at least partly *daf-16*-independent.[10] However, the somatic gonad appears to promote longevity in a *daf-2*-dependent manner,[101] which raises the possibility that olfactory neurons might influence lifespan by also regulating the release of insulin-like peptides that, in this case, act independently of *daf-16*. Indeed, at least one of the worm insulin-like genes

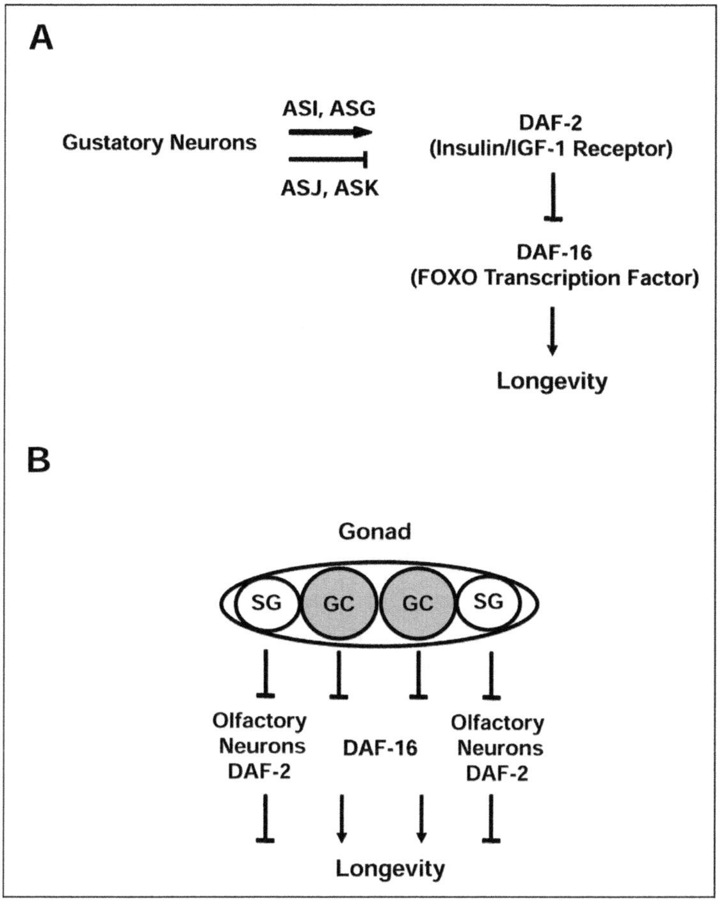

Figure 2. Specific gustatory and olfactory neurons influence *C. elegans* lifespan through hormonal signaling pathways. A) Certain gustatory neurons (ASI, ASG) shorten lifespan, whereas other neurons (ASJ, ASK) lengthen lifespan. Both classes of gustatory neurons appear to modulate insulin/IGF-1 signaling to affect worm lifespan. B) A schematic diagram of the worm's gonad precursor cells showing that the germline cells (GC, gray circles) inhibit longevity in a *daf-16*-dependent manner. On the other hand, the somatic gonad (SG, white circles) promotes longevity in an olfactory neuron-dependent and *daf-2*-dependent manner.

is expressed in a pair of olfactory neurons.[102] So far, it is unclear whether olfactory neurons release a signal that modulates DAF-2 activity or whether DAF-2 blocks the release of a signal from the olfactory neurons to inhibit longevity.

The olfactory influence on lifespan has also been reported in *Drosophila*.[11] Flies that lack the atypical olfactory receptor *Or83b*, which is required for the proper subcellular localization of many of the olfactory receptors,[103] not only have severe olfactory deficits but also live long.[11] Conversely, olfaction has been shown to shorten the lifespan of dietary-restricted flies.[11]

Dietary restriction (DR) is a treatment that can extend the lifespan of many species, ranging from yeast to mammals.[92,104-106] DR, by its very nature, indicates a role for the environment in influencing lifespan. Since the lifespan increase seen in dietary-restricted flies can be partly suppressed when the flies are exposed to food-associated odors alone,[11] it appears that sensory perception itself can trigger a physiological change that can antagonize the DR response. In long-lived *Or83b* mutant flies, the mRNA levels of most of the insulin-like genes are not significantly downregulated, with the exception of one member of this family.[11,107] Since the insulin pathway has also been shown to affect fly lifespan,[108-112] it is possible that the change in expression of one of the insulin-like genes mediates the olfaction-induced physiological change that affects lifespan. Although the mechanism behind this sensory influence on lifespan currently remains unknown, it is not surprising that the olfactory system, which can signal the levels or quality of food in the environment, would mediate, at least in part, the effects of DR on lifespan extension.

Conclusion: Connections Between the Sensory Influence on Homeostasis and Lifespan?

The sensory influence on both homeostasis and lifespan involves the modulation of activities of hormonal signaling pathways. One of these pathways, the insulin/IGF-1 pathway, has been shown to regulate the expression of metabolic, immune-response and stress-response genes that not only regulate animal homeostasis but also affect lifespan.[113,114] Together these observations are consistent with the possibility that specific sensory cues can redefine set points in homeostasis bidirectionally, thus leading to a longer or shorter lifespan. This hypothesis predicts that the effects of a given sensory cue on lifespan will be abolished when either the specific sensory pathway or its modulation of the set point(s) is disrupted. Indeed, sensory neurons have been shown to control lipid homeostasis[115] and a change in lipid homeostasis can also lead to changes in lifespan.[116] Thus, the future identification of the set points affected by the lifespan-influencing sensory cues should yield insight into the mechanisms that regulate the rate and/or onset of aging.

Finally, homeostasis is characterized by a number of feedback mechanisms. Since hormones have also been proposed to modulate the sensory system,[117] this raises another intriguing possibility that the decline in sensory function observed during aging[118] may result from a disruption in feedback regulation between the animal's sensory system and its different homeostatic mechanisms.

Acknowledgements

We apologize to the authors whose work we were unable to cite due to space constraints. We would also like to thank P. Caroni, R. Friedrich, A. Matus, F. Meins, M. Noll and M. Raff for comments on a previous version of this manuscript. This work was supported by the Novartis Research Foundation (J. A. and W. M.), a Deutsche Forschungsgemeinschaft Postdoctoral Fellowship (MA-3995/1; W. M.) and an RCUK Fellowship (Q.C.).

References

1. Nelson DL, Cox MM. Lehninger: Principles of Biochemistry. 5th ed. New York: W.H. Freeman and Company, 2008.
2. Newsholme P, Bender K, Kiely A et al. Amino acid metabolism, insulin secretion and diabetes. Biochem Soc Trans 2007; 035:1180-1186.
3. German MS. Glucose sensing in pancreatic islet beta cells: The key role of glucokinase and the glycolytic intermediates. Proc Natl Acad Sci USA 1993; 90:1781-1785.
4. Drucker DJ. The biology of incretin hormones. Cell Metab 2006; 3:153-165.
5. Murphy KG, Bloom SR. Gut hormones and the regulation of energy homeostasis. Nature 2006; 444:854-859.
6. Nolan CJ, Prentki M. The islet β-cell: fuel responsive and vulnerable. Trends Endocrinol Metab 2008; 19:285-291.
7. Bargmann CI, Hartwieg E, Horvitz HR. Odorant-selective genes and neurons mediate olfaction in C. elegans. Cell 1993; 74:515-527.
8. Challet E, Caldelas I, Graff C et al. Synchronization of the molecular clockwork by light- and food-related cues in mammals. Biol Chem 2003; 384:711-719.
9. Apfeld J, Kenyon C. Regulation of lifespan by sensory perception in Caenorhabditis elegans. Nature 1999; 402:804-809.
10. Alcedo J, Kenyon C. Regulation of C. elegans longevity by specific gustatory and olfactory neurons. Neuron 2004; 41:45-55.
11. Libert S, Zwiener J, Chu X et al. Regulation of Drosophila life span by olfaction and food-derived odors. Science 2007; 315:1133-1137.

12. Li W, Kennedy SG, Ruvkun G. daf-28 encodes a C. elegans insulin superfamily member that is regulated by environmental cues and acts in the DAF-2 signaling pathway. Genes Dev 2003; 17:844-858.
13. Prosser HM, Bradley A, Chesham JE et al. Prokineticin receptor 2 (Prokr2) is essential for the regulation of circadian behavior by the suprachiasmatic nuclei. Proc Natl Acad Sci USA 2007; 104:648-653.
14. Kramer A, Yang F-C, Kraves S et al. A screen for secreted factors of the suprachiasmatic nucleus. Methods Enzymol 2005; 393:645-663.
15. Berthoud HR, Trimble ER, Siegel EG et al. Cephalic-phase insulin secretion in normal and pancreatic islet-transplanted rats. Am J Physiol 1980; 238:E336-E340.
16. Secchi A, Caldara R, Caumo A et al. Cephalic-phase insulin and glucagon release in normal subjects and in patients receiving pancreas transplantation. Metabolism 1995; 44:1153-1158.
17. Barrera NP, Edwardson JM. The subunit arrangement and assembly of ionotropic receptors. Trends Neurosci 2008; 31:569-576.
18. Pierce KL, Premont RT, Lefkowitz RJ. Seven-transmembrane receptors. Nat Rev Mol Cell Biol 2002; 3:639-650.
19. Simon MI, Strathmann MP, Gautam N. Diversity of G proteins in signal transduction. Science 1991; 252:802-808.
20. Oldham WM, Hamm HE. Heterotrimeric G protein activation by G-protein-coupled receptors. Nat Rev Mol Cell Biol 2008; 9:60-71.
21. Pierce KL, Lefkowitz RJ. Classical and new roles of β-arrestins in the regulation of G-PROTEIN-COUPLED receptors. Nat Rev Neurosci 2001; 2:727-533.
22. Edwards RH. Neurotransmitter release: variations on a theme. Curr Biol 1998; 8:R883-885.
23. Xu T, Xu P. Searching for molecular players differentially involved in neurotransmitter and neuropeptide release. Neurochem Res 2008; 33:1915-1919.
24. Südhof TC. The synaptic vesicle cycle. Annu Rev Neurosci 2004; 27:509-547.
25. Südhof TC, Rothman JE. Membrane fusion: grappling with SNARE and SM proteins. Science 2009; 323:474-477.
26. Wojcik SM, Brose N. Regulation of membrane fusion in synaptic excitation-secretion coupling: speed and accuracy matter. Neuron 2007; 55:11-24.
27. Neves SR, Ram PT, Iyengar R. G protein pathways. Science 2002; 296:1636-1639.
28. Jiang M, Bajpayee NS. Molecular mechanisms of Go signaling. Neurosignals 2009; 17:23-41.
29. Mizuno N, Itoh H. Functions and regulatory mechanisms of Gq-signaling pathways. Neurosignals 2009; 17:42-54.
30. Sieburth D, Madison JM, Kaplan JM. PKC-1 regulates secretion of neuropeptides. Nat Neurosci 2007; 10:49-57.
31. Betz A, Ashery U, Rickmann M et al. Munc13-1 is a presynaptic phorbol ester receptor that enhances neurotransmitter release. Neuron 1998; 21:123-136.
32. Rhee JS, Betz A, Pyott S et al. β phorbol ester- and diacylglycerol-induced augmentation of transmitter release is mediated by Munc13s and not by PKCs. Cell 2002; 108:121-133.
33. Madison JM, Nurrish S, Kaplan JM. UNC-13 interaction with syntaxin is required for synaptic transmission. Curr Biol 2005; 15:2236-2242.
34. Morgan A, Burgoyne RD, Barclay JW et al. Regulation of exocytosis by protein kinase C. Biochem Soc Trans 2005; 33:1341-1344.
35. Yarfitz S, Hurley JB. Transduction mechanisms of vertebrate and invertebrate photoreceptors. J Biol Chem 1994; 269:14329-14332.
36. Squire LR, Bloom FE, Roberts JL et al. Fundamental Neuroscience. 2nd ed. USA: Elsevier Science, 2003.
37. Baba T, Sakisaka T, Mochida S et al. PKA-catalyzed phosphorylation of tomosyn and its implication in Ca^{2+}-dependent exocytosis of neurotransmitter. J Cell Biol 2005; 170:1113-1125.
38. Seino S, Shibasaki T. PKA-dependent and PKA-independent pathways for cAMP-regulated exocytosis. Physiol Rev 2005; 85:1303-1342.
39. Ronnett GV, Moon C. G proteins and olfactory signal transduction. Annu Rev Physiol 2002; 64:189-222.
40. Hasin-Brumshtein Y, Lancet D, Olender T. Human olfaction: from genomic variation to phenotypic diversity. Trends Genet 2009; 25:178-184.
41. Nurrish S, Segalat L, Kaplan JM. Serotonin inhibition of synaptic transmission: $G\alpha_o$ decreases the abundance of UNC-13 at release sites. Neuron 1999; 24:231-242.
42. Ch'ng Q, Sieburth D, Kaplan JM. Profiling synaptic proteins identifies regulators of insulin secretion and lifespan. PLoS Genet 2008; 4:e1000283.
43. Hajdu-Cronin YM, Chen WJ, Patikoglou G et al. Antagonism between $G_o\alpha$ and $G_q\alpha$ in Caenorhabditis elegans: the RGS protein EAT-16 is necessary for G_o signaling and regulates $G_q\alpha$ activity. Genes Dev 1999; 13:1780-1793.

44. Winzell MS, Ahren B. G-protein-coupled receptors and islet function-implications for treatment of type 2 diabetes. Pharmacol Ther 2007; 116:437-448.
45. Bargmann CI. Comparative chemosensation from receptors to ecology. Nature 2006; 444:295-301.
46. Bargmann CI. Chemosensation in C. elegans. In: T.C.e.R. Community. WormBook. http://www.wormbook.org, 2006.
47. Sato K, Pellegrino M, Nakagawa T et al. Insect olfactory receptors are heteromeric ligand-gated ion channels. Nature 2008; 452:1002-1006.
48. Wicher D, Schafer R, Bauernfeind R et al. Drosophila odorant receptors are both ligand-gated and cyclic-nucleotide-activated cation channels. Nature 2008; 452:1007-1011.
49. Benton R, Vannice KS, Gomez-Diaz C et al. Variant ionotropic glutamate receptors as chemosensory receptors in Drosophila. Cell 2009; 136:149-162.
50. Chandrashekar J, Hoon MA, Ryba NJ et al. The receptors and cells for mammalian taste. Nature 2006; 444:288-294.
51. Scott K. Taste recognition: food for thought. Neuron 2005; 48:455-464.
52. Benton R. Chemical sensing in Drosophila. Curr Opin Neurobiol 2008; 18:357-363.
53. Hukema R, Rademakers S, Dekkers M et al. Antagonistic sensory cues generate gustatory plasticity in Caenorhabditis elegans. EMBO J 2006; 25:312-322.
54. Welsh DK, Logothetis DE, Meister M et al. Individual neurons dissociated from rat suprachiasmatic nucleus express independently phased circadian firing rhythms. Neuron 1995; 14:697-706.
55. Sakamoto K, Nagase T, Fukui H et al. Multitissue circadian expression of rat period homolog (rPer2) mRNA is governed by the mammalian circadian clock, the suprachiasmatic nucleus in the brain. J Biol Chem 1998; 273:27039-27042.
56. Yamazaki S, Numano R, Abe M et al. Resetting central and peripheral circadian oscillators in transgenic rats. Science 2000; 288:682-685.
57. Dunlap JC. Molecular bases for circadian clocks. Cell 1999; 96:271-290.
58. Hattar S, Liao HW, Takao M et al. Melanopsin-containing retinal ganglion cells: architecture, projections and intrinsic photosensitivity. Science 2002; 295:1065-1070.
59. Provencio I, Rollag MD, Castrucci AM. Anatomy: photoreceptive net in the mammalian retina. Nature 2002; 415:493-493.
60. Ding JM, Chen D, Weber ET et al. Resetting the biological clock: mediation of nocturnal circadian shifts by glutamate and NO. Science 1994; 266:1713-1717.
61 Watts A. The efferent projections of the suprachiasmatic nucleus: anatomical insights into the control of circadian rhythms. In: Klein D, Moore R, Reppert S.M, eds. The Suprachiasmatic Nucleus—The Mind's Clock. New York: Oxford, 1991:77-106.
62 Ha E, Yim S-V, Chung J-H et al. Melatonin stimulates glucose transport via insulin receptor substrate-1/phosphatidylinositol 3-kinase pathway in C_2C_{12} murine skeletal muscle cells. J Pineal Res 2006; 41:67-72.
63. la Fleur SE, Kalsbeek A, Wortel J et al. Role for the pineal and melatonin in glucose homeostasis: pinealectomy increases night-time glucose concentrations. J Neuroendocrinol 2001; 13:1025-1032.
64. Wurtman RJ, Axelrod J, Phillips LS. Melatonin synthesis in the pineal gland: control by light. Science 1963; 142:1071-1073.
65. Wurtman RJ, Axelrod J, Fischer JE. Melatonin synthesis in the pineal gland: effect of light mediated by the sympathetic nervous system. Science 1964; 143:1328-1329.
66. Bargmann CI, Horvitz HR. Control of larval development by chemosensory neurons in Caenorhabditis elegans. Science 1991; 251:1243-1246.
67. Schackwitz WS, Inoue T, Thomas JH. Chemosensory neurons function in parallel to mediate a pheromone response in C. elegans. Neuron 1996; 17:719-728.
68. Butcher RA, Fujita M, Schroeder FC et al. Small-molecule pheromones that control dauer development in Caenorhabditis elegans. Nat Chem Biol 2007; 3:420-422.
69. Jeong PY, Jung M, Yim YH et al. Chemical structure and biological activity of the Caenorhabditis elegans dauer-inducing pheromone. Nature 2005; 433:541-545.
70. Riddle DL, Swanson MM, Albert PS. Interacting genes in nematode dauer larva formation. Nature 1981; 290:668-671.
71. Golden JW, Riddle DL. The Caenorhabditis elegans dauer larva: developmental effects of pheromone, food and temperature. Dev Biol 1984; 102:368-378.
72. Vowels JJ, Thomas JH. Genetic analysis of chemosensory control of dauer formation in Caenorhabditis elegans. Genetics 1992; 130:105-123.
73. Kimura KD, Tissenbaum HA, Liu Y et al. daf-2, an insulin receptor-like gene that regulates longevity and diapause in Caenorhabditis elegans. Science 1997; 277:942-946.
74. Wang J, Kim SK. Global analysis of dauer gene expression in Caenorhabditis elegans. Development 2003; 130:1621-1634.

75. Narbonne P, Roy R. Caenorhabditis elegans dauers need LKB1/AMPK to ration lipid reserves and ensure long-term survival. Nature 2009; 457:210-214.
76. White JG, Southgate E, Thomson JN et al. The structure of the nervous system of the nematode Caenorhabditis elegans. Philos Trans R Soc Lond B Biol Sci 1986; 314.
77. Vosshall LB, Stocker RF. Molecular architecture of smell and taste in Drosophila. Annu Rev Neurosci 2007; 30:505-533.
78. Melcher C, Pankratz MJ. Candidate gustatory interneurons modulating feeding behavior in the Drosophila brain. PLoS Biol 2005; 3:e305.
79. Scott K, Brady JR, Cravchik A et al. A chemosensory gene family encoding candidate gustatory and olfactory receptors in Drosophila. Cell 2001; 104:661-673.
80. Joiner WJ, Crocker A, White BH et al. Sleep in Drosophila is regulated by adult mushroom bodies. Nature 2006; 441:757-760.
81. Pitman JL, McGill JJ, Keegan KP et al. A dynamic role for the mushroom bodies in promoting sleep in Drosophila. Nature 2006; 441:753-756.
82. Zafra MA, Molina F, Puerto A. The neural/cephalic phase reflexes in the physiology of nutrition. Neurosci Biobehav Rev 2006; 30:1032-1044.
83. Saper CB, Chou TC, Elmquist JK. The need to feed: homeostatic and hedonic control of eating. Neuron 2002; 36:199-211.
84. Richardson CT, Feldman M. Salivary response to food in humans and its effect on gastric acid secretion. Am J Physiol 1986; 250:G85-G91.
85. Feldman M, Richardson CT. Role of thought, sight, smell and taste of food in the cephalic phase of gastric acid secretion in humans. Gastroenterology 1986; 90:428-433.
86. Konturek SJ, Konturek JW. Cephalic phase of pancreatic secretion. Appetite 2000; 34:197-205.
87. Lundy Jr RF, Norgren R. Gustatory system. In: Paxinos G, ed. The Rat Nervous System. USA: Elsevier, 2004:891-921.
88. Reed R. After the holy grail: establishing a molecular basis for mammalian olfaction. Cell 2004; 116:329-336.
89. Yoon H, Enquist LW, Dulac C. Olfactory inputs to hypothalamic neurons controlling reproduction and fertility. Cell 2005; 123:669-682.
90. Merchenthaler I, Sétáló G, Petrusz P et al. Identification of hypophysiotropic luteinizing hormone-releasing hormone (LHRH) neurons by combined retrograde labeling and immunocytochemistry. Exp Clin Endocrinol 1989; 94:133-140.
91. Witkin JW, Paden CM, Silverman AJ. The luteinizing hormone-releasing hormone (LHRH) systems in the rat brain. Neuroendocrinology 1982; 35:429-438.
92. Klass MR. Aging in the nematode Caenorhabditis elegans: major biological and environmental factors influencing life span. Mech Ageing Dev 1977; 6:413-429.
93. Kenyon C, Chang J, Gensch E et al. A C. elegans mutant that lives twice as long as wild type. Nature 1993; 366:461-464.
94. Larsen P, Albert PS, Riddle DL. Genes that regulate both development and longevity in Caenorhabditis elegans. Genetics 1995; 139:1567-1583.
95. Lin K, Dorman JB, Rodan A et al. daf-16: An HNF-3/forkhead family member that can function to double the life-span of Caenorhabditis elegans. Science 1997; 278:1319-1322.
96. Ogg S, Paradis S, Gottlieb S et al. The Fork head transcription factor DAF-16 transduces insulin-like metabolic and longevity signals in C. elegans. Nature 1997; 389:994-999.
97. Henderson ST, Johnson TE. daf-16 integrates developmental and environmental inputs to mediate aging in the nematode Caenorhabditis elegans. Curr Biol 2001; 11:1975-1980.
98. Lin K, Hsin H, Libina N et al. Regulation of the Caenorhabditis elegans longevity protein DAF-16 by insulin/IGF-1 and germline signaling. Nat Genet 2001; 28:139-145.
99. Lee RYN, Hench J, Ruvkun G. Regulation of C. elegans DAF-16 and its human ortholog FKHRL1 by the daf-2 insulin-like signaling pathway. Curr Biol 2001; 11:1950-1957.
100. Pierce SB, Costa M, Wisotzkey R et al. Regulation of DAF-2 receptor signaling by human insulin and ins-1, a member of the unusually large and diverse C. elegans insulin gene family. Genes Dev 2001; 15:672-686.
101. Hsin H, Kenyon C. Signals from the reproductive system regulate the lifespan of C. elegans. Nature 1999; 399:362-366.
102. Kodama E, Kuhara A, Mohri-Shiomi A et al. Insulin-like signaling and the neural circuit for integrative behavior in C. elegans. Genes Dev 2006; 20:2955-2960.
103. Larsson MC, Domingos AI, Jones WD et al. Or83b encodes a broadly expressed odorant receptor essential for Drosophila olfaction. Neuron 2004; 43:703-714.
104. Lin SJ, Defossez PA, Guarente LS. Requirement of NAD and SIR2 for life-span extension by calorie restriction in Saccharomyces cerevisiae. Science 2000; 289:2126-2128.

105. Clancy DJ, Gems D, Hafen E et al. Dietary restriction in long-lived dwarf flies. Science 2002; 296:319.
106. McCay CM, Cromwell MF, Maynard LA. The effect of retarded growth upon the length of life span and ultimate body size. J Nutr 1935; 10:63-79.
107. Brogiolo W, Stocker H, Ikeya T et al. An evolutionarily conserved function of the Drosophila insulin receptor and insulin-like peptides in growth control. Curr Biol 2001; 11:213-221.
108. Tatar M, Kopelman A, Epstein D et al. A mutant Drosophila insulin receptor homolog that extends life-span and impairs neuroendocrine function. Science 2001; 292:107-110.
109. Clancy DJ, Gems D, Harshman LG et al. Extension of life-span by loss of CHICO, a Drosophila insulin receptor substrate protein. Science 2001; 292:104-106.
110. Broughton SJ, Piper MD, Ikeya T et al. Longer lifespan, altered metabolism and stress resistance in Drosophila from ablation of cells making insulin-like ligands. Proc Natl Acad Sci USA 2005; 102:3105-3110.
111. Giannakou ME, Goss M, Jünger MA et al. Long-lived Drosophila with overexpressed dFOXO in adult fat body. Science 2004; 305:361.
112. Hwangbo DS, Gersham B, Tu M-P et al. Drosophila dFOXO controls lifespan and regulates insulin signaling in brain and fat body. Nature 2004; 429:562-566.
113. Lee SS, Kennedy S, Tolonen AC et al. DAF-16 target genes that control C. elegans life-span and metabolism. Science 2003; 300: 644-647.
114. Murphy CT, McCarroll SA, Bargmann CI et al. Genes that act downstream of DAF-16 to influence the lifespan of Caenorhabditis elegans. Nature 2003; 424:277-283.
115. Mak HY, Nelson LS, Basson M et al. Polygenic control of Caenorhabditis elegans fat storage. Nat Genet 2006; 38:363-368.
116. Wang MC, O'Rourke EJ, Ruvkun G. Fat metabolism links germline stem cells and longevity in C. elegans. Science 2008; 322:957-960.
117. Martin B, Maudsley S, White CM et al. Hormones in the naso-oropharynx: endocrine modulation of taste and smell. Trends Endocrinol Metab 2009; 20:163-170.
118. Chauhan J, Hawrysh ZJ, Gee M et al. Age-related olfactory and taste changes and interrelationships between taste and nutrition. J Am Diet Assoc 1987; 87:1543-1550.

Chapter 15

Regulation of Muscle Atrophy in Aging and Disease

Manlio Vinciguerra,* Antonio Musaro and Nadia Rosenthal

Abstract

Muscle aging is characterized by a decline in functional performance and restriction of adaptability, due to progressive loss of muscle tissue coupled with a decrease in strength and force output. Together with selective activation of apoptotic pathways, a hallmark of age-related muscle loss or sarcopenia is the progressive incapacity of regeneration machinery to replace damaged muscle. These characteristics are shared by pathologies involving muscle wasting, such as muscular dystrophies or amyotrophic lateral sclerosis, cancer and AIDS, all characterized by alterations in metabolic and physiological parameters, progressive weakness in specific muscle groups. Modulation of extracellular agonists, receptors, protein kinases, intermediate molecules, transcription factors and tissue-specific gene expression collectively compromise the functionality of skeletal muscle tissue, leading to muscle degeneration and persistent protein degradation through activation of proteolytic systems, such as calpain, ubiquitin-proteasome and caspase. Additional decrements in muscle growth factors compromise skeletal muscle growth, differentiation, survival and regeneration. A better understanding of the mechanisms underlying the pathogenesis of muscle atrophy and wasting associated with different diseases has been the objective of numerous studies and represents an important first step for the development of therapeutic approaches.

Among these, insulin-like growth factor-1 (IGF-1) has emerged as a growth factor with a remarkably wide range of actions and a tremendous potential as a therapeutic in attenuating the atrophy and frailty associated with muscle aging and diseases. In this chapter we provide an overview of current concepts in muscle atrophy, focusing specifically on the molecular basis of IGF-1 action and survey current gene and cell therapeutic approaches to rescue muscle atrophy in aging and disease.

Muscle Atrophy and Wasting Diseases

The most remarkable feature of skeletal muscle is the capacity to adapt its morphological, biochemical and molecular properties in response to several factors. Nonetheless, under pathological conditions skeletal muscle loses its adaptability, leading to atrophy or wasting. In particular, atrophy represents a rapid reversible process resulting from disuse, denervation (defined as the loss of nerve supply) or disease, whereas wasting involves an irreversible loss in muscle mass, associated with pathologies like cancer, acquired immune deficiency syndrome (AIDS) or chronic heart failure.

In all these cases, alterations in the metabolic mechanisms represent one of the principal components of the skeletal muscle pathogenesis. The continual synthesis and degradation of cell proteins

*Corresponding Author: Manlio Vinciguerra—Mouse Biology Unit, European Molecular Biology Laboratory (EMBL), Via Ramarini, 32, 00015 Monterotondo (Rome), Italy. Email: Manlio.Vinciguerra@embl.it

Protein Metabolism and Homeostasis in Aging, edited by Nektarios Tavernarakis.

is the result of normal intracellular metabolism and represents an important homeostatic function of muscle tissue. Muscle wasting, in contrast, is a process in which the delicate balance between anabolic and catabolic process is impaired.[1-3] Several pathologies such as muscular dystrophies, fasting, cancer, metabolic acidosis, denervation, AIDS, heart failure, diabetes, sepsis, burns, hyperthyroidism and excess glucocorticoids involve a general protein breakdown, promoting muscle atrophy, alteration in anabolic pathways and muscle wasting.

Molecular Mechanisms of Muscle Atrophy and Wasting

There are several proteolytic systems that are responsible for the protein breakdown and therefore muscle wasting. Calpain-, caspase- and ubiquitin- mediated protein degradation and autophagy/lysosome system are the principal pathways activated in several pathologies, leading to myofiber degeneration and impaired muscle regeneration.

The Calpain Pathway

Calpains are calcium-activated cysteine proteases that participate in various intracellular signal transduction pathways mediated by Ca^{2+}.[4] Calpains consist of two subunits, an 80kDa large subunit, which contains protease activity and a 30kDa smaller subunit, functioning as a regulator of calpain activity. Calpains are activated by several stimuli in which intracellular Ca^{2+} homeostasis is affected, causing disruption of the contractile tissue,[5] mitochondrial swelling, sarcoplasmic reticulum vacuolization[6] and sarcomeric alterations.[7]

Two ubiquitous calpain isoforms, ʃ and m, are well characterized by their relative Ca^{2+} concentration dependence and several other isoforms are selectively expressed in mammalian tissues. Among calpains, calpain 3 is predominantly expressed in skeletal muscle and maintains proteolytic activity at physiological Ca^{2+}.[8]

Calpains are preferentially localized in the Z disk of the sarcomere, where they initiate the proteolytic cleavage of muscle protein, causing the complete disassembly of myofibrils and loss of the Z disk.[9,10] The activity of calpains is further regulated by the endogenous inhibitor calpastatin, which prevents both enzyme activation and expression of catalytic activity. Noteworthy, the interactions occurring between calpastatin and calpains are finely modulated by intracellular Ca^{2+}concentration.

The calpain isozymes are associated with organelles and molecules involved in metabolic and signal transduction pathways, such as glycolytic enzymes, protein kinase A and C, phospholipase C and cytoskeletal protein, such as desmin, vimentin, integrin, cadherin, α-actinin.[11] The non-erythroid spectrin protein known as fodrin, which is a multifunctional protein and a major component of the cortical cytoskeleton of most eukaryotic cells including muscle cells, is another molecular target of calpains.[9,12] Cleavage of fodrin accompanies apoptosis induced by treatment of cells with staurosporine, glucocorticoid, or synthetic ceramide,[13,14] all factors that affect muscle physiology and myofibers survival.

In contrast to other proteolytic systems, calpains cleave target proteins at specific sites, leaving large polypeptide fragments with altered physiological properties. In addition, post-translational modifications provide a mechanism of "marking" specific proteins for calpain degradation. For instance, phosphorylation of Troponin T and I by protein kinase A and C dramatically alters the sensitivity of the protein to calpain degradation.[15]

The activity of calpains is also associated with neutrophil accumulation during exercise,[16,17] suggesting a role for calpains in muscle injury. Growing evidence revealed that the proteolytic activity of calpain is also associated with the pathogenesis of muscle diseases. Proteolysis by calpain causes muscle fiber degradation in Duchenne and Becker muscular dystrophies (DMD/BMD).[18,19] Moreover, alteration in the function of calpain3 was found associated with limb-girdle muscular dystrophy Type 2A,[20] suggesting that calpain3 is fundamental for normal muscle function.

Insights into the roles of Calpain 3 in the skeletal muscle came from studies using transgenic mice. Overexpression of full-length Calpain 3 in mice, under the control of human skeletal α-actin promoter is not toxic to muscle, whereas overexpression of Calpain 3 isoforms lacking exon 6 or exon 15 produces abnormal skeletal muscle due to defects in myoblast fusion, or defects in maturation after fusion.[21] Notably, the expression of a calpastatin transgene, the endogenous inhibitor of calpains, attenuates muscle wasting associated to muscle disuse and completely prevents the shift in fiber type from slow to fast, which normally occurs in muscle unloading.[22] Collectively these studies demonstrate that calpains play a key role in muscle homeostasis and implicate an additional role for alterations in intracellular Ca^{2+} concentration. Future efforts will be needed to test the feasibility of targeting Calpains in gene therapeutic applications and to identify their in vivo substrates.

The Caspase Pathway

The central component of the apoptotic machinery in several tissues is a proteolytic system involving a family of cysteine proteases called caspases.[23] Activation of caspase pathway is related to several diseases, including cancer and muscular dystrophies. Caspases are all expressed as inactive proenzymes with a molecular weight between 50 and 33 kDa and are activated after cleavage at the aspartate residues, generating the active product with lower molecular weight.

Caspases are classified in two major groups: the initiator caspases that initiate the proteolytic cascade and the effector caspases that cleave target protein triggering the apoptotic pathway. A key role of caspases is to inactivate proteins that protect living cells from apoptosis. It has been reported that the cleavage of ICAD,[24-26] an inhibitor of the caspase-activated deoxyribonuclease (CAD), leads to DNA fragmentation. In non-apoptotic cells, CAD is present as an inactive complex with ICAD. Under apoptotic stimuli, caspases inactivate ICAD leaving CAD free to fragment DNA.

In addition, caspases cleave structural proteins of the nucleus and cytoskeleton. Proteolysis of lamins triggers the destruction of nuclear lamina and allows chromatin condensation.[27-29] Desmin is specifically cleaved by caspase 6, generating a cleavage (N-desmin) product which functions as dominant negative inhibitor unable to assemble into intermediate of filament and activating apoptotic pathways.[30] Similarly, the caspase-dependent cleavage of gelsolin, a widely distributed actin binding protein involved in controlling cell morphology, motility, signaling and apoptosis, generates a fragment constitutively active and promoting apoptosis.[31]

Caspase-mediated apoptosis has been associated different muscular diseases. Caspase 3 expression and activity correlates with apoptosis in Duchenne and facioscapulo-humeral dystrophy.[32] In addition, cleavage of β-catenin and focal adesion kinase (FAK) family may interrupt cell-cell contacts and cell-matrix focal adhesions. Moreover, cleavage of regulatory and effector proteins, such as proteins involved in DNA repair, mRNA splicing and DNA replication,[33,34] leads to alteration in cellular functions and activation of muscle wasting mechanisms. Caspases are also involved in the catabolic processes of skeletal muscle associated with cancer cachexia.[35]

The Ubiquitin-Proteasome and the Autophagy-Lysosome Systems

The ubiquitin-proteasome and the autophagy-lysosome pathways play a key role in the turnover of muscle protein and are activated in several catabolic processes leading to muscle wasting.[36] These two systems are coordinately regulated to preserve a normal composition of proteins and organelles in atrophying cells.[37-39] The two pathways are believed to serve separate functions. Proteasomes degrade myofibrillar and short-lived proteins,[40-43] whereas autophagy-lysosomes are believed to control long-lived proteins and organelles.[44,45] Both systems require ATP to activate small molecules: ubiquitin and ubiquitin-like proteins. Different classes of enzymes catalyze the reaction of activation (E1 protein) and the transfer of the small ubiquitins to the conjugation system (E2 proteins). In the ubiquitin-proteasome system, the final step, regulated by enzymes called E3 or ubiquitin ligases, is the transfer of ubiquitin from the conjugation system to the protein leading to polyubiquitination and targeting to the proteasome for degradation.[46]

By contrast in the autophagy/lysosome system, small ubiquitin-like molecules (LC3 and others) are transferred from the conjugation system to membranes for their growth and commitment to become a double membrane vesicle (autophagosome) that engulfs portions of cytoplasm.[44,45] This reaction requires the recruitment and assembly of different components of the autophagy machinery on phospholipids but only the ubiquitin-like components, such as LC3, are covalently bound to phosphatidylethanolamine.[47,48] This covalent bound occurs both on the outer and inner membranes of the autophagosome. Sequestered organelles and proteins are then docked to the lysosomes for their degradation. The fusion of the outer membrane of the autophagosome with the lysosomal membrane also triggers the degradation of the inner membrane and of the associated proteins. Thus, the main difference between the two systems is related to the fate of the ubiquitin and ubiquitin-like proteins. While the ubiquitin proteasome pathway recycles ubiquitin molecules, the autophagy-lysosome system progressively loses the ubiquitin-like proteins, forcing the cell to replenish them in order to maintain the autophagic flux.

Modulators of Muscle Catabolic Pathways

Key components of the ubiquitin and autophagy systems are transcriptionally upregulated during muscle wasting and their induction precedes muscle loss. However, it is unclear whether autophagy is detrimental, causing atrophy, or beneficial, promoting survival during catabolic conditions. Activation of the ubiquitin-proteasome pathway induces muscle atrophy associated with several pathologies. Several recent findings also suggest that excessive autophagy promotes severe wasting during denervation. Moreover, autophagy-lysosome and ubiquitin proteasome systems are coordinately regulated during muscle wasting,[37,39] in part through FoxO transcription factors. Transcription of autophagy related genes, such as LC3B and Bnip3, is mediated by FoxO3. In particular, Bnip3 induces autophagosome formation and is responsible for the induction of autophagy by FoxO3. FoxO3 activates both the ubiquitin-proteasomal system and the autophagic/lysosomal pathway, but is able to control the two pathways independently. Protein breakdown via both the proteasome and lysosome is likely to proceed in a coordinated manner in the activation of muscle atrophy, in which the degradation of myofibrillar proteins via the proteasome proceeds in parallel with the disposal of mitochondria and sarcoplasmic reticulum membranes via the autophagic/lysosomal pathway. Trancriptional targets of FoxO are the atrophy-related ubiquitin ligases atrogin-1/MAFbx and MuRF-1. MuRF1 interacts with titin, a giant sarcomeric protein.[49] Overexpression of MuRF1 disrupts the integrity of sarcomeric structure and alters the thick filament components, whereas overexpression of atrogin-1/MAFbx in myotubes produces atrophy. Notably, mice deficient in either MAFbx or MuRF1 are resistant to atrophy,[50] suggesting that MuRF1 and atrogin-1/MAFbx are potential targets for the development of appropriate strategies to reduces muscle atrophy and degeneration associated with the activation of ubiquitin-proteasome pathway.

Other studies have shown that Cathepsin-L, a lysosomal protease that represents a critical marker of the catabolic pathway, is upregulated in different models of muscle wasting.[51,52] The role of Cathepsin-L induction is still unclear, but recent evidence suggests that it may have a role in degradation that is independent of lysosomal function.[53] Other components of the autophagy machinery are induced during muscle wasting and are involved in regulation of autophagy. Thus, under atrophic conditions, MuRF1 promotes degradation of myofibrillar proteins, whereas the autophagic pathway affects cellular organelles.

Is the activation of autophagy machinery always detrimental? It has been recently demonstrated that inhibition of autophagy induces atrophy and myopathy in the skeletal muscle,[54] raising the hypothesis that the inhibition of autophagy might cause the aggregation of altered proteins and accumulation of abnormal organelles, which in turn would activate detrimental signals. These results suggest that the regulation of the proteolytic systems must be finely balanced.

Another mechanism that is often altered in different atrophic and wasting conditions is the anti-oxidative defense through enzymatic destruction of accumulated reactive oxygen species (ROS).[55] A direct role for oxidative stress on muscle catabolism was demonstrated in a novel transgenic mouse model expressing a mutant form of the SOD1 gene (SOD1G93A) under the transcriptional control of muscle-specific regulatory elements (MLC).[56] The mice developed progressive muscle atrophy, associated with a significant reduction in muscle strength, alterations in the contractile apparatus and mitochondrial dysfunction. ROS accumulation in this model signaled the initiation of autophagy, one of the major intracellular degradation mechanisms and

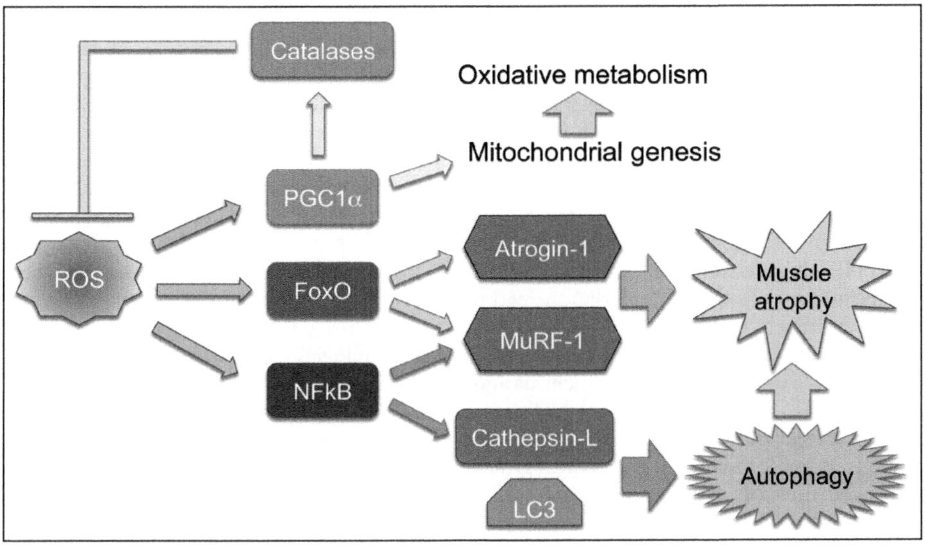

Figure 1. Model of muscle atrophy through ROS production. ROS activates FoxO, NFκB and autophagic signalling. While FoxO modulates both atrogin-1/MAFbx and MuRF1, NFkB acts selectively on MuRF1. The action of FoxO is partly inhibited by PGC-1α, which also induces and coordinates gene expression that stimulates fiber-type switching and metabolic pathways, reflected in the shift from glycolytic to oxidative metabolism in atrophic muscle fibers. The NFκB pathway also increases cathepsin L that together with LC3 represents a critical marker of autophagy.

a key determinant for the induction of muscle atrophy associated with SOD1 mutation, disclosing a novel molecular mechanism whereby oxidative stress triggers muscle atrophy. Accumulation of ROS also activated FoxO, which modulates both atrogin and MuRF1 and is partly inhibited by PGC-1α, a factor that is induced by oxidative stressors and that represents an important regulator of intracellular ROS levels (Fig. 1). In addition, ROS accumulation activated NFkB, which acts selectively on MuRF1 and also increases Cathepsin L. A critical role for autophagy in the promotion of muscle atrophy in this mouse model was disclosed by genetic manipulation of LC3; in vivo electroporation of siRNA against LC3 gene rescued the atrophic phenotype. These results pinpoint modulation of the autophagy pathway is a potential therapeutic intervention to counteract muscle atrophy associated with oxidative stress.

Systemic Effectors of Muscle Atrophy

Among hormones and cytokines that affect muscle homeostasis, glucocorticoids play a pivotal role in muscle degeneration by stimulating proteolysis and inhibiting protein synthesis and are important factors in the development of muscle atrophy in various catabolic conditions.[57] Glucocorticoids also induce an increase in ubiquitin transcript levels[58] and enhance the conversion of amino acids to glucose in the liver. Glucocorticoid-stimulated muscle protein breakdown is therefore primarily mediated by ubiquitin-proteasome-dependent proteolysis, although calcium-dependent protein degradation may also be involved. In certain catabolic conditions, including sepsis, an interaction between glucocorticoids and proinflammatory cytokines is also implicated in the exacerbation of muscle protein breakdown.

Several studies suggest that catabolism is mediated by factors that are connected with systemic inflammation. Many of the perturbations associated with cancer cachexia can be reproduced by the administration of proinflammatory cytokines, such as interleukin-1 (IL-1), interleukin-6 (IL-6), tumor necrosis factor-α (TNF-α) and interferon-γ (IFN−γ).[2,59,60] Cytokines infused or injected into animals decrease skeletal muscle protein mass through increased rates of protein degradation and decreased protein synthesis.[60-62] In addition, TNF-α plays a pivotal role in inhibiting skeletal muscle differentiation, regeneration and inducing muscle wasting.[63-65] One hallmark of TNF signaling is the activation of NFκB, ubiquitous transcription factor normally inactive and sequestered in the cytoplasm through association with IκB. A variety of stimuli, including TNF exposure, lead to the degradation of IκB, allowing NFκB translocation to the nucleus.[66] In parallel with the NFκB system, the transcription factor PW1 participates in the TNF-α signal transduction pathway.[67,68] More recent studies show that TNF-α inhibits muscle differentiation and induces muscle wasting through the activation of caspases and that these effects of TNF-α are dependent upon the presence of PW1 expression.[63-65,69] Taken together, these observations suggest that systemic activation of specific catabolic factors underlies the pathogenesis of several skeletal muscle diseases, leading to muscle atrophy and wasting.

IL-6 is an important cytokine involved in several pathologies characterized by a progressive loss of muscle tissue such as aging, cachexia and muscular dystrophy. Recent studies have demonstrated that IL-6 local infusion into tibialis anterior muscle of healthy rodents results in a decrease of muscle total and myofibrillar fraction protein content, contributing to the development of a strongly atrophic phenotype.[70] Interestingly, high circulating levels of IL-6 causes growth defects associated with a decrease in circulating IGF-I levels, indicating a possible interference with the IGF-1-anabolic pathway.[71]

Muscle Cytoskeletal Disorders and Atrophy

The majority of muscle pathologies are characterized by the progressive loss of muscle tissue due to chronic degeneration combined with the inability of regeneration machinery to replace damaged myofibers. The molecular organization of muscle structure depends on cytoskeletal proteins, which play important roles not only in the maintenance of muscle integrity but also in the modulation of interacellular signal transduction pathways.

The intricate molecular organization of cytoskeleton reflects the complexity of muscle types and the intimate relationship between structural integrity and mechanical function. The muscle cytoskeletal proteins can be subdivided in four major groups: the contractile sarcomeric, the intra-sarcomeric, the peri-sarcomeric and the sub-sarcolemmal proteins. The contractile sarcomeric cytoskeleton is composed of actin and myosin myofilaments and represents the functional core of muscle contraction. The intra-sarcomeric cystoskeleton comprises titin, nebulin, tropomyosin and actinin; the peri-sarcomeric cytoskeleton is composed of desmin-intermediate filament, whereas the sub-sarcolemmal cytoskeleton includes sarcolemma-associated protein, such as dystrophin, vinculin, integrins, α-actinin and ankyrin. Among the proteins that are associated with the plasma membrane, the dystrophin-glycoprotein complex (DGC) plays a critical role in the connection of the cytoskeleton of a muscle fiber to its surrounding extracellular matrix.

Among pathologies leading to muscle wasting, muscular dystrophies are the ones that most severely compromise the functional performance of skeletal muscle fibers, by progressive weakening in specific muscle groups, by persistent protein degradation and by cell-based alterations in regenerative capacity.[72,73] Mutations in genes encoding proteins of the dystrophin-glycoprotein complex (DGC) lead to alteration in muscle structure and cause muscular dystrophy.[72-76] The most common form of the disease is Duchenne muscular dystrophy (DMD) in which dystrophin gene is absent. Dystrophin is part of the DGC that connects the cytoskeleton of a muscle fiber to the surrounding extracellular matrix through the cell membrane. Without dystrophin, the DGC is unstable, leading to increase in muscle damage. Alterations of other components of the DGC and of the cytoskeleton are also linked to muscular dystrophy. For instance, mutations in the sarcoglycan complex,[77] laminin-2[78] and α7-integrin[79,80] genes display several phenotypic traits of muscular dystrophy, suggesting that loss of the link between extracellular matrix and cytoskeleton represents the critical parameter for the maintenance of the structural integrity of skeletal muscle.

Recently, a new convergence between the biology of muscular dystrophy and muscle atrophy has been reported. It is known that Nitric Oxide (NO) signaling is dysregulated in muscular dystrophy due to the disruption of the dystrophin glycoprotein complex (DGC), which anchors neuronal nitric oxide synthase (nNOS) to the plasma membrane: it has been recently shown that nNOS is similarly perturbed in a setting of skeletal muscle atrophy.[81] When it is not tethered from the DGC, as for example in hind-limb suspension that causes atrophy, the nNOS is free to enter the cellular compartment and to potentiate FoxO-mediated transcription and thus up regulate expression of the E3 ligases MAFbx and MuRF-1.[81] Thus, muscular dystrophy is also associated with muscle atrophy and this could clarify why dystrophin is lost from the cell membrane under cancer associated cachexia, leading to atrophy.[82]

In addition to their structural role, the cytoskeletal proteins function as mediators of several signal transduction pathways, including those modulated by calcium. Calcium (Ca^{2+}) is an important intracellular messenger in muscle, controlling numerous cellular process including proliferation, cell growth, differentiation and apoptosis.[83,84] All muscle fibers use Ca^{2+} as their main regulatory and signaling molecule.[85] However in DMD the Ca^{2+} homeostasis results altered, suggesting that dystrophin, which contributes to muscle structural integrity, also regulates intracellular Ca^{2+} levels.[86] In the absence of dystrophin, alterations in intracellular Ca^{2+} may contribute to an imbalance between muscle protein synthesis and protein degradation, culminating in necrosis, fibrosis and shift in fiber content.

Notably, alteration of Ca^{2+} signal may render skeletal muscle more susceptible to free radical-induced injury, exacerbating the pathogenesis of different muscle diseases, such as muscular dystrophy and familiar amyotrophic lateral sclerosis (FALS).[22,56,87-95]

Cellular and Molecular Bases of Muscle Repair

The lack of regeneration of the muscle tissue is another serious complication in several pathologies, leading to the progression of muscle disease. In particular, the imbalance between muscle damage and muscle repair leads to a loss of muscle fibers and an increase in the amount of fibrosis, which compromises the functional and structural organization of skeletal muscle.

Regeneration is a coordinate process in which several factors are sequentially activated to maintain and preserve tissue structure and function upon injured stimuli. Striated muscle is a postmitotic tissue and this limits its ability to repair or to regenerate. Although adult skeletal muscle is composed of fully differentiated fibers, it retains the capacity to regenerate in response to injury and to modify its contractile and metabolic properties in response to changing demand.[96] Regeneration is therefore an important homeostatic process, which guarantees the maintenance of muscle integrity and plasticity.

Muscle regeneration following injury involves three different phases: a local inflammatory response, infiltration of the injured area by blood borne cells and activation of satellite cells. The events that lead to an inflammatory response are characterized by recognition of the site of injury by inflammatory cells, specific recruitment of subpopulations of leukocytes into tissue, removal of the necrotic cells and repair of the site of injury with attempts to reestablish normal parenchymal, stromal and extracellular matrix relationships. It has been shown that chronic inflammatory response dominates the skeletal muscle molecular signature in dystrophin-deficient mdx mice.[97] In fact the inflammatory response contributes conspicuously to dystrophinopathies and exacerbates the clinical disease.[87,98-101] Yet inflammation represents one of the early responses following normal muscle injury, leading to activation of muscle regeneration process (reviewed in ref. 102).

Skeletal muscle retains a resident population of quiescent satellite cells, which are the major source of myogenic precursors in mammalian muscle regeneration. Satellite cells are of mesodermic origin and are found underneath the basal lamina of muscle fibers, closely juxtaposed to the plasma membrane. They represent 2-6% of all nuclei of a given muscle fiber. Satellite cells reside between the basal lamina and plasmalemma and are rapidly activated in response to appropriate stimuli.[103] Once activated, satellite cells proliferate and up-regulate genes involved in the progression of cell cycle including DNA replication factors and in the commitment of the myogenic program such as Pax7 and MyoD.[103] Transition from cell proliferation to differentiation is characterized by expression of several genes induced during skeletal muscle differentiation, including regulatory genes myogenin and MRF4 and contractile proteins such as MyHC, actin, troponin and tropomyosins.[104] However, their number decreases substantially during aging and in muscular dystrophy,[103,105-107] leading to altered regenerative capacity of skeletal muscle. What it appeared clear in the last years is that in order to massively activate satellite cells the muscle should be injured by necrotic and not apoptotic stimuli. Indeed, a dystrophic muscle is able to activate satellite cells, although this activation is defective in the maturation process. In contrast, atrophic muscle, as for example that associated with aging, cachexia, ALS, heart failure, is not able to activate satellite cells, unless a local IGF-1 isoform is expressed,[108] as discussed further, or the Notch pathway is activated.[109]

Although in recent years considerable evidence has been accumulated regarding muscle physiopathology, it is still an open question what molecular mechanisms regulate the phenotypic changes leading to the pathological pattern of skeletal muscle atrophy. It is generally accepted that the primary cause of functional impairment in muscle is a cumulative failure to repair damage related to an overall decrease in anabolic processes. Aging, cancer, AIDS, heart failure and genetic myopathies are all characterized by alterations in metabolic and physiological parameters and the inability to regenerate and repair the injured muscle represents a serious complication in such pathologies.

Despite theses limitations in our knowledge of basic mechanism, cumulative evidence suggests that manipulating appropriate pathways comprise promising therapeutic approaches in staving off advancing muscle weakness associated with aging, muscle pathologies and several other diseases. In this context several factors have been reported to have an effective role in the maintenance of muscle structure and in the promotion of muscle regeneration in different pathological conditions. Programs of muscle regeneration are naturally mediated through growth factors, hormones and cytokines, which are released from injured tissues. Hepatocyte growth factor (HGF) transforming-growth factor-β (TGF-β) and fibroblast growth factor (FGF) each play important roles in the induction and modulation of the myogenic program during muscle regeneration.[110]

More recently, one of the members of the TGF-β family, myostatin, has been described as a negative regulator of skeletal muscle mass and regeneration and is also involved in muscle wasting associated with muscular dystrophy.[111-115] It has been reported that inhibition of the myostatin product, a member of the TGF-β family, ameliorates the dystrophic phenotype in the mdx mouse model of DMD[114,116,117] and the sarcopenia in a mouse model of aging.[118] Administration of a myostatin-neutralizing antibody for three months resulted in an increase in muscle mass and strength with reduction in muscle degeneration.[116] Although with some disease-specific controversies,[119,120] these studies support the concept that prolonging of the regenerative potential of pathological muscle by treating the secondary symptoms of the disease, using agents that circumvent the gene replacement dilemma, represents an alternative strategy.

Recent studies have also implicated NF-κB, a major pleiotropic transcription factor modulating immune, inflammatory, cell survival and proliferative responses, in the modulation of muscle atrophy.[121] Targeted deletion of the activating kinase inhibitor of NF-κB kinase 2 (IKK2) resulted muscle-restricted NF-κB inhibition in mice, which shifted muscle fiber distribution and improved muscle force. In response to denervation, IKK2 depletion protected against atrophy, maintaining fiber type, size and strength, increasing protein synthesis and decreasing protein degradation. In response to muscle damage, IKK2 depletion facilitated skeletal muscle regeneration through enhanced satellite cell activation and reduced fibrosis. As NF-κB signaling is an important modulator of muscle homeostasis, the potential benefits of NF-κB blockade in augmenting muscle performance raise exciting possibilities for therapeutic approaches in the treatment of muscular disorders and for the development of strategies to attenuate muscle wasting.[122]

Other early events involved in the repair of injured muscle are triggered by cytokines. Localization of leukemia inhibitory factor (LIF) and inteleukin-6 (IL-6) are produced by diseases and regenerating skeletal muscle.[123-127] In LIF knockout mice muscle regeneration is significantly reduced compared to control littermates.[123,126,127] Interleukin-15 (IL-15) has been found increased in the skeletal muscle during aging and unloading:[128] IL-15 induction was interpreted as an attempt to counteract muscle mass loss in skeletal muscles upon atrophic stimuli.[128] Interestingly, administration of the IL-10 gene by electrotransfer of naked DNA has anti-inflammatory effect that is applicable to the treatment of rheumatoid arthritis.[129] Conversely, nitric oxide (NO) can function as an anti-inflammatory and cytoprotective molecule, preventing muscle membrane injury and accumulation of creatine kinase concentrations. Expression of a nitric oxide synthase transgene (NOS) ameliorated muscular dystrophy in mdx mice[130] and elegant studies have unvealed the molecular targets of NO at the chromatin level in the frame of DMD.[131,132]

Another signal that could be a potent stimulus for protein degradation by the ubiquitin-proteasome and autophagy-lysosome pathways is a decreased response to growth factors involved in the activation of survival pathways. Insulin-like growth factor-1 (IGF-1) has been implicated in many anabolic pathways in skeletal muscle, where it plays a central role during muscle regeneration.[108,133-135] IGF-1 expression declines with age, which is associated with osteopenic and sarcopenic disorders.[136-138] This decrement is likely to be causally linked to the progress of muscle atrophy in senescence, with upregulation of atrogin-1/MAFbx and MuRF1 levels,[139] thereby limiting the ability of skeletal muscle tissue to sustain regeneration and repair. As a result, considerable effort has been focused on IGF-1 as a promising target for therapeutic intervention. The following section reviews our current understanding of the IGF-1 signaling system, its role in muscle atrophy, wasting and regeneration.

IGF-1 and Prevention of Muscle Atrophy

Over three decades ago, it was discovered that growth hormone (GH) did not directly stimulate the incorporation of sulfate into cartilage, but rather acted through a serum factor, which is predominately IGF-1 in adults. IGF-1 shares 62% homology with pro-insulin but mature IGF-1 differs further from insulin by the retention of the C-domain and the extension of the A-domain to include an extra D-domain (Fig. 2). The IGF-1 gene also encodes for multiple precursor proteins with N- and C-terminal variability.

IGF-1 is a secreted protein, with a blood half-life of less than 10 minutes and is usually stabilized by forming complexes with carrier proteins, the IGFBPs, which impose a complex level of regulation of IGF-1 bioavailability. IGF-1 actions are mediated through binding to its receptor, a heterotetrameric transmembrane glycoprotein complex belonging to the receptor tyrosine kinase family. The IGF-1 receptor is closely related to the insulin receptor but each has significantly different affinities for its cognate ligand.[140]

Studies on mice lacking the IGF-1 gene have shown that normal IGF-1 expression is critical for normal growth and tissue development: both pre and postnatally IGF-1 contributes more to the total body weight then GH alone (35% versus 14%). Mice deficient for IGF-1 have a birthweight of only 60% of that of their wild-type (wt) littermates and most of them die around birth due to respiratory failure. Depending on the genetic background only a small percentage survives until adulthood and adult IGF-1 KO mice are infertile and severely growth retarded, reaching only 30% of the normal adult body weight.[134,141] When the IGF-1R is genetically removed, mice are even more profoundly affected: the birthweight of IGF-1R KO animals is only 45% of that of their wt littermates and they all die immediately after birth because of respiratory failure. The essential role IGF-1 plays in somatic growth is also illustrated by the finding that one single-nucleotide polymorphism in the IGF-1 gene is responsible for the size difference between strains of small and large dogs.[142] While these results firmly establish the importance of IGF-1 in somatic growth and its interplay with GH in this process, it remains unclear how the participation of endocrine, liver-derived IGF-1 was in the promotion of growth differs from the effects of locally produced, IGF-1 presumably acting in an auto- and paracrine fashion on tissues such as skeletal muscle.

The Molecular Complexities of IGF-1 Transcription

An impressive body of knowledge has been accumulated since the IGF-1 locus was first described, but surprisingly the potential diversity of roles played by different IGF-1 isoforms has only recently been appreciated. Although the IGF-1 gene is highly conserved in numerous species, its relatively large size (>70 kb) and its complex transcriptional and splicing pattern, have complicated analysis of its function. As its name implies, the IGF-1 mature protein is similar to insulin in structure, sharing a 50% amino acid identity. However, unlike the insulin gene, the single-copy

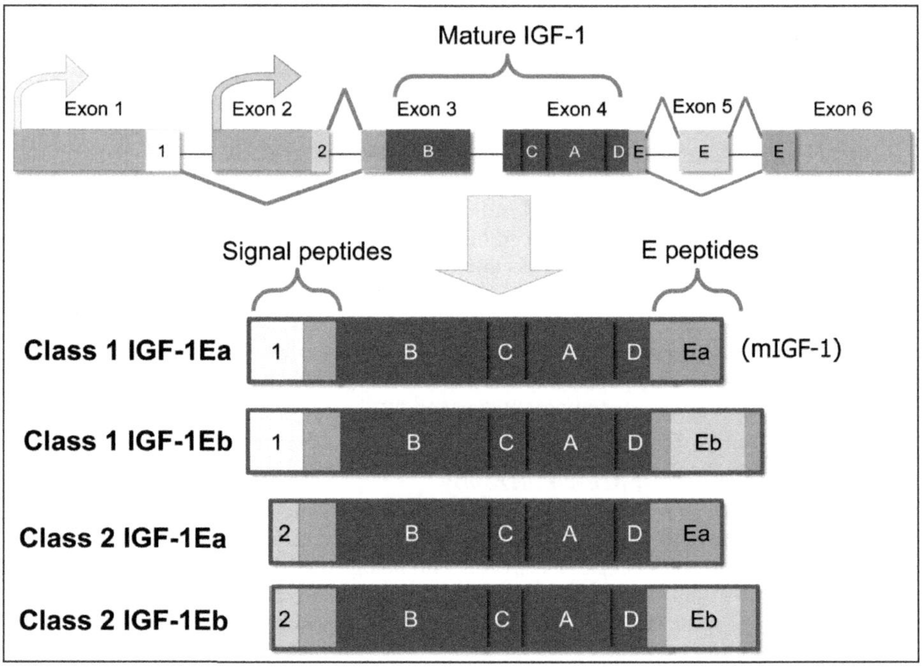

Figure 2. Structure of the mammalian IGF-1 gene locus. The IGF-1 gene contains six exons that encode multiple isoforms, all of which include the core IGF-1 protein body (dark red boxes). Both exons 1 and 2 contain multiple transcription start sites (horizontal arrows). Exons 1, 2 and 3 code for the signal peptide of precursor IGF-1 (orange boxes). Exons 5 and 6 each encode distinct portions of the E-peptides (pink and green boxes). Class 1 IGF-1Ea is also termed mIGF-1. A color version of this figure is available at www.landesbioscience.com/curie.

IGF-1 gene locus encodes multiple propeptides with variable amino- and carboxy-terminal amino acid sequences. The amino acid sequence of mature IGF-1 differs from that of insulin by retention of the C peptide, by a short extension of the A chain to include a novel domain D and by the presence of variable C-terminal E peptides. In mammals, these are encoded by six exons, separated by five introns (Fig. 2). Exons 1 and 2 encode distinct 5′UTRs, as well as variable parts of the signal peptide, whereas Exon 3 encodes 27 amino acids that are part of the signal peptide and common to all isoforms, as well as part of the mature IGF-1 peptide. Exon 4 encodes the rest of the mature peptide and 16 amino acids of the amino-terminal region of the E-peptide, which is also common to all IGF-1 mRNAs. Exons 5 and 6 encode two distinct carboxy-terminal E-peptides and the 3′UTR (Fig. 2).

Although IGF-1 transcripts are not exclusively tissue-restricted, those that initiate at Exon 2 predominate in the liver, are highly growth hormone responsive and as such are considered major endocrine effectors of GH.[143] By contrast, transcripts initiating at Exon 1 are widely expressed in all tissues and according to some reports are less affected by circulating growth hormone levels, presumably performing autocrine or paracrine functions as well. The alternate splicing at the 5′ ends of these two IGF-1 transcripts generates different signal peptides, which purportedly affects the precise N-terminal pro-peptide cleavage site.[143] The functions of the different signal peptides encoded by these transcripts are widely debated.[144] Recent results from our laboratory suggest that proteins encoded by transcripts originating with Exon 2 are dispensable for maintaining normal circulating IGF-1 levels in vivo, as their depletion results in a concomitant upregulation of liver transcription from Exon1 (Temmerman and Rosenthal, ms. submitted).

Elucidation of IGF-1 isoform function is also complicated by alternate splicing at the 3′ end of IGF-1 transcripts. This produces variability in the length and amino acid sequence of the E peptide and in the length and base sequence of the 3'UTR. To date, two prominent splice patterns have been documented in rodents (Fig. 2). Each generates E peptides with a common N-terminal 16 aa sequence and alternate C-terminal sequences.[145] If Exon 4 splices to Exon 6 (the predominant pattern), the length of the 3'UTR is highly variable, but in all cases the Ea peptide is generated with 19 additional amino acids. If Exon 4 splices to Exon 5 and 6, a variant known as Eb is encoded, which is frameshifted relative to Exon 6 and therefore a different 25 aa sequence is added to the common 16 aa encoded by Exon 4. Although E peptide choice appears to be independent of promoter use, Eb-containing transcripts are more prevalent in liver than in other tissues, whereas Ea-containing transcripts are abundant and widespread. In the mouse, two N-linked glycosylation sites occur only in the Ea peptide, suggesting that this post-translational modification is involved in a biological action of the IGF-1 isoform.[146]

The determination of E peptide function and fate is still under investigation. It is unclear when or whether E peptides are cleaved from the mature IGF-1 protein. Recent cell culture studies where cells were transfected with IGF-1 isoforms lacking or including an E peptide suggest that whereas E peptides are not required for IGF-1 secretion, they may modulate cell entry of IGF-1 from the media.[147] After injury to skeletal muscle, the IGF-1Eb isoform is expressed initially, followed by an upsurge of the IGF-1Ea isoforms at later time points; the peak of IGF-1Eb transcripts coincides with satellite cell and myoblast proliferation, whereas of IGF-IEa mRNA is correlated with differentiation to mature myofibers (reviewed in refs. 145,148). Due to the increase in IGF-1Eb transcripts during muscle remodeling, IGF-1Eb gene products have been collectively referred to as mechano-growth factor (MGF). A synthetically manufactured peptide corresponding to the 24 C-terminal residues of the Eb peptide has been prepared to promote cellular proliferation and survival, however, since an analogous peptide product has not been identified in cultured muscle cells or conditioned medium, nor animal tissues or biological fluids, the relevance of these findings in vivo remains unsubstantiated (reviewed in ref. 149). Notably, E peptide splicing patterns are different in the human gene,[145] an anomaly that will need to be considered in the future when translating the results of animal research into clinical applications.

IGF-1 in Skeletal Muscle Growth, Disease and Wasting

The fact that IGF-1 can act either as a circulating hormone or as a local growth factor requires careful experimental design to discriminate between these two roles. As an endocrine factor, the binding of IGF-1 to its receptor controls the hierarchy between the signals which mediate hypertrophy and those which mediate atrophy; IGF-1 signaling prevents expression of muscle atrophy-induced ubiquitin ligases atrogin 1/MAFbx and MuRF1[150] through the canonical PI3K/Akt/mTOR pathway, which is known to attenuate atrophy.[50]

Investigations into the paracrine functions of IGF-1 in skeletal muscle have been facilitated by generating transgenic mice expressing IGF-1 propeptides driven by skeletal muscle-restricted regulatory elements (MLC).[108,144] These transgenes express as early as E9.5 days in embryonic mouse development and persist in the fastest Type IIb fibers. Transgenic animals expressing the IGF-1Ea isoform (termed mIGF-1) exhibited marked skeletal muscle hypertrophy with no undesirable side effects such as tumor formation. Over-expression of this isoform did not lead to an increase in circulating IGF-1 (in contrast to other transgenic mouse lines overexpressing mature IGF-1; 151). The increased muscle mass in mIGF-1 transgenic mice was associated with augmented force generation compared to age-matched wild type littermates.[108] Examination of two year-old animals revealed that whereas wild type mice underwent characteristic muscle atrophy, expression of the mIGF-1 transgene was protective against normal loss of muscle mass during senescence. These studies implicate mIGF-1 as a powerful enhancer of muscle repair, modulating the inflammatory response,[135] which plays an important role in the regenerative process (reviewed in ref. 102).

Studies on the downstream effectors of IGF-1 transgenes have also implicated a naturally occurring splicing variant of the calcineurin A(CnA) catalytic subunit (CnA1) in which the autoinhibitory domain that controls enzyme activation is replaced with a unique C-terminal region. The CnA1 enzyme is constitutively active and dephosphorylates its NFAT target in a cyclosporine-resistant manner.[152] CnA1 is highly expressed in proliferating myoblasts and regenerating skeletal muscle fibers and is activated by both IGF-1Ea (mIGF-1) and IGF-1Eb isoforms. In myoblasts, CnA1 knockdown results in the activation of FoxO-regulated genes, reduces proliferation and induces myoblast differentiation. Conversely, CnA1 overexpression inhibits FoxO and prevents myotube atrophy. Supplemental CnAβ1 transgene expression in skeletal muscle led to enhanced regeneration, reduced scar formation and accelerated resolution of inflammation.[152] This unique mode of action distinguishes the CnAβ1 isoform as potent effector of IGF-1 action and a candidate for interventional strategies in muscle wasting treatment.

In accelerating the repair of injured skeletal muscle in transgenic mice, expression of muscle-specific IGF-1 transgenes preserves the regenerative capacity of muscle tissues by stimulating both the activity of satellite cells and the recruitment of circulating stem cells.[108,153] In addition, local expression of mIGF-1 increases recruitment of proliferating bone marrow cells to injured muscles, accompanied by elevated bone marrow stem cell production in response to distal trauma. Correspondingly, increased cell populations expressing stem cell markers were found in regenerating mIGF-1 transgenic muscles, which exhibited accelerated myogenic differentiation, expressed markers of regeneration and readily converted cocultured bone marrow to muscle.[153]

To test whether mIGF-1 can prevent or diminish muscle loss associated with diseases, mIGF-1 was introduced into the mdx dystrophic animals by crossing the two lines (mdx/mIGF-1).[154] Measurement of muscle morphology and function in transgenic mdx/mIGF-1 mice revealed significant improvement in muscle mass and strength, a decrease in myonecrosis and a reduction in fibrosis in aged diaphragms. In addition, signaling pathways associated with muscle regeneration and protection against apoptosis were significantly elevated in mIGF-1/mdx mouse muscles. Importantly, a net decrease in fibrosis in diaphragms of the mdx/mIGF-1 mice was observed, suggesting that the efficient and rapid repair of the mdx/mIGF-1 muscles prevents the establishment of an environment into which the fibroblasts migrate. This is of particular relevance to the human dystrophic condition where virtually all skeletal muscles succumb to fibrosis.

Elevated muscle wasting is a common complication of advanced congestive heart failure and an important predictor of poor outcome in heart failure patients. Metabolic abnormalities that develop during congestive heart failure lead to progressive catabolism with decrements in the skeletal musculature that result in muscle atrophy. In experimental mouse model of chronic left ventricular dysfunction, skeletal muscle atrophy developed in wildtype mice 12 weeks following myocardial infarction accompanied by an increase in total protein ubiquitination and enhanced proteasome activity, activation of FoxO transcription factors and robust induction of atrogin-1/MAFbx.[155] Further studies identified skeletal muscle myosin as a specific target of ubiquitin-mediated degradation in muscle atrophy. When heart failure was induced in an MLC/mIGF-1 transgenic background, muscle atrophy was blocked, accompanied by skeletal muscle de-activation primarily of FoxO4 and reduced expression of atrogin-1/MAFbx. Thus mIGF-1 supplementation acts as a blockade of the ubiquitin-proteasome pathway in skeletal muscle atrophy that is produced secondarily in response to chronic disease.

Heart failure can be experimentally induced in mice by angiotensin II administration. Targeting a critical downstream signal pathway involved in muscle wasting, angiotensin II involves suppression of IGF-1 signaling via the Akt/mTOR/p70S6K pathway that leads to muscle proteolysis via caspase-3 activation, actin cleavage, stimulation of ubiquitinization and increased apoptosis. These changes were abrogated in angiotensin II-infused MLC/mIGF-1 transgenic mice, suggesting that angiotensin II downregulation of IGF-1 in wildtype skeletal muscle is causally related to angiotensin II–induced wasting.[156] Because the renin-angiotensin system is activated in many catabolic conditions, these findings have broad implications for understanding mechanisms of skeletal muscle wasting and provide new avenues for therapeutic approaches.

IGF-1 Isoforms in Therapeutic Applications

The importance of IGF-1 isoform structure is further underscored by comparative studies of muscle-specific expression of different isoforms in transgenic mice. Analysis of mouse lines generated with an IGF-1 cDNA lacking an E peptide did not display muscle hypertrophy or increased regenerative capacity but had increased levels of circulating IGF-1[151] suggesting that E peptides may be critical for local sequestration and action of IGF-1 expressed in extrahepatic tissues. This conclusion was confirmed in a more recent study[157] in which viral administration of IGF-1Ea, IGF-1Eb and mature IGF-1 lacking an E-peptide extension to young murine muscles revealed gene expression changes that were driven by increased IGF-I regardless of the presence of E-peptide. In contrast, distinct expression patterns were observed after viral delivery of IGF-1 constructs containing an E peptide, which also produced a hypertrophic response lacking in the mature IGF-1 infection. Notably, although both prominent IGF-1 isoforms are capable of enhancing regeneration as muscle-specific transgenes, only the IGF-1Ea isoform (mIGF-1) can induce hypertrophy and protect against atrophy.[144] Furthermore, mIGF-1 induces phosphorylation of phosphoinositide-dependent protein kinase 1 (PDK1) rather than the canonical Akt,[158] whereas increased Akt phosporylation was noted in transgenic mice expressing IGF-1 isoforms including the Eb extension peptide. A schematic representation summarizing current information on IGF-1 isoforms action during muscle regeneration is shown in Figure 3.

Taken together, these studies show that the choice of isoform is critical to the design of gene therapeutic strategies employing IGF-1 for countering muscle atrophy and suggest that at least the IGF-1Ea (mIGF-1) isoform is capable of promoting muscle regenerative capacity and as well as preventing muscle necrosis and could therefore be an effective treatment for muscle loss in aging and disease.

IGF-1: Regenerator or Tumour Promoter?

Substantial evidence supports the involvement of IGF-1 in mitogenesis and neoplastic transformation as well as in tissue repair,[159] suggesting that this growth factor plays an important role in the process of tumor promotion. The MAP kinase pathway is considered to be the main pathway involved in IGF-1 mediated stimulation of cell proliferation, mediated by the Type 1 IGF-receptor, which sequentially activates mitogen-activated-protein kinase kinase (MAPKK) and a whole family of MAP kinases among which extracellular regulated kinases 1 and 2 (Erk1/2).[160] MAP kinases are thought to phosphorylate transcription factors that can induce expression of genes involved in cell cycle progression.

Although the neoplastic potential of mature IGF-1 is an obvious concern to be taken into account when designing therapeutic strategies for human muscle pathologies, the lack of tumors in any of the muscle-specific transgenic IGF-1 mouse lines to date suggest that delivery of isoform propeptides specifically to skeletal muscle is likely to activate intracellular different pathways that promote survival and cell replacement rather than uncontrolled proliferation. Indeed, increased phosphorylation of (PDK1), the mammalian that sits both upstream and downstream of rapamycin (mTOR) and p70 S6 kinase (p70S6K) were documented in the hypertrophic muscle of MLC/mIGF-1 mice, without a significant change in Akt phosphorylation[158] (Fig. 3). Reduced phosphorylation of MAP kinase kinase 1 and 2 (MEK1/2) and mitogen-activated protein kinase kinases 3 and 6 (MKK3/6) were also observed in muscle from transgenic mice, suggesting that the hypertrophic and mitogenic effects of IGF-I are mediated via distinct signaling pathways in skeletal muscle and that inhibition of the MAPkinase pathway may be required for the IGF-I–induced hypertrophic effect.[158] This is an important distinction that may encourage more widespread clinical applications of IGF-1, many of which have been proposed but few have proceeded to human trials.

Gene and Cell Therapies to Rescue Muscle Atrophy and Wasting

The complexity of muscle types, the intimate relationship between structural integrity and mechanical function and the sensitivity of skeletal muscle to metabolic perturbations have impeded rapid progress in successful clinical interventions in muscle atrophy. Gene and stem cell therapy

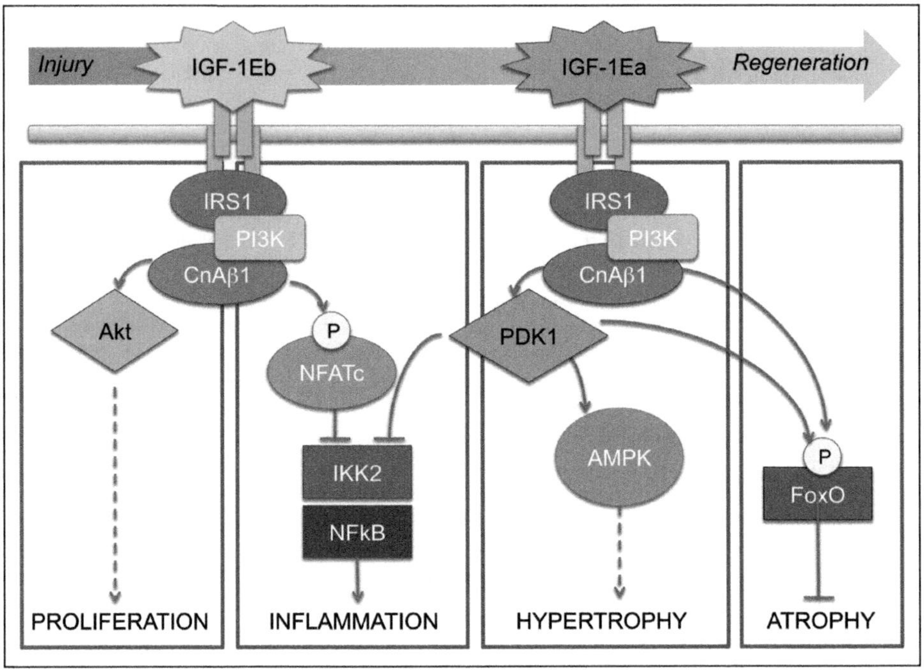

Figure 3. Schematic representation of proposed signaling pathways induced by IGF-1 isoforms during muscle regeneration. In response to traumatic muscle injury, IGF-Eb isoforms are transiently induced, followed by increase in IGF-1Ea isoform levels after 2-3 days. Downregulation of inflammatory signals (such as NF-kB) through the action of CnAb1 is accompanied by a shift from Akt- to PDK1-mediated signaling. IGF-1Ea isoforms also activate AMPK, a fuel gauge for the cell and downregulate inducers of muscle atrophy through phosphorylation of FoxO factors. Supporting data discussed in the text.

represent two promising tools to attenuate the genetic and somatic diseases leading to muscle wasting, although both involve significant hurdles still to overcome. Gene therapy relies on either viral or nonviral vectors to deliver target genes into somatic adult cells. Systemic viral infection of the skeletal muscle system has been achieved in mouse models, but has yet to be translated to human system and nonviral vectors present limitations due to vector instability once delivered into the recipient cells. Cell therapy, which relies on either committed cells or stem cells, also suffers from insufficient persistence of cells in the foreign milieu. Here we will briefly review current gene and stem cell therapies strategies aiming at counteracting muscle atrophy and wasting.

Gene Therapy

The effective delivery of genes encoding diffusible or paracrine effectors that participate in the prevention of muscle catabolism is a promising avenue for the attenuation of muscle wasting in a clinical setting. In order to be effective, vector-mediated gene therapy must satisfy four major parameters: the vectors should have capacity to carry large target genes, must be cheaply produced in large quantity, must produce low immune-response and must be stably expressed. For muscle pathologies as DMD the effort to progress toward an effective gene therapy has been limited by the restriction of viral vectors that can carry the full-length dystrophin gene and delivery into all muscles of the body.[161-163]

Among several viral vectors currently being tested, adenovirus has been used for several pioneer studies since its carrying capacity is large enough to hold the 14-kb dystrophin gene and it has the ability to infect nondividing cells. Nevertheless, adenovirus presents limitations for an efficient gene therapy since it does not integrate into the host DNA and therefore it is not stable and can elicit severe or lethal immune response in humans. In contrast, retrovirus and adeno-associated-virus (AAV) vectors have the ability to integrate the genetic material into the DNA of infected cell. While retroviral vectors can infect only cycling cells, AAV can transduce postmitotic cells, such as muscle fibers, although it cannot package the full-length dystrophin gene. Lentiviral vectors have been increasingly used to efficiently deliver genes to postmitotic cell types offering long-term expression, can be generated in high titers and do not induce immunological complications.[164] For instance, in a mouse model of spinal muscular atrophy (SMA), a recessive autosomal disorder caused by mutations or deletion of the telomeric copy of the survival motor neuron (SMN) gene, it was shown that injections of a lentiviral vector expressing SMN in various muscles of SMA mice restored SMN to motor neurons and reduced motor neuron death.[165]

A significant obstacle to designing effective vector mediated replacement gene therapy for myopathies is the necessity to infect the entire musculature, a problem that cannot easily be overcome unless more systemic delivery systems are developed. Alternative strategies to attenuate the progressive skeletal muscle wasting associated with DMD, cancer, AIDS are actively being pursued. Non viral-vectors currently present an attractive alternative, reducing or eliminating immune response and recent advances in gene delivery to muscle promise to overcome limitations due to the instability of naked DNA once delivered into the recipient cells.[166,167]

The anti-atrophic actions of IGF-1 in skeletal muscle have also been explored in a therapeutically relevant setting. Local delivery of mIGF-1 to individual 27 month old mouse muscles by AAV-mediated gene transfer reversed the age-related reduction in force production and loss of fast fibers, all of which are typical of aging skeletal muscle.[168] Young adult muscles receiving virally delivered IGF-1 exhibited muscle hypertrophy, while the functional properties of old muscles injected with the IGF-1 virus exhibited the same mechanical force as legs of younger mice. Changes in muscle mass and cross sectional area of mIGF-1 injected muscles also translated into increased force production. This further demonstrates that the anti-atrophic action of mIGF-1 is not dependent on life-long expression and as such represents a potentially effective gene therapeutic strategy to combat muscle wasting.

How do IGF-1 isoforms differ in their capacity to promote muscle growth and repair? Infections of mouse skeletal muscle with AAV-based IGF-1 expression vectors have revealed that in young animals, IGF-1Eb produced more protein than IGF-1Ea (mIGF-1) yet comparable levels of hypertrophy were observed with both isoforms.[148] The situation changed in older animals (6 months), where only IGF-1Ea expressing viruses produced significant increases in muscle size, despite equal expression of both isoforms. Whether the bioavailability of the IGF-1Eb isoform or its receptor affinity diminishes with age remains to be determined, but this distinction is particularly relevant to therapies directed towards the prevention of age-related loss of muscle size and strength.

Cell Therapy

A promising line of research to ameliorate muscle tissue regeneration focuses on the characterization of quiescent satellite cells. Substantial effort has been done to optimize, the isolation, the activation and the transplantation potential of SC cells and encouraging results have been obtained in the mdx mouse model and in several preclinical studies.[103] On the therapeutic side, a major hurdle remains that SC cells do not cross the endothelial wall making systemic delivery impossible.

Myoblast cell therapy has also been extensively explored as a promising alternative to correct genetic diseases by contributing to tissue regeneration. Replacement of diseased muscles with healthy and functional muscle fibers, has long been a major therapeutic strategy for muscular dystrophies.[169-171] However, the failure of injected committed cells to survive in the recipient animals and successfully engraft within their target organs has proven disappointing. The poor survival of injected cells (less than 1%), minimal migration from injection site (1 mm) and rapid senescence of the surviving population, has failed to produce satisfactory protocols of muscle regeneration that might be considered for therapeutic purposes.

Several lines of research have been involved to increase the survival of injected myoblasts. One of these implicates immunosuppression of the host: myoblast death observed at 3 days after transplantation was significantly reduced when the hosts were irradiated, suggesting that modification of the host cell environment were contributing to this phenomenon.[172] In addition, efficient myoblast transplantation was achieved in mice that were immunosuppressed with monoclonal antibodies against CD8, CD4 and CTLA4 Ig.[172] The authors postulated that neutrophils mediate myoblast mortality by a leucocyte function-associated antigen-1 (LFA-1)-dependent mechanism. Other studies reported that immunosuppression with FK506 insures good success of myoblast transplantation in mdx mice and in nonhuman primates.[173-176] In the last two decades, several clinical trials of myoblast transplantation have been performed in DMD patients, in presence or not of immunosuppressive drugs and with rather positive outcomes.[103] In general, modulation of the inflammatory reaction to foreign cells is emerging as a necessary prerequisite for effective clinical applications of myoblast transplantation. Thus, integrating gene and cell therapy approaches may circumvent the major problem associated with the survival of transplanted cells, enhancing cell engraftment and improving muscle regeneration.

The identification of multi-potent stem cells residing in extra-hematopoietic adult tissues has offered new perspectives in cell-mediated therapy for genetic diseases and muscle regeneration. Different subpopulations of haematopoietic $CD133^+$ stem cells (HSC), reside in the muscle beds of non-injured animals[177] and more migrate into sites of injury, suggesting a mechanism by which damaged tissues are repaired.[178,179] However this seems a rare event and presents limitations for efficient tissue repair. It is likely that poor recruitment of HSC into the dystrophic muscle of the mdx mouse and the low frequency of the myogenic potency of HSC may be the major obstacle for muscle regeneration and rescue of the genetic disease.[103,180]

Mobilization of HSC from the bone marrow into the circulation is increasingly used for other clinical applications; however HSC participation in normal regeneration processes is not yet proven. Local commitment of these cells to myogenesis and replenishment the satellite cell compartment has been proposed as a potential route to improved response to injury;[181] alternatively, they may fuse directly into regenerating muscle fibers.[181-184] However, in all animal studies to date, it has been necessary to replace host bone marrow with marked progenitor cells to prove their provenance. This experimental manipulation inevitably involves lethal irradiation of the host animal, a necessary prerequisite for bone marrow engraftment into injured muscle.[185] In any case, the total number of bone marrow stem cells recruited to a muscle fate in these studies appears still insufficient to be of therapeutic benefit.[103,180]

Conclusion

In this chapter we review various aspects of myocyte growth, regeneration and survival tactics, which may prove useful for devising therapeutic strategies to combat muscle atrophy. Although stem cell therapy has not yet solved the major problem related to cell transplantation, namely the capacity to survive and to improve muscle regeneration, recent studies are beginning to elucidate the signals and mechanisms whereby regenerating muscle recruits circulating cells to sites of injury or degeneration. These cells need not be stem cells as long as they maintain sufficient plasticity to participate in muscle repair, either by rebuilding the damaged tissue or by instructing resident precursors. Research is now directed at identifying the optimal progenitor cell type for this purpose, at enhancing the recruitment of the patient's own reserves or transplanted cells to damaged muscle. Such cells could also be used to deliver genes that replace missing functions or counter degenerative processes, a promising avenue for designing future therapies of muscle wasting.

Blocking proteolytic pathways or inducing muscle compensatory growth or stem cell incorporation with pharmacological intervention will require an ever more detailed knowledge of the signaling mechanisms involved. The extent to which blocking or enhancing these pathways will be detrimental to other physiological parameters such as motor innervation remains to be seen. Although many pathway components constitute attractive therapies for the muscle atrophic phenotype in animal models, these applications cannot yet be directly applied to human clinical trials. Precautions must be taken before these agents are used to manipulate signaling in different muscle conditions, to ensure the specificity of the existing molecules. In addition, long-range pharmacological stimulation or inhibition of signaling raises serious concerns about the effects on other body systems. Different muscle wasting conditions might require specific manipulations in a more cell-type- restricted way. Finally, the potential feedback between skeletal muscle and other tissue types such as heart, adipose tissue and bone is a major aspect of physiology that merits consideration and which will undoubtedly play an important role in preventing muscle atrophy in the context of the clinic.

References

1. Baracos V. Hypercatabolism and hypermetabolism in wasting states. Curr Opin Clin Nutr Metab Care 2002; 5(3):237-39.
2. Baracos V. Cancer-associated cachexia and underlying biological mechanisms. Ann Rev Nutr 2006; 26:435-61.
3. Saini A, Faulkner S, Al-Shanti N et al. Powerful signals for weak muscles. Ageing Res Rev 2009; 8(4):251-67.
4. Croall D, Ersfeld K. The calpains: modular designs and functional diversity. Genome Biol 2007; 8(6):218.
5. Cannon J, Meydani SN, Fielding RA et al. Acute phase response in exercise. II. Associations between vitamin E, cytokines and muscle proteolysis. Am J Physiol 1991; 260:R1235-40.
6. Armstrong R, Warren GL, Warren JA. Mechanisms of exercise-induced muscle fibre injury. Sports Med 1991; 12(3):184-207.
7. Murphy R. Calpains, skeletal muscle function and exercise. Clin Exp Pharmacol Physiol 2009. [Epub ahead of print].
8. Sorimachi H, Imajoh-Ohmi S, Emori Y et al. Molecular cloning of a novel mammalian calcium-dependent protease distinct from both m- and mu-types. Specific expression of the mRNA in skeletal muscle. J Biol Chem 1989; 264:20106-11.
9. Huang J, Forsberg NE. Role of calpain in skeletal-muscle protein degradation. Proc Natl Acad Sci USA 1998; 95:12100-005.
10. Kumamoto T, Kleese WC, Cong JY et al. Localization of the Ca(2+)-dependent proteinases and their inhibitor in normal, fasted and denervated rat skeletal muscle. Anat Rec 1992; 232:60-77.
11. Saido T, Sorimachi H, Suzuki K. Calpain: new perspectives in molecular diversity and physiological-pathological involvement. FASEB J 1994; 8:814-22.
12. Takamure M, Murata KY, Tamada Y et al. Calpain-dependent alpha-fodrin cleavage at the sarcolemma in muscle diseases. Muscle Nerve 2005; 32(3):303-9.
13. Martin S, O'Brien GA, Nishioka WK et al. Proteolysis of fodrin (non-erythroid spectrin) during apoptosis. J Biol Chem 1995; 270:6425-8.
14. Schakman O, Gilson H, Thissen JP. Mechanisms of glucocorticoid-induced myopathy. J Endocrinol 2008; 197(1):1-10.
15. Di Lisa F, De Tullio R, Salamino F et al. Specific degradation of troponin T and I by mu-calpain and its modulation by substrate phosphorylation. Biochem J 1995; 308:57-61.
16. Noma H, Kato T, Fujita H et al. Calpain inhibition induces activation of the distinct signalling pathways and cell migration in human monocytes. Immunology 2009; 128(1 Suppl):e487-96.
17. Stupka N, Tarnopolsky MA, Yardley NJ et al. Cellular adaptation to repeated eccentric exercise-induced muscle damage. J Appl Physiol 2001; 91(4):1669-78.
18. Kumamoto T, Fujimoto S, Ito T et al. Proteasome expression in the skeletal muscles of patients with muscular dystrophy. Acta Neuropathol 2000; 100(6):595-602.
19. Richard I, Roudaut C, Marchand S et al. Loss of calpain 3 proteolytic activity leads to muscular dystrophy and to apoptosis-associated IkappaBalpha/nuclear factor kappaB pathway perturbation in mice. J Cell Biol 2000; 151(7):1583-90.
20. Baghdiguian S, Martin M, Richard I et al. Calpain 3 deficiency is associated with myonuclear apoptosis and profound perturbation of the IkappaB alpha/NF-kappaB pathway in limb-girdle muscular dystrophy type 2A. Nat Med 1999; 5:503-11.
21. Spencer M, Guyon JR, Sorimachi H et al. Stable expression of calpain 3 from a muscle transgene in vivo: immature muscle in transgenic mice suggests a role for calpain 3 in muscle maturation. Proc Natl Acad Sci USA 2002; 99:8874-9.

22. Tidball J, Wehling-Henricks M. The role of free radicals in the pathophysiology of muscular dystrophy. J Appl Physiol 2007; 102(4):1677-86.
23. Li J, Yuan J. Caspases in apoptosis and beyond. Oncogene 2008; 27(48):6194-206.
24. Enari M, Sakahira H, Yokoyama H et al. A caspase-activated DNase that degrades DNA during apoptosis and its inhibitor. Nature 1998; 391:43-50.
25. Lechardeur D, Drzymala L, Sharma M et al. Determinants of the nuclear localization of the heterodimeric DNA fragmentation factor (ICAD/CAD). Oncogene 2000; 150(2):321-34.
26. Liu X, Zou H, Slaughter C et al. DFF, a heterodimeric protein that functions downstream of caspase-3 to trigger DNA fragmentation during apoptosis. Cell 1997; 89:175-84.
27. Kivinen K, Kallajoki M, Taimen P. Caspase-3 is required in the apoptotic disintegration of the nuclear matrix. Exp Cell Res 2005; 15(311(1)):62-73.
28. Orth K, Chinnaiyan AM, Garg M et al. The CED-3/ICE-like protease Mch2 is activated during apoptosis and cleaves the death substrate lamin A. J Biol Chem 1996; 271:16443-6.
29. Takahashi A, Alnemri ES, Lazebnik YA et al. Cleavage of lamin A by Mch2 alpha but not CPP32: multiple interleukin 1 beta-converting enzyme-related proteases with distinct substrate recognition properties are active in apoptosis. Proc Natl Acad Sci USA 1996; 93:8395-400.
30. Chen F, Chang R, Trivedi M et al. Caspase proteolysis of desmin produces a dominant-negative inhibitor of intermediate filaments and promotes apoptosis. J Biol Chem 2003; 278(9):6848-53.
31. Kothakota S, Azuma T, Reinhard C et al. Caspase-3-generated fragment of gelsolin: effector of morphological change in apoptosis. Science 1997; 278(5336):294-8.
32. Sandri M, El Meslemani AH, Sandri C et al. Caspase 3 expression correlates with skeletal muscle apoptosis in Duchenne and facioscapulo human muscular dystrophy. A potential target for pharmacological treatment? J Neuropathol Exp Neurol 2001; 60(3):302-12.
33. Hotti A, Järvinen K, Siivola P et al. Caspases and mitochondria in c-Myc-induced apoptosis: identification of ATM as a new target of caspases. Oncogene 2000; 19(19):2354-62.
34. Rhéaume E, Cohen LY, Uhlmann F et al. The large subunit of replication factor C is a substrate for caspase-3 in vitro and is cleaved by a caspase-3-like protease during Fas-mediated apoptosis. EMBO J 1997; 16(21):6346-54.
35. Belizario J, Lorite MJ, Tisdale MJ. Cleavage of caspases-1, -3, -6, -8 and -9 substrates by proteases in skeletal muscles from mice undergoing cancer cachexia. Br J Cancer 2001; 84:1135-40.
36. Sandri M. Signaling in muscle atrophy and hypertrophy. Physiology (Bethesda) 2008; 23:160-70.
37. Mammucari C, Milan G, Romanello V et al. FoxO3 Controls Autophagy in Skeletal Muscle In Vivo. Cell Metab 2007; 6:458-71.
38. Mammucari C, Schiaffino, S, Sandri M. Downstream of Akt: FoxO3 and mTOR in the regulation of autophagy in skeletal muscle. Autophagy 2008; 4:524-6.
39. Zhao J, Brault JJ, Schild A et al. FoxO3 Coordinately Activates Protein Degradation by the Autophagic/Lysosomal and Proteasomal Pathways in Atrophying Muscle Cells. Cell Metab 2007; 6:472-83.
40. Clarke B, Drujan D, Willis MS et al. The E3 Ligase MuRF1 degrades myosin heavy chain protein in dexamethasone-treated skeletal muscle. Cell Metab 2007; 6:376-85.
41. Cohen S, Brault JJ, Gygi SP et al. During muscle atrophy, thick, but not thin, filament components are degraded by MuRF1-dependent ubiquitylation. J Cell Biol 2009; 185:1083-95.
42. Fielitz J, Kim MS, Shelton JM et al. Myosin accumulation and striated muscle myopathy result from the loss of muscle RING finger 1 and 3. J Clin Invest 2007; 117:2486-95.
43. Solomon V, Goldberg AL. Importance of the ATP-ubiquitin-proteasome pathway in the degradation of soluble and myofibrillar proteins in rabbit muscle extracts. J Biol Chem 1996; 271:26690-7.
44. Levine B, Kroemer G. Autophagy in the pathogenesis of disease. Cell 2008; 132:27-42.
45. Mizushima N, Levine B, Cuervo AM et al. Autophagy fights disease through cellular self-digestion. Nature 2008; 45:1069-75.
46. Lecker S, Goldberg AL, Mitch WE. Protein degradation by the ubiquitin proteasome pathway in normal and disease states. J Am Soc Nephrol 2006; 17:1807-19.
47. Tanida I, Ueno T, Kominami E. Human light chain 3/MAP1LC3B is cleaved at its carboxyl-terminal Met121 to expose Gly120 for lipidation and targeting to autophagosomal membranes. J Biol Chem 2004; 279(47704-47710.).
48. Tanida I, Ueno T, Kominami E. LC3 conjugation system in mammalian autophagy. Int J Biochem Cell Biol 2004; 36:2503-18.
49. McElhinny A, Kakinuma K, Sorimachi H et al. Muscle-specific RING finger-1 interacts with titin to regulate sarcomeric M-line and thick filament structure and may have nuclear functions via its interaction with glucocorticoid modulatory element binding protein-1. J Cell Biol 2002; 157:125-36.
50. Bodine S, Latres E, Baumhueter S et al. Identification of ubiquitin ligases required for skeletal muscle atrophy. Science 2001; 294:1704-8.

51. Deval C, Mordier S, Obled C et al. Identification of cathepsin L as a differentially expressed message associated with skeletal muscle wasting. Biochem J 2001; 360:143-50.
52. Lecker S, Jagoe RT, Gilbert A et al. Multiple types of skeletal muscle atrophy involve a common program of changes in gene expression. FASEB J 2004; 18:39-51.
53. Duncan E, Muratore-Schroeder TL, Cook RG et al. Cathepsin L proteolytically processes histone H3 during mouse embryonic stem cell differentiation. Cell 2008; 135:284-94.
54. Masiero E, Sandri M. Autophagy inhibition induces atrophy and myopathy in adult skeletal muscles. Autophagy 2010; 6(2):[Epub ahead of print].
55. Musaro A, Fulle S, Fano G. Oxidative stress and muscle homeostasis. Curr Opin Clin Nutr Metab Care 2010: Epub ahead of print.
56. Dobrowolny G, Aucello M, Rizzuto E et al. Skeletal muscle is a primary target of SOD1G93A-mediated toxicity. Cell Metab 2008; 8(5):425-36.
57. Mitch W, Goldberg AL. Mechanisms of muscle wasting. The role of the ubiquitin-proteasome pathway. New Engl J Med 1996; 335:1897-905.
58. Auclair D, Garrel DR, Chaouki Zerouala A et al. Activation of the ubiquitin pathway in rat skeletal muscle by catabolic doses of glucocorticoids. Am J Physiol 1997; 272((3 Pt 1)):C1007-16.
59. Llovera M, Garcia-Martinez C, Lopez-Soriano J et al. Role of TNF receptor 1 in protein turnover during cancer cachexia using gene knockout mice. Mol Cell Endocrinol 1998; 142(1-2):183-9.
60. Argiles JM, Lopez-Soriano FJ. Catabolic proinflammatory cytokines. Curr Opin Clin Nutr Metab Care 1998; 1(3):245-51.
61. Tisdale M. Catabolic mediators of cancer cachexia. Curr Opin Support Palliat Care 2008; 2(4):256-61.
62. Tisdale M. Mechanisms of cancer cachexia. Physiol Rev 2009; 89(2):381-410.
63. Moresi V, Pristerà A, Scicchitano BM et al. Tumor necrosis factor-alpha inhibition of skeletal muscle regeneration is mediated by a caspase-dependent stem cell response. Stem Cells 2008; 26(4):997-1008.
64. Moresi V, Garcia-Alvarez G, Pristerà A et al. Modulation of caspase activity regulates skeletal muscle regeneration and function in response to vasopressin and tumor necrosis factor. PLoS One 2009; 4(5):e5570.
65. Schwarzkopf M, Coletti D, Sassoon D et al. Muscle cachexia is regulated by a p53-PW1/Peg3-dependent pathway. Genes Dev 2006; 20(24):3440-52.
66. Ghosh S, Karin M. Missing pieces in the NF-kappaB puzzle. Cell 2002; 109 Suppl:S81-96.
67. Relaix F, Weng X, Marazzi G et al. Sassoon D. Pw1, a novel zinc finger gene implicated in the myogenic and neuronal lineages. Dev Biol 1996; 177(2):383-96.
68. Relaix F, Wei XJ, Wu X et al. Peg3/Pw1 is an imprinted gene involved in the TNF-NFkappaB signal transduction pathway. Nat Genet 1998; 18(3):287-91.
69. Coletti D, Yang E, Marazzi G et al. TNFalpha inhibits skeletal myogenesis through a PW1-dependent pathway by recruitment of caspase pathways. EMBO J 1997; 21(4):631-42.
70. Haddad F, Zaldivar F, Cooper DM et al. IL-6-induced skeletal muscle atrophy. J Appl Physiol 2005; 98(3):911-7.
71. De Benedetti F, Alonzi T, Moretta A et al. Interleukin 6 causes growth impairment in transgenic mice through a decrease in insulin-like growth factor-I. A model for stunted growth in children with chronic inflammation. J Clin Invest 1997; 99(4):643-50.
72. Jimenez-Mallebrera C, Brown SC, Sewry CA et al. Congenital muscular dystrophy: molecular and cellular aspects. Cell Mol Life Sci 2005; 62(7-8):809-23.
73. Machuca-Tzili L, Brook D, Hilton-Jones D. Clinical and molecular aspects of the myotonic dystrophies: a review. Muscle Nerve 2005; 32(1):1-18.
74. Batchelor C, Winder SJ. Sparks, signals and shock absorbers: how dystrophin loss causes muscular dystrophy. Trends Cell Biol 2006; 16(4):198-205.
75. Lapidos K, Kakkar R, McNally EM. The dystrophin glycoprotein complex: signaling strength and integrity for the sarcolemma. Circ Res 2004; 94(8):1023-31.
76. Sciandra F, Gawlik KI, Brancaccio A et al. Dystroglycan: a possible mediator for reducing congenital muscular dystrophy? Trends Biotechnol 2007; 25(6):262-8.
77. Sandonà D, Betto R. Sarcoglycanopathies: molecular pathogenesis and therapeutic prospects. Expert Rev Mol Med 2009; 11:e28.
78. Patton B, Wang B, Tarumi YS et al. A single point mutation in the LN domain of LAMA2 causes muscular dystrophy and peripheral amyelination. J Cell Sci 2008; 121(10):1593-604.
79. Mayer U, Saher G, Fassler R et al. Absence of integrin alpha 7 causes a novel form of muscular dystrophy. Nat Genet 1997; 17:318-23.
80. Rooney J, Welser JV, Dechert MA et al. Severe muscular dystrophy in mice that lack dystrophin and alpha7 integrin. J Cell Sci 2006; 119(11):2185-95.

81. Suzuki N, Motohashi N, Uezumi A et al. NO production results in suspension-induced muscle atrophy through dislocation of neuronal NOS. J Clin Invest 2007; 117(9):2468-76.
82. Acharyya S, Butchbach ME, Sahenk Z et al. Dystrophin glycoprotein complex dysfunction: a regulatory link between muscular dystrophy and cancer cachexia. Cancer Cell 2005; 8(5):421-32.
83. Berridge M, Lipp P, Bootman MD. The versatility and universality of calcium signalling. Nat Rev Mol Cell Biol 2000; 1(1):11-21.
84. Berridge M, Bootman MD, Roderick HL. Calcium signalling: dynamics, homeostasis and remodelling. Nat Rev Mol Cell Biol 2003; 4(7):517-29.
85. Berchtold M, Brinkmeier H, Müntener M. Calcium ion in skeletal muscle: its crucial role for muscle function, plasticity and disease. Physiol Rev 2000 80(3):1215-65.
86. Hopf F, Turner PR, Steinhardt RA. Calcium misregulation and the pathogenesis of muscular dystrophy. Subcell Biochem 2007; 45:429-64.
87. Acharyya S, Villalta SA, Bakkar N et al. Interplay of IKK/NF-kappaB signaling in macrophages and myofibers promotes muscle degeneration in Duchenne muscular dystrophy. J Clin Invest 2007; 117(4):889-901.
88. Blake D, Weir A, Newey SE et al. Function and genetics of dystrophin and dystrophin-related proteins in muscle. Physiol Rev 2002; 82:291-329.
89. Dudley R, Danialou G, Govindaraju K et al. Sarcolemmal damage in dystrophin deficiency is modulated by synergistic interactions between mechanical and oxidative/nitrosative stresses. Am J Pathol 2006; 168(4):1276-87.
90. Hand C, Rouleau GA. Familial amyotrophic lateral sclerosis. Muscle Nerve 2002; 25(2):135-59.
91. Robert V, Massimino ML, Tosello V et al. Alteration in calcium handling at the subcellular level in mdx myotubes. J Biol Chem 2001; 276(7):4647-51.
92. Shkryl V, Martins AS, Ullrich ND et al. Reciprocal amplification of ROS and Ca(2+) signals in stressed mdx dystrophic skeletal muscle fibers. Pflugers Arch 2009; 458(5):915-28.
93. Turner P, Westwood T, Regen CM et al. Increased protein degradation results from elevated free calcium levels found in muscle from mdx mice. Nature 1988; 335:735-8.
94. Valentine J, Doucette PA, Zittin Potter S. Copper-zinc superoxide dismutase and amyotrophic lateral sclerosis. Annu Rev Biochem 2005; 74:563-93.
95. von Lewinski F, Keller BU. Ca2+, mitochondria and selective motoneuron vulnerability: implications for ALS. Trends Neurosci 2005; 28(9):494-500.
96. Schiaffino S, Sandri M, Murgia M. Activity-dependent signaling pathways controlling muscle diversity and plasticity. Physiology (Bethesda) 2007; 22:269-78.
97. Porter J, Khanna S, Kaminski HJ et al. A chronic inflammatory response dominates the skeletal muscle molecular signature in dystrophin-deficient mdx mice. Hum Mol Genet 2002; 11(3):263-72.
98. Evans N, Misyak SA, Robertson JL et al. Dysregulated intracellular signaling and inflammatory gene expression during initial disease onset in Duchenne muscular dystrophy. Am J Phys Med Rehabil 2009; 88(6):502-22.
99. Haslett J, Sanoudou D, Kho AT et al. Gene expression comparison of biopsies from Duchenne muscular dystrophy (DMD) and normal skeletal muscle. Proc Natl Acad Sci USA 2002; 99(23):15000-5.
100. Peterson J, Guttridge DC. Skeletal muscle diseases, inflammation and NF-kappaB signaling: insights and opportunities for therapeutic intervention. Int Rev Immunol 2008; 27(5):375-87.
101. Vetrone S, Montecino-Rodriguez E, Kudryashova E et al. Osteopontin promotes fibrosis in dystrophic mouse muscle by modulating immune cell subsets and intramuscular TGF-beta. J Clin Invest 2009; 119(6):1583-94.
102. Mourkioti F, Rosenthal N. IGF-1, inflammation and stem cells: interactions during muscle regeneration. Trends Immunol 2005; 26(10):535-42.
103. Tedesco F, Dellavalle A, Diaz-Manera J et al. Repairing skeletal muscle: regenerative potential of skeletal muscle stem cells. J Clin Invest 2010; 120(1):doi: 10.1172/JCI40373.
104. Yan Z, Choi S, Liu X et al. Highly coordinated gene regulation in mouse skeletal muscle regeneration. J Biol Chem 2003; 278:8826-36.
105. Carosio S, Berardinelli MG, Aucello M et al. Impact of ageing on muscle cell regeneration. Ageing Res Rev 2009. [Epub ahead of print].
106. Cohn R, Henry MD, Michele DE et al. Disruption of DAG1 in differentiated skeletal muscle reveals a role for dystroglycan in muscle regeneration. Cell 2002; 110:639-48.
107. Shi X, Garry DJ. Muscle stem cells in development, regeneration and disease. Genes Dev 2006; 20(13):1692-708.
108. Musaro A, McCullagh K, Paul A et al. Localized Igf-1 transgene expression sustains hypertrophy and regeneration in senescent skeletal muscle. Nat Genet 2001; 27(2):195-200.
109. Conboy IM, Conboy MJ, Wagers AJ et al. Rejuvenation of aged progenitor cells by exposure to a young systemic environment. Nature 2005; 433(7027):760-4.

110. Karalaki M, Fili S, Philippou A et al. Muscle regeneration: cellular and molecular events. In Vivo 2009; 23(5):779-96.
111. Joulia-Ekaza D, Cabello G. The myostatin gene: physiology and pharmacological relevance. Curr Opin Pharmacol 2007; 7(3):310-5.
112. Kollias H, McDermott JC. Transforming growth factor-beta and myostatin signaling in skeletal muscle. J Appl Physiol 2008; 104(3):579-87.
113. Mozzetta C, Minetti G, Puri PL. Regenerative pharmacology in the treatment of genetic diseases: the paradigm of muscular dystrophy. Int J Biochem Cell Biol 2009; 41(4):701-10.
114. Patel K, Amthor H. The function of Myostatin and strategies of Myostatin blockade-new hope for therapies aimed at promoting growth of skeletal muscle. Neuromuscul Disord 2005; 15(2):117-26.
115. Roth S, Walsh S. Myostatin: a therapeutic target for skeletal muscle wasting. Curr Opin Clin Nutr Metab Care 2004; 7(3):259-63.
116. Bogdanovich S, Krag TO, Barton ER et al. Functional improvement of dystrophic muscle by myostatin blockade. Nature 2002; 420(6914):418-21.
117. Nakatani M, Takehara Y, Sugino H et al. Transgenic expression of a myostatin inhibitor derived from follistatin increases skeletal muscle mass and ameliorates dystrophic pathology in mdx mice. FASEB J 2008; 22(2):477-87.
118. Siriett V, Salerno MS, Berry C et al. Antagonism of myostatin enhances muscle regeneration during sarcopenia. Mol Ther 2007; 15(8):1463-70.
119. Bogdanovich S, McNally EM, Khurana TS. Myostatin blockade improves function but not histopathology in a murine model of limb-girdle muscular dystrophy 2C. Muscle Nerve 2008; 37(3):308-16.
120. Sumner C, Wee CD, Warsing LC et al. Inhibition of myostatin does not ameliorate disease features of severe spinal muscular atrophy mice. Hum Mol Genet 2009; 18(17):3145-52.
121. Mourkioti F, Kratsios P, Luedde T et al. Targeted ablation of IKK2 improves skeletal muscle strength, maintains mass and promotes regeneration. J Clin Invest 2006; 116(11):2945-54.
122. Mourkioti F, Rosenthal N. NF-kappaB signaling in skeletal muscle: prospects for intervention in muscle diseases. J Mol Med 2008; 86(7):747-59.
123. Gregorevic P, Hayes A, Lynch GS et al. Functional properties of regenerating skeletal muscle following LIF administration. Muscle Nerve 2000; 23(10):1586-8.
124. Pedersen B, Steensberg A, Keller P et al. Muscle-derived interleukin-6: lipolytic, anti-inflammatory and immune regulatory effects. Pflugers Arch 2003; 446(1):9-16.
125. Pedersen B. IL-6 signalling in exercise and disease. Biochem Soc Trans 2007; 35(5):1295-7.
126. White J, Bower JJ, Kurek JB et al. Leukemia inhibitory factor enhances regeneration in skeletal muscles after myoblast transplantation. Muscle Nerve 2001; 24(5):695-7.
127. White J, Davies M, McGeachie J et al. An evaluation of leukaemia inhibitory factor as a potential therapeutic agent in the treatment of muscle disease. Neuromuscul Disord 2002; 12(10):909-16.
128. Pistilli E, Siu PM, Alway SE. Interleukin-15 responses to aging and unloading-induced skeletal muscle atrophy. Am J Physiol Cell Physiol 2007; 292(4):C1298-304.
129. Saidenberg-Kermanac'h N, Bessis N, Deleuze V et al. Efficacy of interleukin-10 gene electrotransfer into skeletal muscle in mice with collagen-induced arthritis. J Gene Med 2003; 5:164-71.
130. Wehling M, Spencer MJ, Tidball JG. A nitric oxide synthase transgene ameliorates muscular dystrophy in mdx mice. J Cell Biol 2001; 155(1):123-31.
131. Colussi C, Mozzetta C, Gurtner A et al. HDAC2 blockade by nitric oxide and histone deacetylase inhibitors reveals a common target in Duchenne muscular dystrophy treatment. Proc Natl Acad Sci USA 2008; 105(49):19183-7.
132. Colussi C, Mozzetta C, Gurtner A et al. Nitric oxide deficiency determines global chromatin changes in Duchenne muscular dystrophy. FASEB J 2009; 23(7):2131-41.
133. Coleman M, DeMayo F, Yin KC et al. Myogenic vector expression of insulin-like growth factor I stimulates muscle cell differentiation and myofiber hypertrophy in transgenic mice. J Biol Chem 1995; 270:12109-16.
134. Liu JP, Baker J, Perkins AS et al. Mice carrying null mutations of the genes encoding insulin-like growth factor I (Igf-1) and type 1 IGF receptor (Igf1r). Cell 1993; 75(1):59-72.
135. Pelosi L, Giacinti C, Nardis C et al. Local expression of IGF-1 accelerates muscle regeneration by rapidly modulating inflammatory cytokines and chemokines. FASEB J 2007; 21(7):1393-402.
136. Grounds M. Reasons for the degeneration of ageing skeletal muscle: a central role for IGF-1 signalling. Biogerontology 2002; 3:19-24.
137. Rudman D, Kutner MH, Rogers CM et al. Impaired growth hormone secretion in the adult population: relation to age and adiposity. J Clin Invest 1981; 67:1361-69.
138. Vanitallie T. Frailty in the elderly: contributions of sarcopenia and visceral protein depletion. Metabolism 2003; 52(10 Suppl 2):22-6.

139. Clavel S, Coldefy AS, Kurkdjian E et al. Atrophy-related ubiquitin ligases, atrogin-1 and MuRF1 are up-regulated in aged rat Tibialis Anterior muscle. Mech Ageing Dev 2006; 127(10):794-801.
140. Belfiore A, Frasca F, Pandini G et al. Insulin receptor isoforms and insulin receptor/insulin-like growth factor receptor hybrids in physiology and disease. Endocr Rev 2009; 30(6):586-623.
141. Powell-Braxton L, Hollingshead P, Warburton C et al. IGF-I is required for normal embryonic growth in mice. Genes Dev 1993; 7(12B):2609-17.
142. Sutter N, Bustamante CD, Chase K et al. A single IGF1 allele is a major determinant of small size in dogs. Science 2007; 316(5821):112-5.
143. LeRoith D, Roberts, CT Jr. Insulin-like growth factor I (IGF-I): a molecular basis for endocrine versus local action? Mol Cell Endocrinol 1991; 77:C57-61.
144. Winn N, Paul A, Musaro A et al. Insulin-like growth factor isoforms in skeletal muscle aging, regeneration and disease. Cold Spring Harb Symp Quant Biol 2002; 67:507-18.
145. Shavlakadze T, Winn N, Rosenthal N et al. Reconciling data from transgenic mice that overexpress IGF-I specifically in skeletal muscle. Growth Horm IGF Res 2005; 15(1):4-18.
146. Bach M, Roberts CT Jr, Smith EP et al. Alternative splicing produces messenger RNAs encoding insulin-like growth factor-I prohormones that are differentially glycosylated in vitro. Mol Endocrinol 1990; 4:899-904.
147. Pfeffer L, Brisson BK, Lei H et al. The insulin-like growth factor (IGF)-I E-peptides modulate cell entry of the mature IGF-I protein. Mol Biol Cell 2009; 20(17):3810-7.
148. Barton E. Viral expression of insulin-like growth factor-I isoforms promotes different responses in skeletal muscle. J Appl Physiol 2006; 100(6):1778-84.
149. Matheny RJ, Nindl BC, Adamo ML. Minireview: Mechano-Growth Factor: A Putative Product of IGF-I Gene Expression Involved in Tissue Repair and Regeneration. Endocrinology 2010. [Epub ahead of print].
150. Stitt TN, Drujan D, Clarke BA et al. The IGF-1/PI3K/Akt pathway prevents expression of muscle atrophy-induced ubiquitin ligases by inhibiting FOXO transcription factors. Mol Cell 2004; 14(3):395-403.
151. Shavlakadze T, Boswell JM, Burt DW et al. Rskalpha-actin/hIGF-1 transgenic mice with increased IGF-I in skeletal muscle and blood: impact on regeneration, denervation and muscular dystrophy. Growth Horm IGF Res 2006; 16(3):157-73.
152. Lara-Pezzi E, Winn N, Paul A et al. A naturally occurring calcineurin variant inhibits FoxO activity and enhances skeletal muscle regeneration. J Cell Biol 2007; 179(6):1205-18.
153. Musarò A, Giacinti C, Borsellino G et al. Stem cell-mediated muscle regeneration is enhanced by local isoform of insulin-like growth factor 1. Proc Natl Acad Sci USA 2004; 101(5):1206-10.
154. Barton E, Morris L, Musaro A et al. Muscle-specific expression of insulin-like growth factor I counters muscle decline in mdx mice. J Cell Biol 2002; 157(1):137-48.
155. Schulze P, Fang J, Kassik KA et al. Transgenic overexpression of locally acting insulin-like growth factor-1 inhibits ubiquitin-mediated muscle atrophy in chronic left-ventricular dysfunction. Circ Res 2005; 97(5):418-26.
156. Song Y, Li Y, Du J et al. Muscle-specific expression of IGF-1 blocks angiotensin II-induced skeletal muscle wasting. J Clin Invest 2005; 115(2):451-8.
157. Barton E, Demeo J, Lei H. The insulin-like growth factor (IGF)-1 E-peptides are required for isoform specific gene expression and muscle hypertrophy after local IGF-1 production. J Appl Physiol 2010.
158. Song Y, Godard M, Li Y et al. Insulin-like growth factor I-mediated skeletal muscle hypertrophy is characterized by increased mTOR-p70S6K signaling without increased Akt phosphorylation. J Investig Med 2005; 53(3):135-42.
159. Pollak M. Insulin and insulin-like growth factor signalling in neoplasia. Nat Rev Cancer 2008; 8:915-28.
160. Florini J, Ewton DZ, Coolican SA. Growth hormone and the insulin-like growth factor system in myogenesis. Endocr Rev 1996; 17(5):481-517.
161. Nelson S, Crosbie RH, Miceli MC et al. Emerging genetic therapies to treat Duchenne muscular dystrophy. Curr Opin Neurol 2009; 22(5):532-8.
162. Tang Y, Cummins J, Huard J et al. AAV-directed muscular dystrophy gene therapy. Expert Opin Biol Ther 2010; 10(3):395-408.
163. Trollet C, Athanasopoulos T, Popplewell L et al. Gene therapy for muscular dystrophy: current progress and future prospects. Expert Opin Biol Ther 2009; 9(7):849-66.
164. Nanou A, Azzouz M. Gene therapy for neurodegenerative diseases based on lentiviral vectors. Prog Brain Res 2009; 175:187-200.
165. Azzouz M, Le T, Ralph GS et al. Lentivector-mediated SMN replacement in a mouse model of spinal muscular atrophy. J Clin Invest 2004; 114(12):1726-31.

166. Hojman P, Gissel H, Andre F et al. Physiological effects of high and low voltage pulse combinations for gene electrotransfer in muscle. Hum Gen Ther 2008. [Epub ahead of print].
167. Zampaglione I, Arcuri M, Cappelletti M et al. In vivo DNA gene electro-transfer: a systematic analysis of different electrical parameters. J Gene Med 2005; 7(11):1475-81.
168. Barton-Davis E, Shoturma DI, Musaro A et al. Viral mediated expression of insulin-like growth factor I blocks the aging-related loss of skeletal muscle function. Proc Natl Acad Sci USA 1998; 95(26):15603-7.
169. Liu Z, Wu Y, Chen BG. Myoblast therapy: from bench to bedside. Cell Transplant 2006; 15(6):455-62.
170. Menasché P. Skeletal myoblasts and cardiac repair. J Mol Cell Cardiol 2008; 45(4):545-53.
171. Smythe G, Hodgetts SI, Grounds MD. Problems and solutions in myoblast transfer therapy. J Cell Mol Med 2001; 5(1):33-47.
172. Skuk D, Tremblay JP. Myoblast transplantation: the current status of a potential therapeutic tool for myopathies. J Muscle Res Cell Motil 2003; 24(4-6):285-300.
173. Camirand G, Caron NJ, Asselin I et al. Combined immunosuppression of mycophenolate mofetil and FK506 for myoblast transplantation in mdx mice. Transplantation 2001; 72(1):38-44.
174. Kinoshita I, Roy R, Dugré FJ et al. Myoblast transplantation in monkeys: control of immune response by FK506. J Neuropathol Exp Neurol 1996; 55(6):687-97.
175. Skuk D, Goulet M, Roy B et al. Myoblast transplantation in whole muscle of nonhuman primates. J Neuropathol Exp Neurol 2000; 59(3):197-206.
176. Tremblay J, Skuk D, Palmieri B et al. A case for immunosuppression for myoblast transplantation in duchenne muscular dystrophy. Mol Ther 2009; 17(7):1122-4.
177. McKinney-Freeman SL, Jackson KA, Camargo FD et al. Muscle-derived hematopoietic stem cells are hematopoietic in origin. Proc Natl Acad Sci USA 2002; 99:1341-6.
178. Bhagavati S. Stem cell based therapy for skeletal muscle diseases. Curr Stem Cell Res Ther 2008; 3(3):219-28.
179. Ferrari G, Cusella-De Angelis G, Coletta M et al. Muscle regeneration by bone marrow-derived myogenic progenitors. Science 1998; 279(5356):1528-30.
180. Ferrari G, Stornaiuolo A, Mavilio F. Failure to correct murine muscular dystrophy. Nature 2001; 411:1014-5.
181. LaBarge M, Blau HM. Biological progression from adult bone marrow to mononucleate muscle stem cell to multinucleate muscle fiber in response to injury. Cell 2002; 111:589-601.
182. Blau H, Brazelton TR, Weimann JM. The evolving concept of a stem cell: entity or function? Cell 2001; 105(7):829-41.
183. Corbel S, Lee A, Yi L et al. Contribution of hematopoietic stem cells to skeletal muscle. Nat Med 2003; 9(12):1528-32.
184. Rudnicki M. Marrow to muscle, fission versus fusion. Nat Med 2003; 9(12):1461-2.
185. Morgan JE, Gross JG, Pagel CN et al. Myogenic cell proliferation and generation of a reversible tumorigenic phenotype are triggered by preirradiation of the recipient site. J Cell Biol 2002; 157:693-702.

Chapter 16

Confronting Cellular Heterogeneity in Studies of Protein Metabolism and Homeostasis in Aging Research

Louise Boisen and Peter Kristensen*

Abstract

In this chapter we review the different technologies that can be applied in the analysis of protein homeostasis and metabolism in aging research. Special focus will be on technologies with a potential to circumvent the problems associated with cell heterogeneity in biomarker discovery. Often studies aimed at increasing our understanding of cellular senescence take advantage of model systems. This can be in the form of cell culture, where specific cell lines are cultivated, thus undergoing cellular senescence according to the Hayflick phenomenon. [1] Alternatively, model organisms can be included, such as yeast, nematodes and zebra fish.[2-5] Even though such model systems allow the researcher to control many parameters of the system, it is well established that even in a simple cell culture system the individual cells are morphologically and functionally different.

The heterogeneity observed even within the least complex systems, makes it a difficult task to identify biomarkers of cellular senescence and to fully understand the various molecular networks. The complexity is further increased when taking the step from model systems of aging to human aging.

Analysing Aging

The aim of many studies of cellular senescence is to increase our understanding of human aging. However, it still remains to be determined to what extent cellular senescence will lead to human aging. While the number of senescent cells increase, in the human as a consequence of advanced age, the number of senescent cells is still limited even at extreme age.[6]

While the influence of cellular senescence on human aging is still not fully understood, it is evident that the physiological signs of aging occur at all levels of the organism and no part of the body can be said to be exempt from the effects of aging. The aging phenotype is evident in the greying of hair, wrinkling of skin, redistribution of body fat, fragility of bones, general slowness of movement and increase in frailty, which is often accompanied by various disorders like arthritis, hypertension, dementia and cancers. The aging phenotype is the sum of many individual contributions from the different organs and tissues, but, ultimately, aging of the body originates from the cells and molecules that make up the organism and their interplay with the surroundings.

Based on this, it becomes evident that an examination of the molecular building blocks will lead to an understanding of the changes that eventually result in the aging phenotype.

*Corresponding Author: Peter Kristensen—Department of Molecular Biology, University of Aarhus. Email: pk@mb.au.dk

Protein Metabolism and Homeostasis in Aging, edited by Nektarios Tavernarakis.

The central dogma of molecular biology states that the flow of information is carried from the heritable material to the active components of the cells via a messenger system.[7] Although simple at first sight, today we know that complexity is the underlying theme in biological systems.

Today large initiatives aim to establish what role the genetic component plays in healthy human aging.[8] While the genome and the regulation of individual genes certainly play a great role in many human diseases and we can learn a great deal about the possible age related effects of a given cell by studying the genome, we need to complement such studies by studies of the functional components of the cell, the proteins. Again the complexity of the aging process and the effects on protein homeostasis and metabolism necessitates large scale initiatives.

A single gene can give rise to several alternatively spliced mRNAs that can be translated into similar, but distinct, proteins. A single mRNA can be translated numerous times or perhaps just once, depending in part on its stability. After translation, the protein product can be further modified in a number of ways. Therefore, the linkage between the genome and cellular function is complex and multiple genome- or transcriptome- based analyses have revealed that genes that are completely turned on or off during aging are exceedingly rare. This fact negates the value of genomic and transcriptomic approaches in this context (Kristensen et al, 2000). Conversely, proteomics is the study of the protein universe of the cell. It is by studying protein function and interactions we can come to a real understanding of what goes on in the cell.

Traditional proteomic approaches have relied heavily on 2D-PAGE[9,10] (two-dimensional polyacrylamide gel electrophoresis) and more recently 2D-DIGE[11] (two-dimensional fluorescence difference gel electrophoresis) analysis platforms to investigate changes occurring to the proteome in different circumstances, such as aging. In 2D-PAGE, proteins are initially separated in a pH-gradient according to their isoelectric point, followed by separation according to their molecular weight. This technique makes it possible to separate thousands of proteins on one gel, thus greatly enhancing the possible resolution of a protein sample. The separated proteins can be excised from the gel and identified using mass spectrometry (MS), which identifies proteins based on their specific breakdown patterns. Typically, the protein to be identified is enzymatically digested in the gel and the resulting peptides ionised and analysed using techniques such as electrospray ionisation (ESI) or matrix-assisted laser desorption ionisation time of flight (MALDI-TOF). The identified peptides are then matched with a database and the protein identified.[12] Theoretically, this approach can lead to the identification of thousands of proteins, although in practice the number is much lower.

Several research groups have used 2D-PAGE to analyse proteomic changes happening with age in various organs and cultured cells. For example, it was reported that the hearts of young and old mice showed changes in protein expression and that several of these changes were found in the mitochondria, which may suggest that dysfunctional mitochondria could contribute to cardiac aging.[13] Altered mitochondrial protein expression was identified in skeletal muscle of aging rats as well, but closer examination revealed that the observed changes most likely represented an adaptation to the increasing oxidative stress observed during aging rather than a failure of the mitochondria.[14] Accordingly, several proteins associated with oxidative stress were found to be up-regulated with age in skin biopsies from young and old human donors.[15] Also proteomic profiling of cultured cells has revealed changes in protein expression with senescence, which for human umbilical vein endothelial cells have included alterations in proteins involved in structural integrity of the cytoskeleton and nucleus, as well as proteins involved in protein processing and cellular repair processes.[16,17] Even though 2D-gel based approaches can lead to the identification of numerous proteins and changes in protein expression, certain limitations exist, most notably the inability to reach a single cell level of analysis when confronted with sample heterogeneity.

Improved sensitivity can be achieved using new techniques that bypass 2D-PAGE and today it is possible to analyse an entire, complex sample with the help of sophisticated chromatography-coupled MS machinery.[18] Advances in software development and technology are continuous and, most likely, gel-free and label-free differential proteomic analyses will play a greater role in the discovery of biomarkers in the future.[19-21]

The traditional 2D-gel and MS based proteomics approaches are very powerful in the large-scale identification of possible changes in abundance or modification pattern of proteins. However, typically the sensitivity of the methods, especially 2D-PAGE, dictates that a large number of cells is sampled together. As for, example, senescent cells only constitute a small fraction of a given sample, this most probably results in interesting biomarkers being overlooked.

The traditional proteomic approaches are unsurpassed in their ability to provide global overviews of the proteomic profile but deciphering the functional significance of the results can be problematic. Conversely, phage display is a very good tool for the analysis of protein function but the entire pattern of changes when comparing two samples will seldom be achievable. Consequently, when embarking on a biomarker quest one should consider whether functional significance is more important than complete coverage or vice versa. Below we will describe various applications of the phage display technology for biomarker identification. As this is a fairly new field in the study of protein metabolism and homestatis in aging research, most examples are taken from the cancer field; however, the methods described should be applicable in aging research.

Antibody Phage Display as a Discovery Tool

Antibody phage display was originally described by John McCafferty and coworkers.[22] The technology involves fusing libraries of antibody fragments[23] to the surface of filamentous bacteriophage, which allows for selection of specific antibodies against a target of interest. Traditionally, antibody phage technology has been used to select antibodies against well-defined targets that are non-immunogenic, self-antigens, toxic or otherwise problematic to raise traditional monoclonal antibodies against. Compared to hybridoma technology, the phage antibody technology provides some unique possibilities for generating antibodies, as will be described below. While traditional antibody technology often provides better reagents for biochemical analysis, the phage display technology is superior with respect to being applied as a discovery tool, for example to find new biomarkers for certain diseases or in the field of aging.

Different formats of antibodies can be displayed on phage. Full-size immunoglobulins (Ig) are very large and complex and difficult to express in bacteria and, thus, difficult to display on phage. Therefore, various fragmented antibody formats have been developed and used for fusion to the different phage coat proteins.[24]

Antibody Libraries

Just as researchers can choose a specific antibody format to work with, there is also a choice between different types of antibody libraries. Generally speaking, libraries can be immune, naïve or synthetic. Immune libraries were the first to be described and are created using B-cells isolated from various lymphoid sources from an immunised donor. The rearranged antibody genes are cloned into an appropriate vector creating a library with antibody specificities biased towards the immunising agent.[25] Immunised libraries can harbour high affinity antibodies against the desired target, but the use of them requires a new library to be created for each new target and the target must be immunogenic but not acutely toxic. Additionally, the use of human donors raises several ethical concerns and limitations as the active immunisation of a human being is ethically impossible. However, immune libraries can be constructed from passively immunised donors. To obviate the need for the construction of new libraries to each experiment, libraries from non-immunised donors have been created. These repertoires are the so-called naïve libraries and are cloned from the IgM mRNA isolated from B-cells of non-immunized donors.[26] These libraries are not biased against any particular antigen and it is even possible to isolate antibodies against self-antigens. The final class of libraries are the synthetic libraries, which are constructed entirely in vitro. These libraries are constructed by artificial combination of V genes with D and J segments randomised to varying degrees.[27] Problems with nonfunctional antibodies can be minimised by carefully designing which gene segments should be included and where and to what extent randomisation should occur, as described in the design of the HuCAL library principle.[28]

Antibody Selection against Complex Antigens

Traditional selection of phage-displayed antibodies against a defined target involves coating of immunotubes with the purified target protein. Through one or more rounds of affinity selection, or bio-panning, antibodies binding to the immobilised target are retained, while unbound phage is washed away. This procedure is fairly simple and works well with most purified proteins. Multiple alternatives for passive absorption of the antigen to the plastic surface of an immunotube have been devised in the past 20 years. By avoiding adsorption, normally the antigen is kept in a more native like conformation and often better control of the actual amount of antigen used in selection is obtained.[29-31] In the following sections, we will summarise some of the strategies used when working with complex antigens, especially in the form of whole cells.

Selection of Antibodies against Low Abundance Proteins

When phage display of antibodies is applied as a discovery tool to find differentially expressed or differentially modified antigens, the selection has to be performed on a complex proteins mixture either presented on the cell surface of whole intact cells or on extracts of proteins from, for example, the cytoplasm. In such situations the protein composition varies greatly and not all the components will be present in equal amounts, some will be dominating and others will be present only in minute quantities. This complexity causes difficulties for the discovery process, where low-abundant proteins may be at least as interesting in the search for novel targets as more dominating proteins in the mix.

The traditional, multiple rounds of bio-panning procedure will always enrich high-affinity antibodies against abundant proteins at the expense of antibodies of moderate affinity as well as antibodies against non-abundant proteins, which will be removed during washing steps or possibly in the amplification procedure. This loss of diversity presents a major problem in the discovery process and there have been different attempts to minimise the problem. One of the most popular strategies has been to use depletion techniques, where the phage library is pre-incubated with the unwanted antigen(s) or an irrelevant cell type.[32] In general, pre-adsorption is going to be most efficient when high-affinity antibodies against abundant antigens are to be removed. Because of binding kinetics, the chances of removing low affinity antibodies and antibodies against antigens that are difficult to access (either because of low copy number on the cell surface or because of steric hindrance) are quite low.[33] Additionally, there will be a risk of actually depleting the library of desired binders because of unspecific binding to the adsorbing cells.[34] Furthermore, if a phagemid library rescued by a wild-type helper phage is used, a large proportion of the phages in the library at any one time will not display any fusion protein and will, therefore, not be specifically depleted.[35-37] As such, nondisplaying phages are not depleted but will be propagated and in the next round of selection they will present antibodies, which optimally should have been depleted in the initial round. This renders the preselection fruitless, unless it is performed in every single round of selection just before the actual selection (i.e., no amplification step in between).

An alternative approach, which in addition will assure a high diversity of binders to be retained, is to minimise the number of selection rounds, preferably to one. One of the most significant contributing factors that previously made selection in one round problematic is the fact that the phage particle is a very large and extremely sticky assembly of proteins, therefore nonspecific binding is often observed. To diminish unspecific background, specific elution has been used, where the phages binding the antigen of interest were eluted using an excess of antigen or another antibody directed against the target.[38,39] The obvious drawback of this approach is the necessity of knowing the identity of the antigen, making specific elution problematic in a discovery setting. Another option is to render the majority of background phage non-infective, which can be achieved by the use of a protease cleavable helper phage, such as the one developed by Kristensen and Winter.[36] In the protease cleavable helper phage KM13, the gene which encodes protein III of the helper phage is modified in such a way that a protease cleavable site is located between domain 2 and 3 of protein III. Previously, it has been established that domains 1 and 2 areis needed for infection.[40,41] Therefore, removal of domains 1 and 2 of the KM13 encoded protein renders the helper phage, or phage only presenting protein III from the helper phage, non-infective. During the selection, phage is eluted with trypsin treatment, which will cleave pIII derived from the helper phage leaving only fusion-protein displaying phage infective. Thus, a large proportion (90-99.9%) of the background is eliminated, making it possible to identify specific phage antibodies after a single round of selection.[42]

Cell Surface Selection

We now turn to applications of phage antibody technology to find specific biomarker on cell surfaces.. The cell surface is extremely complex with various types of membrane spanning and heavily glycosylated proteins and selection strategies must be carefully designed if they are to result in the selection of antibodies with the desired properties. The cells used for cell surface selections can be chosen according to the design of the experiment and can be cultured cells, freshly isolated cells, cells from tissue sections, or even cells still in the body. Cultured cells are typically the easier choice, but cells in culture often show quite different protein expression patterns when compared to cells in vivo.

Simple Selection on Freshly Isolated or Cultured Cells

The use of cultured or freshly isolated cells as the basis for selection has some advantages. Firstly, the cells are in a controlled environment and the presence of irrelevant, contaminating cells can be minimised. Secondly, the cells can be manipulated before selection is performed, e.g., starving the cells, stimulating them with specific agents, transfecting the cells etc. In addition subsequent screening, especially when working with cells in culture, is much easier. Consequently, the majority of cell surface selections described to date have taken place on either cultured or freshly isolated cells.

The first cell surface selection described was performed on red blood cells isolated from the blood of human donors.[33] The cells were kept in suspension and incubated with a naïve single chain variable fragment library. A few rounds of direct selection resulted in the isolation of a few blood-group specific scFvs. However, some of the selected antibodies had a more ubiquitous specificity in that they recognised dominating antigens present on several different blood groups. In an attempt to weed out antibodies against more common antigens, the researchers decided to include a negative selection, or pre-absorption, step where the antibody library was first incubated with irrelevant cells containing the unwanted antigens before unbound phage was incubated with the target cells. In this way they were able to select two new scFvs, one of which was specific to the cell type selected on. The other scFv was specific towards an antigen present on the absorbing cells as well as the selection cells, demonstrating that pre-absorption was not 100% effective. Another design using subtractive selection was described by the Genentech laboratories.[43] In this study, tumour and normal cell lines were used for selection in order to identify novel tumour-associated antigens. The cells were detached from their culture flasks with EDTA to preserve the integrity of cell surface proteins and the subsequent selection was performed in solution on the unfixed cells. The naïve scFv library was incubated twice with a normal cell line before incubation with the tumour cell line. Again, the results indicated that the counter-selection procedure was only partly successful, as antibodies recognising antigens found on both cell lines were selected. Finnern and colleagues succeeded in selecting scFvs against a well-known epithelial tumour marker, MUC-1, which is overexpressed and under-glycosylated in most breast cancers. Three rounds of panning on live monolayers of the breast carcinoma cell line MCF-7 resulted in a number of clones, which were tested for binding to synthetic MUC-1 peptides. Twenty-five percent of the cell-binding scFvs were shown to bind MUC-1 peptides as well as other MUC-1 expressing cancer cell lines.[44] Direct selection on freshly isolated, fixed cells from primary colorectal carcinoma did not result in the isolation of antibodies uniquely specific for colorectal carcinoma but rather antibodies recognising tumour-associated antigens were selected.[45] However, the epitopes recognised were all formed in the fixation step, therefore the antibodies did not recognise cells in vivo. This situation once again underlines the importance of designing the selection procedure to meet the expectations of the results.

Several other experimental designs have been used in the simple selection of antibodies against cells but in general the conclusions are similar.[42,46] To recapitulate, selections can be performed either on freshly isolated or on cultivated cells. Freshly isolated cells may be the better choice but are not always available. If they are available, they could be fixed to preserve their structure and overall integrity. However, cell surface antigens may be modified by the fixation process and selected antibodies may be directed towards the modified epitopes, thus limiting their potential use. Additionally, subtraction techniques could be employed but the pitfalls of this approach should be kept in mind. If using cultivated cells, the paramount problem will be the selection of phages binding specifically to serum proteins, which could potently mask any binders to the cells. Again, subtractive selection could be a way to diminish this problem, but also using serum to block the library may be effective. Generally, serum problems will also be minimised by selecting on cells in suspension rather than monolayers but the way the cells are detached should also be considered. Traditional detachment using trypsin will cleave many of the cell surface proteins that are of interest and therefore a nonproteolytic detachment agent such as EDTA (ethylenediamine-N,N,N′,N′-tetraacetic acid) should be used. The fixation and negative selection comments made above are also applicable to cultured cells. The final thing that should be considered when using cultured cells is that all culture conditions are abnormal and that the protein expression pattern of cultivated cells may differ markedly from cells in vivo.

Selection Using Cell Sorting Techniques, Such as MACS or FACS

With the simple selection techniques described in the previous section a number of antibodies to the different target cells has been selected. However, when compared to the number of antibodies often isolated with traditional bio-panning on purified targets, the cell selections are often found to be inadequate.[34] In recent years improved selections schemes haves been introduced using new cell separation techniques.

Exploiting the fact that different cell types have different characteristics and display different surface molecules, de Kruif et al selected scFvs against different cell types from a venous blood sample using fluorescence activated cell sorting (FACS).[47] The nucleated cells from the blood sample were labelled with fluorescent antibodies against specific cell surface markers (CD3 and CD20) after incubation with the phage library and the cells were then sorted according to their labelling. Because a large number of the total cell population was not CD3 or CD20 positive, the authors reasoned that the nonlabelled cells would serve to absorb a lot of the unspecific binding. Since this first report, several other studies have employed the FACS based selection to isolate antibodies against, for example, dendritic cells,[48] cells from patients with acute myeloid leukaemia[49,50] and against predefined antigens overexpressed on engineered cell lines.[51]

Despite the relative success of the FACS based selection procedure, the conditions the cells are subjected to are quite harsh and sorting large numbers of cells is very time-consuming.[34] This has led to the development of a different sorting based selection strategy. From a mixed population of freshly isolated murine thymic cells, Palmer et al isolated antibodies against MHC class II positive thymic epithelial cells using magnetically activated cell separation (MACS).[52] The desired major histocompatibility complex (MHC) class II positive cells were positively selected from the diverse cell population by incubating the cell-phage suspension with magnetic beads labelled with an anti MHC class II antibody. The exposure to a magnet allowed for the nonlabelled cells to be removed along with unbound and unspecifically bound phage particles—yet again allowing the nonlabelled cells (98%) to act as absorber cells. After six rounds of selection, 50-80% of the selected antibodies bound specifically to mouse thymic epithelium.. Using a positive-negative selection principle McWhirther et al succeeded in selecting antibodies against a few up regulated cell surface receptors on CLL cells from human donors using a chronic lymphocytic leukaemia (CLL) immune library from immunised mice.[53]

In summary, the idea of simultaneous positive and negative selection by having specifically labelled target cells in a large pool of un-labelled absorber cells seems very promising and it is definitely an approach that is becoming more widespread with the development of more sensitive sorting equipment and the increased knowledge of different specific and unspecific labelling techniques.

Selection on Cells from Tissues Subjected to Laser Capture Microdissection

In the previous sections, various strategies for selecting antibodies against cells removed from their original environment were described. However, cells in the body are intricately influenced by their environment, which is composed of multiple different cell types and extracellular components. Although isolation of the desired cell type from its environment immediately prior to selection should alleviate most of the problems caused by the removal of the natural milieu not all cell types can be easily isolated. To circumvent this problem, selection on tissue sections was examined as a means of obtaining specific antibodies.[54] However, tissue sections are obviously composed of many different cell types and while this is desirable in terms of keeping the interesting cells in the appropriate environment, it makes the analysis of selected antibodies very complicated.[55] Nonetheless, recent technological advances have made it possible to select phage antibodies or, more commonly, peptides against tissue sections while only retrieving phage bound to the cell type of interest. The technique that makes this possible is called laser capture microdissection (LCM).[56]

In order to screen tumours in their natural environment, Lu et al incubated tissue sections of human tumours zenografted in mice with a peptide phage library. Phage-bearing tumour cells were then dissected from the tissue slide using LCM and bound phages were allowed to infect *E. coli*. After three rounds of selection, peptides able to bind and penetrate tumour tissues were identified, thus demonstrating the feasibility of this technique. A few years later, the same approach was used to select internalising antibodies against slides of prostate cancer tissue.[57] In this study, an antibody against an already known tumour marker was selected, again proving the relevance of the technique.

The combination of LCM and phage display is still very new and has mainly been applied to tissue sections of various tumours.[55-58] However, the potential of the technique is great and in the future might be applicable in the search for biomarkers of aging by selection on senescent cells identified in tissue sections.

In Vivo Selection

Although the laser capture microdissection approach described above may result in selection of antibodies which recognise specific cells in tissue sections, such as senescent cells, there is no guarantee that the relevant epitopes will have the same conformation when presented in a live human being. To ensure recognition of specific cells types in the native in vivo conformtation the group of Rouslathi established in vivo selections. Here a peptide library injected into mice resulted in organ-specific retention of phage.[59] This report demonstrated that the vasculature is heterogeneous and that it is possible to create compounds that specifically can target one area while leaving surrounding areas largely unaffected. Since the first report, a small number of groups have utilised the in vivo strategy to select peptides homing to specific parts of the vascular endothelium in mice or rats.[60-63] The majority of in vivo phage display selections have used peptide libraries but a few reports on the use of antibody phage display in vivo have been published.[64-67] Generally, the in vivo selection schemes involve injection of the phage library into the vasculature of the subject. After a suitable circulation period, the subject is sacrificed and the target organ harvested. Unbound phage is washed away and specifically bound phage eluted.

Regardless of the objective of the in vivo selection, the subject used for the selection is normally sacrificed in order to access the target organ. Obviously, this poses severe limitations on the exploitation of the technique in aging research to study human aging, as only animal subjects can be used. In summary, in vivo phage display has mainly been used to target the vascular endothelium, either with the objective of isolating specific targeting agents to be used for therapeutic purposes or to increase the understanding of the complexity of the vascular endothelium.

Conclusion

Normal human somatic cells have a limited capacity to proliferate before entering a state of permanent cell cycle arrest known as replicative senescence. Entry into senescence in many cell types is triggered by telomere shortening producing viable, postmitotic, cells that display radically altered patterns of gene expression. The differences in the transcriptome between these senescent cells and their growing counterparts are as large as those seen when one cell type differentiates into another. It has long been proposed that replicative senescence has the potential to act as a causal mechanism of organismal aging in at least two distinct ways, either through simple loss of the proliferative capacity of mitotic tissues or through alterations in the tissue microenvironment or tissue fluids as a result of the accumulation of senescent cells. At present our understanding of the molecular pathways triggering the processes of senescence within individual cells is far more advanced than our understanding of the role (if any) that such senescent cells play in human aging.

The senescent phenotype itself has only been properly investigated in relatively few cell types. Although it is commonly believed that senescence gives rise to a 'pro-inflammatory' phenotype there is considerable evidence that the situation is more complex than this.[68] For example, senescent human vascular smooth muscle cells secrete pro-inflammatory cytokines such as IL1 and IL8 but also adopt a pro-calcificatory phenotype in blood vessels as a result of partial transdifferentiation into osteoblasts. Senescent ocular keratocytes do not show a pro-inflammatory phenotype but do show a profound dedifferentiation that has the potential to compromise ocular function. This indicates the need to considerably increase our understanding of senescence in many more cell types.

As the field moves into a translational phase it is critically important that studies of this type are improved, broadened and used as a platform for in vivo study design. It is extremely difficult to demonstrate the presence of senescent cells in vivo due to a lack of reliable histological markers capable of distinguishing them from their quiescent counterparts. To study the protein metabolism and homeostasis during aging, phage antibody technology presents an interesting new technology, capable of dealing with the cellular heterogeneity and limited sample material.

Acknowledgements

This work has been supported under the frame of the EU FP6 funded project PROTEOMAGE FP6-518230.

References

1. Hayflick L, Moorhead PS. The serial cultivation of human diploid cell strains. Exp Cell Res 1961; 25(3):585-621.
2. Kaeberlein M, McVey M, Guarente L. Using yeast to discover the fountain of youth. Sci Aging Knowledge Environ 2001; 2001(1):pe1.
3. Gershon H, Gershon D. Caenorhabditis elegans—a paradigm for aging research: advantages and limitations. Mech Ageing Dev 2002; 123(4):261-74.
4. Gerhard GS. Comparative aspects of zebrafish (Danio rerio) as a model for aging research. Exp Gerontol 2003; 38(11-12):1333-41.
5. Partridge L. Some highlights of research on aging with invertebrates, 2009. Aging Cell 2009; 8(5):509-13.
6. de Magalhaes JP. From cells to ageing: a review of models and mechanisms of cellular senescence and their impact on human ageing. Exp Cell Res 2004; 300(1):1-10.
7. Crick F. Central dogma of molecular biology. Nature 1970; 227(5258):561-3.
8. Franceschi C, Bezrukov V, Blanche H et al. Genetics of healthy aging in Europe: the EU-integrated project GEHA (GEnetics of Healthy Aging). Ann N Y Acad Sci 2007; 1100:21-45.
9. O'Farrell PH. High resolution two-dimensional electrophoresis of proteins. J Biol Chem 1975; 250(10):4007-21.
10. Gorg A, Weiss W, Dunn MJ. Current two-dimensional electrophoresis technology for proteomics. Proteomics 2004; 4(12):3665-85.
11. Kondo T, Hirohashi S. Application of highly sensitive fluorescent dyes (CyDye DIGE Fluor saturation dyes) to laser microdissection and two-dimensional difference gel electrophoresis (2D-DIGE) for cancer proteomics. Nat Protoc 2006; 1(6):2940-56.

12. Mann M, Hendrickson RC, Pandey A. Analysis of proteins and proteomes by mass spectrometry. Annu Rev Biochem 2001; 70:437-73.
13. Chakravarti B, Oseguera M, Dalal N et al. Proteomic profiling of aging in the mouse heart: Altered expression of mitochondrial proteins. Arch Biochem Biophys 2008; 474(1):22-31.
14. Lombardi A, Silvestri E, Cioffi F et al. Defining the transcriptomic and proteomic profiles of rat ageing skeletal muscle by the use of a cDNA array, 2D- and Blue native-PAGE approach. J Proteomics 2009; 72(4):708-21.
15. Gromov P, Skovgaard GL, Palsdottir H et al. Protein profiling of the human epidermis from the elderly reveals up-regulation of a signature of interferon-gamma-induced polypeptides that includes manganese-superoxide dismutase and the p85beta subunit of phosphatidylinositol 3-kinase. Mol Cell Proteomics 2003; 2(2):70-84.
16. Kamino H, Hiratsuka M, Toda T et al. Searching for genes involved in arteriosclerosis: Proteomic analysis of cultured human umbilical vein endothelial cells undergoing replicative senescence. Cell Struct Funct 2003; 28(6):495-503.
17. Eman MR, Regan-Klapisz E, Pinkse MWH et al. Protein expression dynamics during replicative senescence of endothelial cells studied by 2-D differences in-gel electrophoresis. Electrophoresis 2006; 27(8):1669-82.
18. Levin Y, Schwarz E, Wang L et al. Label-free LC-MS/MS quantitative proteomics for large-scale biomarker discovery in complex samples. J Sep Sci 2007; 30(14):2198-203.
19. Wiener MC, Sachs JR, Deyanova EG et al. Differential mass spectrometry: A label-free LC-MS method for finding significant differences in complex peptide and protein mixtures. Anal Chem 2004; 76(20):6085-96.
20. Lengqvist J, Andrade J, Yang Y et al. Robustness and accuracy of high speed LC-MS separations for global peptide quantitation and biomarker discovery. J Chromatogr B Analyt Technol Biomed Life Sci 2009; 877(13):1306-16.
21. Kaplan A, Söderström M, Fenyö D et al. An automated method for scanning LC-MS data sets for significant peptides and proteins, including quantitative profiling and interactive confirmation. J Proteome Res 2007; 6(7):2888-95.
22. McCafferty J, Griffiths AD, Winter G et al. Phage antibodies: Filamentous phage displaying antibody variable domains. Nature 1990; 348(6301):552-4.
23. Huse WD, Sastry L, Iverson SA et al. Generation of a large combinatorial library of the immunoglobulin repertoire in phage lambda. Science 1989; 246(4935):1275-81.
24. Holliger P, Hudson PJ. Engineered antibody fragments and the rise of single domains. Nat Biotechnol 2005; 23(9):1126-36.
25. Hoogenboom HR. Selecting and screening recombinant antibody libraries. Nat Biotechnol 2005; 23(9):1105-16.
26. Hoogenboom HR, De Brune AP, Hufton SE et al. Antibody phage display technology and its applications. Immunotechnology 1998; 4(1):1-20.
27. Benhar I. Biotechnological applications of phage and cell display. Biotechnol Adv 2001; 19(1):1-33.
28. Knappik A, Ge L, Honegger A et al. Fully synthetic human combinatorial antibody libraries (HuCAL) based on modular consensus frameworks and CDRs randomized with trinucleotides. J Mol Biol 2000; 296(1):57-86.
29. Parsons HL, Earnshaw JC, Wilton J et al. Directing phage selections towards specific epitopes. Protein Eng 1996; 9(11):1043-9.
30. Hawkins RE, Russell SJ, Winter G. Selection of phage antibodies by binding affinity. Mimicking affinity maturation. J Mol Biol 1992; 226(3):889-96.
31. Griffiths AD, Duncan AR. Strategies for selection of antibodies by phage display. Curr Opin Biotechnol 1998; 9(1):102-8.
32. Jarutat T, Nickels C, Frisch C et al. Selection of vimentin-specific antibodies from the HuCAL phage display library by subtractive panning on formalin-fixed, paraffin-embedded tissue. Biol Chem 2007; 388(6):651-8.
33. Marks JD, Ouwehand WH, Bye JM et al. Human antibody fragments specific for human blood group antigens from a phage display library. Nat Biotechnol 1993; 11(10):1145-9.
34. Siegel DL, Chang TY, Russell SL et al. Isolation of cell surface-specific human monoclonal antibodies using phage display and magnetically-activated cell sorting: Applications in immunohematology. J Immunol Methods 1997; 206(1-2):73-85.
35. Lowman HB, Bass SH, Simpson N et al. Selecting high-affinity binding proteins by monovalent phage display. Biochemistry 1991; 30(45):10832-8.
36. Kristensen P, Winter G. Proteolytic selection for protein folding using filamentous bacteriophages. Fold Des 1998; 3(5):321-8.

37. Pavoni E, Monteriu G, Cianfriglia M et al. New display vector reduces biological bias for expression of antibodies in E. coli. Gene 2007; 391(1-2):120-9.
38. Hoogenboom HR, Lutgerink JT, Pelsers MMAL et al. Selection-dominant and nonaccessible epitopes on cell-surface receptors revealed by cell-panning with a large phage antibody library. Eur J Biochem 1999; 260(3):774-84.
39. Meulemans EV, Slobbe R, Wasterval P et al. Selection of phage-displayed antibodies specific for a cytoskeletal antigen by competitive elution with a monoclonal antibody. J Mol Biol 1994; 244(4):353-60.
40. Holliger P, Riechmann L, Williams RL. Crystal structure of the two N-terminal domains of g3p from filamentous phage fd at 1.9 A: evidence for conformational lability. J Mol Biol 1999; 288(4):649-57.
41. Riechmann L, Holliger P. The C-terminal domain of TolA is the coreceptor for filamentous phage infection of E. coli. Cell 1997; 90(2):351-60.
42. Jensen KB, Jensen ON, Ravn P et al. Identification of keratinocyte-specific markers using phage display and mass spectrometry. Mol Cell Proteomics 2003; 2(2):61-9.
43. Ridgway JBB, Ng E, Kern JA et al. Identification of a human anti-CD55 single-chain Fv by subtractive panning of a phage library using tumor and nontumor cell lines. Cancer Res 1999; 59(11):2718-23.
44. Wong C, Waibel R, Sheets M et al. Human scFv antibody fragments specific for the epithelial tumour marker MUC-1, selected by phage display on living cells. Cancer Immunol Immunother 2001; 50(2):93-101.
45. Roovers RC, Van der Linden E, De Brune AP et al. Identification of colon tumour-associated antigens by phage antibody selections on primary colorectal carcinoma. Eur J Cancer 2001; 37(4):542-9.
46. Mutuberria R, Satijn S, Huijbers A et al. Isolation of human antibodies to tumor-associated endothelial cell markers by in vitro human endothelial cell selection with phage display libraries. J Immunol Methods 2004; 287(1-2):31-47.
47. De Kruif J, Terstappen L, Boel E et al. Rapid selection of cell subpopulation-specific human monoclonal antibodies from a synthetic phage antibody library. Proc Natl Acad Sci USA 1995; 92(9):3938-42.
48. Lekkerkerker A, Logtenberg T. Phage antibodies against human dendritic cell subpopulations obtained by flow cytometry-based selection on freshly isolated cells. J Immunol Methods 1999; 231(1-2):53-63.
49. Geuijen CAW, Bijl N, Smit RCM et al. A proteomic approach to tumour target identification using phage display, affinity purification and mass spectrometry. Eur J Cancer 2005; 41(1):178-87.
50. Bakker ABH, Van Den Oudenrijn S, Bakker AQ et al. C-type lectin-like molecule-1: A novel myeloid cell surface marker associated with acute myeloid leukemia. Cancer Res 2004; 64(22):8443-50.
51. Yuan QA, Robinson MK, Simmons HH et al. Isolation of anti-MISIIR scFv molecules from a phage display library by cell sorter biopanning. Cancer Immunology, Immunotherapy 2008; 57(3):367-78.
52. Palmer DB, George AJT, Ritter MA. Selection of antibodies to cell surface determinants on mouse thymic epithelial cells using a phage display library. Immunology 1997; 91(3):473-8.
53. McWhirter JR, Kretz-Rommel A, Saven A et al. Antibodies selected from combinatorial libraries block a tumor antigen that plays a key role in immunomodulation. Proc Natl Acad Sci USA 2006; 103(4):1041-6.
54. Tordsson J, AbrahmsÃ©n L, Kalland T et al. Efficient selection of scFv antibody phage by adsorption to in situ expressed antigens in tissue sections. J Immunol Methods 1997; 210(1):11-23.
55. Kubo N, Akita N, Shimizu A et al. Identification of oligopeptide binding to colon cancer cells separated from patients using laser capture microdissection. J Drug Target 2008; 16(5):396-404.
56. Lu H, Jin D, Kapila YL. Application of laser capture microdissection to phage display peptide library screening. Oral Surg Oral Med Oral Pathol Oral Radiol Endod 2004; 98(6):692-7.
57. Ruan W, Sassoon A, An F et al. Identification of clinically significant tumor antigens by selecting phage antibody library on tumor cells in situ using laser capture microdissection. Mol Cell Proteomics 2006; 5(12):2364-73.
58. Sun Y, Shukla GS, Weaver D et al. Phage-display selection on tumor histological specimens with laser capture microdissection. J Immunol Methods 2009; 347(1-2):46-53.
59. Pasqualini R, Ruoslahti E. Organ targeting in vivo using phage display peptide libraries. Nature 1996; 380(6572):364-6.
60. Ballard VLT, Sharma A, Duignan I et al. Vascular tenascin-C regulates cardiac endothelial phenotype and neovascularization. FASEB J 2006; 20(6):717-9.
61. Cai D, Holm JM, Duignan IJ et al. BDNF-mediated enhancement of inflammation and injury in the aging heart. Physiol Genomics 2006; 24(3):191-7.
62. Chen Y, Shen Y, Guo X et al. Transdermal protein delivery by a coadministered peptide identified via phage display. Nat Biotechnol 2006; 24(4):455-60.
63. Zhang L, Hoffman JA, Ruoslahti E. Molecular profiling of heart endothelial cells. Circulation 2005; 112(11):1601-11.
64. Johns M, George AJT, Ritter MA. In vivo selection of sFv from phage display libraries. J Immunol Methods 2000; 239(1-2):137-51.

65. Mutuberria R, Hoogenboom HR, Van Der Linden E et al. Model systems to study the parameters determining the success of phage antibody selections on complex antigens. J Immunol Methods 1999; 231(1-2):65-81.
66. Robert R, Jacobin-Valat MJ, Daret D et al. Identification of human scFvs targeting atherosclerotic lesions: Selection by single round in vivo phage display. J Biol Chem 2006; 281(52):40135-43.
67. Valadon P, Garnett JD, Testa JE et al. Screening phage display libraries for organ-specific vascular immunotargeting in vivo. Proc Natl Acad Sci USA 2006; 103(2):407-12.
68. Campisi J, d'Adda di Fagagna F. Cellular senescence: when bad things happen to good cells. Nat Rev Mol Cell Biol 2007; 8(9):729-40.

INDEX

A

Aggregation 6, 14, 21, 31, 39, 70, 83, 96, 100, 101, 108, 109, 111, 113, 118-120, 138, 141-153, 214
Aging 1-9, 14-23, 25, 30-33, 35, 38, 39, 41-44, 47, 48, 52-54, 56, 57, 61, 62, 64-66, 69, 70, 72, 74-78, 81-83, 94, 96-99, 101, 102, 108, 111, 117-120, 134, 138, 139, 141-143, 145, 148-150, 153, 160, 163-168, 172-179, 181-187, 198, 204, 206, 211, 216, 218, 223, 225, 234-236, 240, 241
Alzheimer's disease (AD) 5, 8, 96, 100, 139, 142, 143, 146, 150, 174, 180
AMP-activated protein kinase (AMPK) 49, 51, 55, 56, 65, 75, 224
Antagonistic pleiotropy 30-33, 35
Anti-aging 53, 56, 57, 64, 78, 183
Antibiotic 2, 3, 141, 152
Antibody 6, 129, 131, 218, 226, 236-241
Antioxidant 38, 43, 48, 52, 139, 153
Apoptosis 8, 108, 117-119, 129, 133, 134, 152, 164-166, 175, 183, 212, 213, 217, 222
Atrophy 31, 32, 64, 117, 184, 211, 212, 214-219, 221-227
Autophagy 15, 20-22, 34, 47-51, 53-57, 61-66, 69, 72, 73, 77, 78, 82, 84, 94, 96-101, 108-110, 140, 148, 172-174, 212-215, 219

B

Biomarker 42, 129, 130, 234-236, 238, 240
Breast cancer associated protein 1 (BRCA1) 177, 178, 186

C

Cachexia 213, 215-218
Caenorhabditis elegans 14, 16-21, 31, 32, 33, 47-57, 63, 69, 74-76, 78, 81-101, 116, 142-151, 153, 184, 186, 198, 200-202, 204, 205
Calorie restriction 61, 66, 183
Calpain 211-213
Cancer 23-25, 64, 83, 127, 139, 142, 163, 175, 177, 181, 184, 211-213, 215, 217, 218, 225, 234, 236, 238, 240
Caspase 183, 211-213, 216, 222
Cell
- cycle 2, 4, 8, 16, 61, 85, 127, 129-132, 162, 163, 175-181, 217, 223, 241
- damage 57, 66, 70, 77, 78, 108
- growth 4, 5, 35, 69, 74-76, 83, 160, 164, 165, 217
- selection 239
- senescence 14, 32, 126, 128, 129, 131, 160, 163-168, 172, 174-176, 181, 234

Centenarian 44
Chronological life span (CLS) 16, 76
Circadian clock 202
c-Jun terminal kinase (JNK) 49, 51, 53, 55, 56, 69, 76, 77, 127, 129
Clp protease 111-113, 116

D

DAF-2 48, 49, 51-54, 74, 82, 142, 204, 205
DAF-16 31, 32, 47-49, 51, 54, 55, 75, 76, 82-98, 101, 142, 143, 149, 186, 204, 205
Deamidation 6, 83
Degradation 9, 15, 22, 38-40, 42-44, 47, 48, 51, 53, 61, 62, 69, 70, 72, 73, 77, 78, 83, 96, 97, 100-102, 108, 109, 111-117, 119, 120, 138, 140-143, 148, 150-152, 160, 168, 172-174, 176, 177, 179-186, 211-219, 222
Dietary restriction (DR) 15-17, 19-21, 23-25, 30, 32-34, 47-49, 52-56, 69, 82-85, 94-99, 101, 139, 167, 183, 186, 204-206
Differential translation 20, 34
Disease 1, 5, 8, 9, 14, 21, 23-25, 30, 35, 42, 44, 54, 57, 64, 81, 83, 96, 100, 101, 108, 111, 113, 115, 117, 118, 120, 138, 139, 141-143, 145-153, 167, 172-175, 180, 182, 184, 187, 211-213, 216-218, 221-226, 235, 236

DNA repair 7, 61, 64, 65, 127, 162, 163, 167, 172-174, 176-179, 182, 213
Drosophila 2-4, 6, 14, 31, 32, 41, 51, 52, 54, 74, 75, 96, 98, 100, 139, 140, 143-147, 149-151, 182, 201, 203-205
Dystrophy 24, 139, 211-213, 216-219, 225

E

eIF4F 32, 34
eIF4G 32, 75, 76, 85, 89, 94
Elongation 2-5, 7, 8, 53, 71, 72, 127, 177
Elongation factor (EF) 3, 5, 7, 8, 71
Error 1, 2, 82, 83, 138, 141
  catastrophe 2, 82, 83
Evolution of aging 30

F

Fidelity 3, 82, 141

G

Gcn4 20, 21, 35, 85
Genetic screen 53, 146, 149, 151, 152
Glycation 6, 14, 83, 139
G-protein coupled receptor (GPCR) 198, 199, 201

H

Heat shock 2, 4, 48, 52, 70, 96, 100, 111, 116, 129, 141-143, 148-150, 152, 153, 162
Homeodynamics 4
Homeostasis 5, 20-22, 34, 38, 39, 41-44, 53, 54, 56, 61, 69, 72, 81-84, 95, 97, 101, 109, 138, 140-143, 148-150, 153, 163, 197-204, 206, 212, 213, 215, 217, 218, 234, 235, 241
Hormone 5, 7, 8, 62, 64, 65, 77, 82, 184, 197-204, 206, 215, 218-221
HSF1 77, 96, 97, 142, 143, 148-150, 152, 162
Huntington's disease (HD) 21, 24, 96, 100, 142, 143, 145, 153, 180

I

i-AAA protease 113-117
Initiation 2-5, 8, 15, 17, 18, 20, 23, 25, 32, 34, 35, 50, 53, 71-73, 84-89, 94, 95, 101, 119, 152, 214
Initiation factor 5, 8, 15, 17, 18, 20, 23, 34, 53, 71, 72, 84-89, 94, 95, 119, 152
Inner mitochondrial membrane 111-113, 117, 118
Insulin 4, 5, 8, 14, 21, 23-25, 31-34, 47-49, 51-56, 62-65, 69, 74-77, 81, 82, 84, 94-98, 101, 131, 142, 143, 184, 186, 197-206, 211, 219, 220
Insulin growth factor-1 (IGF-1) 32, 48, 53, 54, 69, 74-77, 81, 82, 84, 95, 96, 101, 184, 204-206, 211, 216, 218-225
Insulin/IGF-1 signaling 74-77, 81, 84, 96, 101, 205
Insulin-like signaling 14, 21, 31, 48, 49, 51, 54-56, 142, 186
Invertebrate model 69, 138, 145, 146, 149, 152, 153
Ionotropic receptor 198, 200, 201

K

Kinase target of rapamycin (kinase TOR) 50, 53, 75, 77, 83, 97

L

Lifespan 2, 4, 6, 7, 15-17, 21, 30-35, 38, 39, 41-44, 69, 75, 81-102, 120, 132, 138, 139, 142, 143, 163, 164, 166, 186, 197, 198, 204-206
  extension 15, 21, 30-34, 38, 43, 81, 83-94, 96, 98-102, 143, 186, 204, 206
Longevity 7, 14-17, 20-23, 25, 32, 33, 38, 39, 42-44, 47-57, 61-65, 69, 72, 74-78, 82-85, 94, 95, 97, 98, 101, 142, 186, 204, 205
  pathway 15, 17, 21, 47-49, 52, 54, 55, 57, 69, 72, 74-78
Lon protease 39, 108, 111, 116-120, 140
Lysosome 9, 39, 50, 53, 54, 62, 64, 69, 72, 73, 100, 102, 108, 109, 181, 183, 212-214, 219

M

m-AAA protease 111-113, 115, 118
Mammal 20, 21, 23, 25, 32, 33, 41, 47, 51-53, 56, 62-64, 70, 74, 76, 81, 83, 96, 100, 113, 115, 118, 161, 167, 201-203, 205, 220
Metabolism 5, 15, 24, 57, 65, 66, 69, 70, 72, 74, 78, 81, 150, 163, 175, 184, 202, 212, 215, 234-236, 241
Methylation 7, 162
Mice 2, 3, 5-7, 15, 21, 23-25, 30, 41, 43, 51, 53, 54, 56, 62-66, 75, 76, 83, 96, 100, 111, 116, 118, 119, 127, 133, 134, 141, 147, 152, 163, 164, 166, 175, 177, 179, 181, 183-186, 212, 214, 215, 217-219, 221-226, 235, 239, 240
Mitochondria 15, 16, 19, 20, 24, 32, 39, 47-49, 52-57, 65, 69, 70, 74, 75, 77, 81-93, 98, 100, 102, 108-120, 130, 140, 148, 173, 181, 183, 184, 212, 214, 235
  chaperone 109-111
  dysfunction 65, 85, 115, 117, 181, 214
  matrix 108, 111-113, 116, 117, 119, 181
  membrane 111-113, 115, 117-119
Mitogen-activated protein kinase (MAPK) 5, 23, 51, 69, 76, 91, 126, 127, 133, 134, 181, 184, 223
Mitophagy 102, 109
Molecular chaperone 72, 96, 97, 100, 101, 116, 138, 141-143, 145, 148, 150-152
Molecular damage 108, 119, 142
mRNA translation 14-17, 19-25, 32-35, 70-72, 74-78, 82, 97
Muscle 3, 6, 7, 24, 41, 52, 54, 81, 96, 98, 100, 119, 129, 144, 146, 147, 149-151, 181, 184, 185, 211-219, 221-227, 235, 241

N

Nematode 2, 3, 6, 14-20, 22, 25, 47, 48, 53, 54, 56, 62, 63, 65, 69, 74, 76, 78, 81-85, 96, 100, 101, 161, 186, 234
Neural circuit 198, 199
Neurodegenerative disease 23, 24, 42, 54, 81, 96, 101, 108, 111, 113, 115, 117, 118, 138, 141, 143, 145, 146, 150, 153, 182
Neuronal cell death 100, 153
NFκB 215, 224

O

Olfaction 203-206
Oncogene 128, 129, 132, 133, 164
Organismal aging 160, 163, 168, 181, 241
Oxidation 4-6, 14, 39, 42, 65, 83, 139
Oxidative damage 6, 39, 57, 61, 82, 83, 117, 139, 140
Oxidative stress 6, 38, 39, 41-43, 52, 56, 57, 84-93, 101, 116, 118, 120, 126-128, 130, 132-134, 139, 140, 164, 175, 183, 186, 214, 215, 235

P

p27 180
$p38^{MAPK}$ 76, 126-134
p53 4, 63, 77, 98, 127, 128, 130-132, 134, 162, 164-167, 175, 177-180
  /pRB 164-166, 175
  signalling pathway 167
Parkinson's disease (PD) 100, 146, 182
Phage display 236, 237, 240
Post-translational modification 1, 2, 4, 7-9, 160, 162, 164, 165, 172, 175, 178, 212, 221
Premature aging 61, 62, 64-66, 118, 163, 164, 177
Progeria 5, 6, 61, 65, 66, 186
Progeroid syndrome 64, 66, 172, 177, 178, 185, 186
Proliferating cell nuclear antigen (PCNA) 162, 163, 178, 180, 182
Protease 8, 9, 39, 72, 73, 108-120, 140, 148, 150, 152, 161, 162, 165, 167, 174, 175, 179, 183, 212-214, 237
Proteasome 9, 38-44, 61, 69, 72, 73, 77, 78, 97, 100, 101, 109, 112, 113, 116, 119, 140, 143, 148, 152, 160, 168, 172-186, 211, 213-215, 219, 222
  activation 42, 43
  function 38, 41, 42, 44, 101
  inhibition 41, 42

Protein
  aggregation 100, 108, 109, 111, 118, 120, 143, 145, 150-152
  damage 39, 43, 70, 72, 77, 78, 83, 101, 108-111, 116, 119, 139, 141, 149, 151, 153
  degradation 9, 15, 38, 39, 42, 43, 48, 61, 69, 70, 72, 73, 78, 83, 96, 97, 113, 140, 150, 168, 172, 173, 182, 183, 184, 211, 212, 215-219
  folding 8, 110, 115, 116, 138, 139-141, 143, 146, 149-151, 153
  homeostasis 20-22, 34, 38, 39, 43, 44, 81-84, 95, 97, 101, 109, 138, 142, 143, 150, 234, 235
  misfolding 14, 138, 141, 142, 145-149, 151, 152
  oxidation 5, 39, 42
  quality control (PQC) 61, 108-112, 115-120, 151, 181
  repair 6, 38, 39, 43
  stability 38, 43, 44, 140, 162, 177
  synthesis 1-5, 7, 8, 14, 17, 21, 22, 24, 30, 32-35, 69, 70, 72, 74-78, 81-85, 94-96, 98, 99, 101, 109, 111, 115, 119, 142, 150, 151, 168, 215-218
  turnover 9, 35, 61, 62, 69, 70, 72, 77, 78, 83, 84, 97, 101, 109, 116, 117, 168
Proteolysis 8, 9, 38, 39, 42, 44, 73, 111, 113, 114, 116, 117, 146, 168, 179, 180, 183, 184, 212, 213, 215, 222
Proteolytic processing 8, 115
Proteome 34, 38, 39, 101, 138, 142, 153, 235
Proteomics 1, 9, 160, 235, 236
Proteostasis 138, 140-143, 145, 148, 149, 151, 152
Proteotoxicity 21, 44, 83, 96, 97, 100, 101, 142, 143, 145, 146, 148, 151, 186

Q

Quality control 53, 61, 100, 108-110, 115, 120, 141, 142, 146, 151, 181

R

Racemization 6, 7, 83
Rapamycin 14, 21, 22, 24, 25, 30, 52, 64, 69, 75, 223
Ras 4, 8, 34, 69, 75-77, 126-128, 131-134, 166, 175, 180, 181, 184
Reactive oxygen species (ROS) 43, 49, 52, 53, 55, 57, 70, 77, 108, 110, 114, 119, 126, 133, 139, 180, 214, 215
Replicative life span (RLS) 16, 17, 19, 20, 23, 76, 164, 181
Replicative senescence 39, 41, 126, 127, 129-131, 134, 163, 164, 166, 167, 172, 174-180, 241
Respiratory chain 75, 77, 98, 112-117, 120
Ribosomal protein RNA interference 85

S

S6 kinase 17, 20, 23-25, 31, 84, 86, 223
Saccharomyces cerevisiae 14, 16, 18, 32, 35, 62, 97, 100, 111, 113, 116, 117, 145, 150
Satellite cell 217, 218, 221, 222, 225, 226
Secretion 119, 129, 130, 197-204, 221
$SNEV^{Prp19/Pso4}$ 179
Sensory system 198, 204, 206
SIR-2.1 20, 51, 86-93, 150
Small molecule 145, 151-153, 200, 213
Small ubiquitin-related modifier (SUMO) 160-168, 174-179, 186, 187
  E3 ligase 160, 165, 177
  isoform 160, 165
  specific peptidase 160, 161
  SUMOylation 160-168, 174, 175, 177-179, 186
Stem cell 16, 126, 167, 222-224, 226
Stress 2, 6, 21, 25, 31-34, 38-43, 51-53, 55-57, 66, 69, 70, 72, 81-98, 101, 111, 116-118, 120, 126-134, 138-143, 148, 149, 151-153, 160, 161, 164, 165, 167, 175, 179, 181, 183, 186, 202, 206, 214, 215, 235
  response 21, 31, 85, 94-96, 101, 133, 138, 142, 143, 148, 151, 152, 160, 164, 165, 175, 206
Substrate binding 116

T

Target of rapamycin (TOR) 14-17, 19-25, 30-34, 47, 49-54, 65, 69, 75-77, 81, 83-95, 97-100
Taste 203
TAX-6 49, 51, 55
Telomere 14, 61, 126, 127, 129, 133, 164, 167, 168, 175-177, 186, 241
Termination 2, 4, 71, 72, 203
TGF-β signaling 132
TOR signaling 15, 17, 19-21, 25, 31, 47, 49, 54, 65, 75-77, 81, 83-85, 94, 95, 97, 98
Translation initiation factor 15, 17, 18, 20, 23, 84-89, 94, 95, 152

U

Ubiquitin (Ub) 9, 39-41, 50, 53, 61, 69, 72, 73, 97, 101, 112, 113, 140, 160-162, 167, 172-187, 211-215, 219, 221, 222
Ubiquitin proteasome system (UPS) 69, 72, 73, 97, 100, 101, 113, 176, 182, 183, 186, 213, 214
Ubiquitinome 172, 173, 175, 182, 186

V

Visual cue 201, 202

W

Wasting 66, 184, 211-214, 216, 218, 219, 221-227

Y

Yeast 3, 14-23, 25, 30, 31, 33, 35, 42, 50, 52-54, 56, 62, 65, 70, 75, 76, 83, 85, 94, 95, 97-99, 111, 113-118, 140, 151, 161-163, 167, 174, 177, 205, 234